Cardiovascular and Metabolic Disease
Scientific Discoveries and New Therapies

RSC Drug Discovery Series

Titles in the Series:

1: Metabolism, Pharmacokinetics and Toxicity of Functional Groups
2: Emerging Drugs and Targets for Alzheimer's Disease; Volume 1
3: Emerging Drugs and Targets for Alzheimer's Disease; Volume 2
4: Accounts in Drug Discovery
5: New Frontiers in Chemical Biology
6: Animal Models for Neurodegenerative Disease
7: Neurodegeneration
8: G Protein-Coupled Receptors
9: Pharmaceutical Process Development
10: Extracellular and Intracellular Signaling
11: New Synthetic Technologies in Medicinal Chemistry
12: New Horizons in Predictive Toxicology
13: Drug Design Strategies: Quantitative Approaches
14: Neglected Diseases and Drug Discovery
15: Biomedical Imaging
16: Pharmaceutical Salts and Cocrystals
17: Polyamine Drug Discovery
18: Proteinases as Drug Targets
19: Kinase Drug Discovery
20: Drug Design Strategies: Computational Techniques and Applications
21: Designing Multi-Target Drugs
22: Nanostructured Biomaterials for Overcoming Biological Barriers
23: Physico-Chemical and Computational Approaches to Drug Discovery
24: Biomarkers for Traumatic Brain Injury
25: Drug Discovery from Natural Products
26: Anti-Inflammatory Drug Discovery
27: New Therapeutic Strategies for Type 2 Diabetes: Small Molecules
28: Drug Discovery for Psychiatric Disorders
29: Organic Chemistry of Drug Degradation
30: Computational Approaches to Nuclear Receptors

31: Traditional Chinese Medicine
32: Successful Strategies for the Discovery of Antiviral Drugs
33: Comprehensive Biomarker Discovery and Validation for Clinical Application
34: Emerging Drugs and Targets for Parkinson's Disease
35: Pain Therapeutics; Current and Future Treatment Paradigms
36: Biotherapeutics: Recent Developments using Chemical and Molecular Biology
37: Inhibitors of Molecular Chaperones as Therapeutic Agents
38: Orphan Drugs and Rare Diseases
39: Ion Channel Drug Discovery
40: Macrocycles in Drug Discovery
41: Human-based Systems for Translational Research
42: Venoms to Drugs: Venom as a Source for the Development of Human Therapeutics
43: Carbohydrates in Drug Design and Discovery
44: Drug Discovery for Schizophrenia
45: Cardiovascular and Metabolic Disease: Scientific Discoveries and New Therapies

How to obtain future titles on publication:
A standing order plan is available for this series. A standing order will bring delivery of each new volume immediately on publication.

For further information please contact:
Book Sales Department, Royal Society of Chemistry, Thomas Graham House, Science Park, Milton Road, Cambridge, CB4 0WF, UK
Telephone: +44 (0)1223 420066, Fax: +44 (0)1223 420247,
Email: booksales@rsc.org
Visit our website at www.rsc.org/books

Cardiovascular and Metabolic Disease
Scientific Discoveries and New Therapies

Edited by

Philip Peplow
University of Otago, Dunedin, New Zealand
Email: phil.peplow@anatomy.otago.ac.nz

James Adams
University of Southern California, Los Angeles, USA
Email: jadams@usc.edu

Tim Young
Vertex Pharmaceuticals, San Diego, USA
Email: Tim_Young@vrtx.com

ROYAL SOCIETY OF CHEMISTRY

THE QUEEN'S AWARDS FOR ENTERPRISE: INTERNATIONAL TRADE 2013

RSC Drug Discovery Series No. 45

Print ISBN: 978-1-78262-046-4
PDF eISBN: 978-1-78262-239-0
ISSN: 2041-3203

A catalogue record for this book is available from the British Library

Published by The Royal Society of Chemistry,
Thomas Graham House, Science Park, Milton Road,
Cambridge CB4 0WF, UK

Registered Charity Number 207890

For further information see our web site at www.rsc.org

Preface

Obesity, hypertension, cardiovascular disease, and diabetes mellitus have been the subject of clinical trials and experimental studies using animal models for many years. There is substantial overlap between these diseases in etiology and disease mechanisms. The increasing effort devoted to understanding the cellular pathways and molecular links that exist between these diseases, together with the design of new medications in both academic and pharmaceutical drug discovery programmes, as well as testing in randomized clinical trials and experimental animal models, has provided the impetus for the construction of this book.

It is the goal of this book to afford a forum for clinical experts to present recent data on biomarkers, risk factors and gene–environment interactions for these diseases. It also brings together scientific experts who have been directly involved in research to discover and develop new drugs for treating these diseases, and provide them a forum to present the state of the art in their field including drugs with exciting potential that have advanced to clinical trials or to the market. It is my hope that the topics covered herein will provide essential background information on the etiology and cellular and biochemical mechanisms for these diseases, and that knowing the common causes and disease mechanisms will allow a more effective and proactive approach in their prevention and treatment.

I wish to express my great appreciation to each of the chapter authors for the time and effort spent to write informative reviews on their respective areas of special clinical and research interest. Also I wish to express my gratitude to Professor James Adams (University of Southern California), Dr Tim Young (Vertex Ltd), and Dr Cara Sutton, Helen Prasad and Rowan Frame at the Royal Society of Chemistry for their help, support and patience in the course of putting together this book.

Philip V. Peplow

RSC Drug Discovery Series No. 45
Cardiovascular and Metabolic Disease: Scientific Discoveries and New Therapies
Edited by Philip Peplow, James Adams and Tim Young

Published by the Royal Society of Chemistry, www.rsc.org

Substantial overlap exists between obesity and hypertension leading to cardiovascular disease and diabetes. There is also evidence linking obesity to cancer. This illustration of linking circles represents the likely overlapping of obesity (yellow circle), hypertension (brown circle), cardiovascular disease (red circle), diabetes (blue circle), and cancer (green circle) in a human subject (purple circle).

Contents

RSC Drug Discovery Series No. 45
Cardiovascular and Metabolic Disease: Scientific Discoveries and New Therapies
Edited by Philip Peplow, James Adams and Tim Young

Published by the Royal Society of Chemistry, www.rsc.org

Introduction

CHAPTER 1

The Relevance of Biomarkers, Risk Factors and Gene–Environment Interactions in Disease: Scientific Developments and Therapeutic Approaches

PHILIP V. PEPLOW*[a] AND JAMES D. ADAMS JR[b]

[a] Department of Anatomy, University of Otago, Dunedin, New Zealand; [b] Department of Pharmacology and Pharmaceutical Sciences, School of Pharmacy, University of Southern California, Los Angeles, CA 90089, USA
*Email: phil.peplow@otago.ac.nz

1.1 Introduction

An increasing prevalence of obesity, hypertension and cardiovascular disease, and diabetes has been reported for children and adults in many developed and developing countries. These conditions are interlinked and are associated with both immediate and long-term health effects.

Childhood obesity has more than doubled in children and tripled in adolescents in the past 30 years. In the United States of America the percentage of children aged 6–11 years who were obese increased from 7% in

RSC Drug Discovery Series No. 45
Cardiovascular and Metabolic Disease: Scientific Discoveries and New Therapies
Edited by Philip Peplow, James Adams and Tim Young

Published by the Royal Society of Chemistry, www.rsc.org

1980 to nearly 18% in 2010. Likewise, the percentage of adolescents aged 12–19 years who were obese increased from 5% to 18% over the same time period.[1,2] In 2010, more than one third of children and adolescents were overweight or obese.[1] Overweightness and obesity are caused by caloric (energy) imbalance with too few calories expended for the amount of calories consumed, and are affected by various behavioral, genetic and environmental factors.[3] Obese youth are more likely to have risk factors for cardiovascular disease such as high cholesterol or high blood pressure.[4] Obese adolescents are more likely to have prediabetes, a condition in which blood glucose levels indicate a high risk of developing diabetes.[5] Data from the National Health and Nutrition Examination Survey (NHANES) on the prevalence and control of hypertension in the USA between 1960 and 2008 have shown that the prevalence is higher in older individuals, non-Hispanic blacks, and women.[6,7] It is unclear why non-Hispanic blacks are more likely to become hypertensive. Both genetic and environmental factors are probably important, one of which may be a greater likelihood of low birth weight, which appears to predict higher blood pressures in adulthood. NHANES data from 2007 to 2008 showed a 28 to 30% prevalence of hypertension in the 18 year and older population in the USA.[7] This equates to approximately 65 million adult hypertensives in the USA, which is markedly higher than the estimated 43 million from the 1988–1991 NHANES survey.[6–8] Thus, there has been a 50% increase in the number of adult hypertensives over a decade. With the increased prevalence of hypertension there has also been a marked increase in obesity and body mass index,[9] and it has been estimated that one-half of the relative increased prevalence of hypertension may be due to an increased weight of the average individual. In surveys similar to the NHANES in Canada and Europe, the age- and sex-adjusted prevalence of hypertension ranged from 20 to 55%.[10] Analysis of world-wide data suggested that 26% of the world adult population had hypertension in 2000.[11] Rates of hypertension appear to be similar in developed and developing countries.[12] Worldwide, hypertension is not adequately controlled, and control is worse in lower income countries.[13]

There is increasing evidence that adult hypertension has its origins during childhood, as childhood blood pressure predicts adult blood pressure. Hypertension in childhood and adolescence may contribute to premature atherosclerosis and the early development of cardiovascular disease. Identifying children with hypertension and successfully treating their hypertension will have an important impact on long-term outcomes of cardiovascular disease. The definition of childhood hypertension is based on the normative distribution of blood pressure in healthy children. This contrasts with adult hypertension which is primarily defined by clinical outcomes (*i.e.* risk of cardiovascular disease and mortality) from large trials of antihypertensive therapy.[14] This clinical definition cannot be applied to children because cardiovascular events other than left ventricular hypertrophy do not typically occur until adulthood. Body size is the most important determinant of blood pressure in children and adolescents.

In 2010, an estimated 19 million persons in the USA had diagnosed diabetes and another 7 million had undiagnosed diabetes.[15] Since 1990, the prevalence of diagnosed diabetes in the USA has risen sharply among all age groups, both sexes, and all racial/ethnic groups for which data are available.[16] The substantial increase in the prevalence of diagnosed diabetes is likely the result of improved survival of persons with diabetes and increasing diabetes incidence. Nationally representative data suggest that mortality among adults with diabetes in the USA declined substantially between 1997 and 2006, and at a faster rate than among adults without diabetes.[17] This decline is paralleled by improvements in the health of persons with diabetes, including lower levels of risk factors for complications (*e.g.* hyperglycemia, uncontrolled blood pressure), decreased rates of complications associated with increased risk of death, and improvements in quality of care and medical treatments.[18]

The main driver of the increase in diabetes prevalence is the increase in the incidence of diabetes in the USA since 1990. Increasing incidence could be the result of many factors, including changes in diagnostic criteria, enhanced detection of undiagnosed diabetes, demographic changes in the population (*e.g.* aging of the population and growth of minority populations who are at greater risk for diabetes), and an increase in the prevalence of risk factors for the development of diabetes (*e.g.* obesity and sedentary lifestyle). The increase in diabetes prevalence coincides with the increase in obesity prevalence in the USA.[16,19] Continued surveillance of diabetes prevalence and incidence, its risk factors, and prevention efforts is important to measure progress of prevention efforts. Strategies that target the entire population and high-risk groups are needed to reverse the trend of increasing diabetes prevalence. Many governments and public health planners still remain largely unaware of the current magnitude and future potential for increases in diabetes and its serious complications in their own countries. In addition to diabetes, the condition of impaired glucose tolerance is a major public health concern, both because of its association with diabetes incidence and its own association with an increased risk of cardiovascular disease.

From the above it is clear that an urgent need exists to identify as early as possible those children and adults at risk of developing obesity, hypertension and cardiovascular disease, and diabetes and to discover new improved therapies for treating these conditions and their associated complications. Complications from diabetes, which include coronary artery and peripheral vascular disease, stroke, diabetic neuropathy, amputations, renal failure, and blindness, are resulting in increasing disability, reduced life expectancy, and enormous health costs for society.

1.2 Biomarkers of Disease

In 2001, the National Institute of Health defined a biomarker as a characteristic that is objectively measured and evaluated as an indicator of normal biological processes, pathogenic processes, or pharmacologic responses to a therapeutic intervention or other health care intervention.[20] The biomarker is either produced by the diseased organ (*e.g.* tumour) or by the body in response to disease. Before diagnosis, biomarkers could be used for screening and risk assessment. During diagnosis, biomarkers can determine staging, grading, and selection of initial therapy. Later, they can be used to monitor therapy, select additional therapy, or monitor recurrent diseases.[20] Thus, identifying biomarkers include all diagnostic tests, imaging technologies, and any other objective measures of a person's health status. Genetics, genomics, proteomics, and modern imaging techniques and other high-throughput technologies allow more biomarkers than before to be measured. In addition, a greater understanding of disease pathways, the targets of intervention, and the pharmacologic consequences of medicine is achieved.[21]

If a biomarker is to be used as a diagnostic test, it should be sensitive and specific and have a high predictive value. A highly sensitive test will be positive in nearly all patients with the disease, but it may also be positive in many patients without the disease. To be of clinical value, a test with high sensitivity should also have high specificity, so that most patients without the disease should have negative test results. For predicting the likelihood of disease on the basis of the test result, rather than the converse, the appropriate measures are positive and negative predictive values. However, the positive predictive value falls as the prevalence of the disease falls, so that tests for rare conditions have many more false positive results than true positive results.

Most biological markers are not simply present or absent but have wide ranges of values that overlap in persons with a disease and in those without it. The risk typically increases progressively with increasing levels; few biomarkers have a threshold at which the risk suddenly rises, so various cut-off points must be evaluated for their ability to detect disease. Cut-off points with high sensitivity producing few false negative results are used when the consequences of missing a potential case are severe, whereas highly specific cut-off points producing few false positive results are used to avoid mislabeling a person who is actually free of the disease. Sensitivity and specificity calculated at various cut-off points generate a receiver operating characteristic curve, which ideally will be highly sensitive throughout the range of specificity. The most useful clinical tests are those with the largest area under the curve.[21]

The *C*-statistic, or area under the receiver operating characteristic curve, is a method used to test model discrimination. *C*-statistic for a multivariable model represents the probability that a case has a higher measure or risk score (or a shorter time to event in survival analyses) than a comparable control. The *C*-statistic measures the concordance of the score and disease state. The value of the *C*-statistic ranges from 0.5 (no discrimination) to 1.0 (perfect discrimination) and for the Framingham coronary heart disease risk score, the *C*-statistic is approximately 0.76.[22]

When considering the efficacy of novel biomarkers in risk stratification, one approach is to determine to what extent entering the candidate biomarker into standard risk prediction models will actually increase the model's *C*-statistic. The value of a biomarker may also be assessed by studying how biomarker information may lead to a reclassification of individuals in low-, medium- and high-risk categories based on traditional risk factors. The ultimate goal of this approach is to refine risk stratification, and it has been emphasized when considering biomarker information that would serve to shift individuals who are in the intermediate-risk groups upwards into the high-risk category or downwards into the low-risk category.

Some screening methods (*e.g.* Pap smears, colonoscopy) have successfully reduced mortality. However the field of early detection is affected by problems of overdiagnosis (*e.g.* prostate specific antigen (PSA), mammographic screening), inadequate specificity of individual markers (*e.g.* carcinoembryonic antigen (CEA), alpha-fetoprotein (AFP)), low compliance (colonoscopy), and a lack of analytical methods for discovering new diagnostic markers. High-throughput technologies have been used to assess genomic data, transcriptional data, and proteomic data and have resulted in DNA, RNA and protein biomarkers in clinical use.

1.3 Risk Factors of Disease

Many factors influence the health of a person with some functioning on an individual level (*e.g.* health behaviours, genetic make-up) while others act at a broader societal level (*e.g.* availability of health services, vaccination programs, a clean and healthy environment). All these influencing factors are known collectively as determinants of health. Health determinants can influence health of an individual in either a positive or negative way. Those determinants affecting health in a negative way are referred to as risk factors. They can increase the likelihood of developing disease or adversely affect the management of existing conditions (*e.g.* hypertension can increase the likelihood of developing cardiovascular disease). Positive determinants of health are known as protective factors and include high-quality nutrition, safe sexual behaviour, maintaining a healthy body weight and exercise.

The prevalence of many chronic diseases is increasing in many countries. There are many reasons for this, and they include early detection and improved treatments for diseases that previously caused premature death, lifestyle behaviours such as smoking or low-nutritional diet, and an aging population. Chronic diseases impact heavily on the use of health services and contribute to major funding pressures on the health care system.

Risk factors affect health with varied levels of severity and measuring this can be quite complex. Burden of disease studies attempt to quantify the impact of selected risk factors, individually and in combination, on individual and population health. Burden of disease is a measure used by epidemiologists, statisticians and health economists to assess and compare the relative impact of different diseases and injuries on people or populations. Risk factors carry different levels of burden and the benefits of reducing them vary for individuals and the community.

The development of chronic diseases is strongly associated with the behavioural risk factors of smoking, physical inactivity, low-nutritional diet, and the harmful use of alcohol. These behaviours can contribute to the development of biomedical risk factors such as hypertension, obesity, and hypercholesterolemia. The top five risk factors for males and females in Australia in 2003 were smoking, hypertension, high body weight, hypercholesterolemia, and physical inactivity.[23] Some chronic conditions themselves are considered to be risk factors for other conditions, *e.g.* diabetes, where there are known links between diabetes and increased likelihood of cardiovascular, eye and kidney disease.

A recent study conducted in Australia[24] revealed that almost all individuals aged 15 years or over have at least one risk factor (99%) and most persons have three. More than one-third of persons have at least one of the risk factors of daily smoking, risky alcohol consumption or obesity. A higher proportion of males than females reported having five or more risk factors. Persons living in the most socioeconomic disadvantaged areas have greater numbers of risk factors than those living in less disadvantaged arcas. Persons living in the major cities have fewer risk factors than those living elsewhere. As the number of risk factors increases, so does the likelihood of reporting certain chronic diseases. Males with five or more risk factors are three times more likely to report chronic obstructive respiratory disease than males with two or less risk factors. Females with five or more risk factors are three times more likely to report stroke, and two-and-half times more likely to report depression than females with two or less risk factors. Studying risk factor combinations that are prevalent in the community complements information about the prevalence of individual risk factors and the number of risk factors. Knowledge about patterns in multiple risk factors can assist health professionals treating persons with chronic disease to develop management regimens and indicate where interventions might best be targeted. Having multiple risk factors can affect the speed at which a condition progresses and develops into a new condition. The combination of hypertension, dyslipidemia and diabetes accelerates atherogenesis, thereby leading to blockage of the artery. Controlling those risk factors can stabilize lesions and slow progression. Some risk factors can be the result of others *e.g.* low-nutritional diet or inadequate physical activity can contribute to obesity.[24]

1.4 Gene–Environment Interactions in Disease

Studies of gene–environment interactions aim to describe how genetic and environmental factors jointly influence the risk of developing a disease. Information is required on both elements of the relationship. Genetic predisposition can be inferred from family history, phenotype (*e.g.* skin colour) or direct analysis of DNA sequence. Environmental and lifestyle factors are measured in epidemiologic studies using self-reported information. This information is obtained by interview or questionnaire, from records or direct measures in individuals (*e.g.* anthropometry) or biomarker-based inference on environmental exposures. Until recently, many studies of genetic predisposition obtained little information on environment and lifestyle. Similarly, many epidemiologic studies of unrelated individuals (known as association studies) did not obtain blood samples or other sources of DNA that would allow direct assessment of genetic variation. More recently, in both family-based and association studies, collection of both genetic and environmental data is becoming more common so that the interaction between the two can be examined.

The studies of gene–environment interactions provide a better estimate of the population-attributable risk for genetic and environmental risk factors by accounting for their joint interaction. They strengthen the associations between environmental factors and disease by examining these factors in genetically susceptible individuals. They help to identify disease mechanisms by using information on susceptibility and resistance genes to focus on the biologic pathways that are most relevant to that disease and the environmental factors that are most relevant to the pathways. The information on biologic pathways is used to design new preventive and therapeutic strategies. The studies allow tailored preventive advice that is based on the knowledge that an individual carries susceptibility or resistance genes.[25]

Visceral obesity alters the activities of many genes, especially the genes that make adipokines and toxic lipids.[26] During visceral obesity, visceral fat, perivascular fat and macrophages make the inflammatory adipokines, visfatin, leptin, tumor necrosis factor-α, interleukin-6, C-reactive protein, resistin, and others. Toxic lipids, such as ceramide and the endocannabinoids, are made in increased amounts and circulate in the blood. Ectopic fat deposits form in muscle, liver, brain and other organs. These ectopic fat deposits, especially in muscle, correlate with insulin resistance and type 2 diabetes. Fibrinogen is released by the liver when fat deposits form in the liver.

Hypertension is caused by many factors.[26] Ceramide delivery to the kidney increases and causes blood pressure to increase. C-reactive protein, interleukin-6 and tumor necrosis factor-α cause the dysfunction of both endothelial nitric oxide synthase (eNOS) and induced nitric oxide synthase (iNOS) which decreases nitric oxide (NO) formation and increases oxygen radical formation. Visfatin and resistin inhibit vasodilation and increase endothelin transcription. Angiotensinogen is released by visceral fat, increases angiotensin II levels and blood pressure.

Atherosclerosis and heart disease are caused by a multifactorial process.[26] Hypertension is induced as described already. Leptin and visfatin induce wall defects in arteries by stimulating oxygen radical formation in the arteries. Platelets, neutrophils and monocytes adhere to the defects due to adhesion molecule synthesis at the sites, as induced by tumor necrosis factor-α, resistin and C-reactive protein. Neutrophils start an inflammatory process.

LDL-cholesterol becomes oxidized due to oxygen radical formation induced by leptin and visfatin. Macrophages become activated and take up the oxidized LDL-cholesterol making foam cells. Arterial smooth muscle cells proliferate and enlarge due to the adipokines, platelet-derived growth factor, angiotensin II and heparin-binding epidermal growth factor-like growth factor. Plaques grow and become unstable due to matrix metalloproteinase induction by C-reactive protein. This leads to thrombosis and embolism.

It should also be mentioned that toxic lifestyles other than visceral obesity are important in heart disease.[26] Alcohol consumption induces sterol regulatory element binding protein that increases visceral fat deposition. Smoking causes damage to arteries since nicotine interacts with non-innervated nicotinic acetylcholine receptors in endothelial cells. This interaction causes oxygen radical formation that damages arteries leading to atherosclerosis.

Insulin resistance is caused by several factors.[26] Endocannabinoids in visceral fat induce the formation of inflammatory adipokines such as visfatin, resistin, tumor necrosis factor-α and others. Ceramide inhibits insulin receptor phosphorylation leading to receptor dysfunction. Ceramide also causes iNOS dysfunction leading to oxygen radical formation and beta cell destruction. Resistin and resistin-like molecules cause insulin resistance. Visfatin, interleukin-6 and tumor necrosis factor-α cause long term insulin receptor dysfunction.

1.5 Drug Development and Targeted Therapies

Recent advances in preclinical and clinical research have resulted in the development of improved treatment strategies. These newer approaches are intended to identify individualized patient benefits of treatment, minimize the risk of drug toxicity and reduce the cost of treatment. In deciding on which type of biomarker to use across the wide range of diseases, the easy accessibility of proteins, peptides or metabolites in body fluids make them candidates for measuring outcomes and studying changes in disease state. Patterns of markers are tending to be used more than a single marker. With increasing knowledge of the molecular pathways underlying the development of various diseases, selection of patients and treatments will be based on molecular profiling or phenotypic expression of target molecules in malignant disease. Drugs will be designed to act on target molecules and shut down a specific molecular pathway or sets of pathways in malignant tissues.[21]

Many preclinical studies on drug development have utilized knockout mice or rats. A knockout mouse or rat is a genetically engineered animal in which an existing gene has been inactivated by replacing it or disrupting it with an artificial piece of DNA. The inactivation of the gene often causes changes in an animal's phenotype, which includes appearance, behaviour and other physical and biochemical characteristics. Humans share many genes with mice and rats, so that by observing the characteristics of knockout mice and rats information can be obtained that enables a better understanding of how a similar gene may cause or contribute to disease in humans. Knockout mice and rats have been used in studying many disease states including obesity, cardiovascular disease, and diabetes. Rats are physiologically more similar to humans than mice, and it is considered that the rat is a better model than the mouse for studying cardiovascular disease and diabetes.[27]

Transgenic animal models have also been used in drug discovery and testing. Transgenic animal models are developed by the insertion of a particular human DNA into fertilized oocytes which are then allowed to develop to term by implantation into the oviducts of pseudopregnant females.[28] Transgenic animal models have been used for studying various diseases including cardiovascular disease and diabetes.[29] Transgenic animals can be used widely in drug development *e.g.* in validation of potential drug targets and in understanding of specificity, efficacy and safety of drugs.

1.6 Conclusion

It is clear that there is an ever increasing global burden of obesity, hypertension and cardiovascular disease, and diabetes with serious consequences and implications for the health status of individuals and populations. Urgent action needs to be taken by governments now to control and lessen the impact of this pandemic. Early detection of disease in children and adults by identifying and measuring reliable biomarkers of disease, as well as identifying risk factors of disease in individuals together with genetic predisposition to disease in families and communities, are essential steps to lowering the prevalence of disease. Modifications to risk factors that are influenced by lifestyle, nutrition, and environment will be imperative. Research directed towards gaining knowledge and understanding of molecular pathways involved in the disease process is necessary to designing new drugs and treatment strategies that have greater efficacy, patient compliance and cost-benefit.

The three main sections of this book will present recent findings on biomarkers, risk factors and gene–environment interactions for each of these diseases, and consider the ways in which they are interlinked. The design of new drugs and therapies for each disease and findings from preclinical and clinical trials will be highlighted.

References

1. C. L. Ogden, M. D. Carroll, B. K. Kit and K. M. Flegal, *JAMA, J. Am. Med. Assoc.*, 2012, **307**, 483.
2. National Center for Health Statistics, Health, United States, 2011: With Special Features on Socioeconomic Status and Health, Hyattsville, MD; U.S. Department of Health and Human Services; 2012.

3. S. R. Daniels, D. K. Arnett, R. H. Eckel, S. S. Gidding, L. L. Hayman, S. Kumanyika, T. N. Robinson, B. J. Scott, S. St Jeor and C. L. Williams, *Circulation*, 2005, **111**, 1999.
4. D. S. Freedman, Z. Mei, S. R. Srinivasan, G. S. Berenson and W. H. Dietz, *J. Pediatr.*, 2007, **150**, 12.
5. C. Li, E. S. Ford, G. Zhao and A. H. Mokdad, *Diabetes Care*, 2009, **32**, 342.
6. V. L. Burt, P. Whelton, E. J. Roccella, C. Brown, J. A. Cutler, M. Higgins, M. J. Horan and D. Labarthe, *Hypertension*, 1995, **25**, 305.
7. B. M. Egan, Y. Zhao and R. N. Axon, *JAMA, J. Am. Med. Assoc.*, 2010, **303**, 2043.
8. L. E. Fields, V. L. Burt, J. A. Cutler, J. Hughes, E. J. Roccella and P. Sorlie, *Hypertension*, 2004, **44**, 398.
9. I. Haijar and T. A. Kotchen, *JAMA, J. Am. Med. Assoc.*, 2003, **290**, 199.
10. M. Joffres, E. Falaschetti, C. Gillespie, C. Robitaille, F. Loustalot, N. Poulter, F. A. McAlister, H. Johansen, O. Baclic and N. Campbell, *BMJ Open*, 2013, **3**, e003423.
11. P. M. Kearney, M. Whelton, K. Reynolds, P. Muntner, P. K. Whelton and J. He, *Lancet*, 2005, **365**, 217.
12. M. Pereira, N. Lunet, A. Azevedo and H. Barros, *J. Hypertens.*, 2009, **27**, 963.
13. C. K. Chow, K. K. Teo, S. Rangarajan, S. Islam, R. Gupta, A. Avezum, A. Bahonar *et al.*, *JAMA, J. Am. Med. Assoc.*, 2013, **310**, 959.
14. A. V. Chobanian, G. L. Bakris, H. R. Black, W. C. Cushman, L. A. Green, J. L. Izzo Jr, D. W. Jones, B. L. Materson, S. Oparil, J. T. Wright Jr and E. J. Rocella, *JAMA, J. Am. Med. Assoc.*, 2003, **289**, 2560.
15. Centers for Disease Control and Prevention. National diabetes fact sheet: national estimates and general information on diabetes and prediabetes in the United States, 2011. Available at http://www.cdc.gov/diabetes/pubs/factsheet11.htm.
16. L. S. Geiss and C. Cowie, Type 2 diabetes and persons at high risk of diabetes, in *Diabetes public health: from data to policy*, ed. K. M. V. Narayan, D. Williams, E. W. Gregg and C. Cowie, Oxford University Press, New York, NY, 2011, 15–32.
17. E. W. Gregg, Y. J. Cheng, S. Daydah, C. Cowie, S. Garfield, L. Geiss and L. Barker, *Diabetes Care*, 2012, **35**, 1252.
18. Y. Li, N. R. Burrows, E. W. Gregg, A. Albright and L. S. Geiss, *Diabetes Care*, 2012, **35**, 273.
19. A. H. Mokdad, E. S. Ford, B. A. Bowman, D. E. Nelson, M. M. Engelgau, F. Vinicor and J. S. Marks, *Diabetes Care*, 2000, **23**, 1278.
20. A. J. Atkinson, W. A. Colburn, V. G. DeGruttola and D. L. Demets, *Clin. Pharmacol. Ther.*, 2001, **69**, 89.
21. M. Kumar and S. K. Sarin, *Curr. Trends Sci.*, 2009, 403.
22. P. W. Wilson, R. B. D'Agostino, D. Levy, A. M. Belanger, H. Silbershatz and W. B. Kannel, *Circulation*, 1998, **97**, 1837.
23. S. Begg, T. Vos, B. Barker, C. Stevenson, L. Stanley and A. Lopez, AIHW Cat. No, PHE 82, 2007.

24. AIHW 2012, available at www.aihw.gov.au/publicationdetail/? id=10737421466&tab=2.
25. D. J. Hunter, *Nat. Rev. Genet.*, 2005, **6**, 287.
26. J. D. Adams and K. K. Parker, *Extracellular and Intracellular Signaling*, Royal Society of Chemistry, London, 2011.
27. A. Abbott, *Nature*, 2004, **428**, 464.
28. A. D. Kandhare, K. S. Raygude, P. Ghosh, T. P. Gosavi and S. L. Bodhankar, *Int. J. Pharm. Biol. Arch.*, 2011, **2**, 1024.
29. T. R. Bagle, R. R. Kunkulol, M. S. Baig and S. Y. More, *Int. J. Med. Res. Health Sci.*, 2012, **2**, 107.

Part 1
Obesity

CHAPTER 2

Molecular Links between Obesity, Hypertension and Metabolic Dysfunction

GEOFFREY A. HEAD,* KYUNGJOON LIM, BENJAMIN BARZEL, SANDRA L. BURKE AND PAMELA J. DAVERN

Neuropharmacology Laboratory, Baker IDI Heart and Diabetes Institute, 75 Commercial Road, Melbourne 3004, Australia
*Email: geoff.head@bakeridi.edu.au

2.1 Introduction

The alarming increase in the prevalence of obesity and the ensuing dyslipidemia, hypertension, insulin resistance, type 2 diabetes and cardiovascular disease occurs both in developed and developing countries. The enormity of the health issues of being overweight or obese is therefore a major worldwide burden of disease that promises to outstrip the capacity of health systems to cope.[1] This is due to the increased incidence of metabolic disease, cardiovascular disease and premature death (Figure 2.1). It is expected that in countries like the USA and Australia the prevalence of being overweight or obese will increase to 75% for females and 83% for males by 2025.[2] At this time and even now, the normal condition of the population will be overweight or obese. The consequences of obesity are a range of metabolic changes that have been incorporated to define a "metabolic syndrome".[3] These include dyslipidemia and insulin resistance, which is the primary feature of type 2 diabetes. The progression from obesity to

RSC Drug Discovery Series No. 45
Cardiovascular and Metabolic Disease: Scientific Discoveries and New Therapies
Edited by Philip Peplow, James Adams and Tim Young

Published by the Royal Society of Chemistry, www.rsc.org

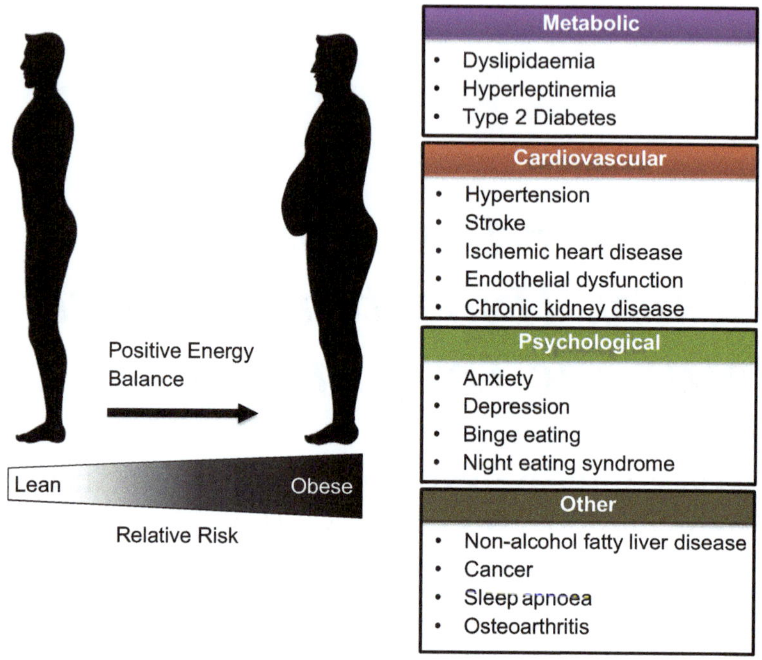

Figure 2.1 Illustration of the deleterious health conditions associated with a long term positive energy balance resulting in the progression from being lean to obese. The consequences include an increased relative risk of metabolic, cardiovascular and psychological disease.

metabolic dysfunction to hypertension and diabetes ultimately occurs unless there are significant interventions. It is thus very important to understand the short term and long term impact on health and wellbeing of being overweight or obese.

There is also a considerable reduction in the quality of life related to reduced mobility, public stigma, negative psychological changes including depression and higher levels of unemployment. While effective weight loss therapies including dieting, pharmacological appetite suppressing agents and even surgery are currently only partly successful and have some issues, this should not diminish our resolve to find solutions. Clearly evidence-based allocation of resources for prevention, reduction and mitigation should be a major priority. However, it should be understood that the long term impact of obesity-related metabolic disease is cardiovascular disease associated with greater rates of atherosclerosis, hypertension, heart disease, vascular disease and cardiovascular-related mortality. The analysis of the Framingham cohort suggests that over two thirds of newly diagnosed cases of hypertension are related to being overweight or obese.[4] While there are many factors contributing to the increase in blood pressure in obese and overweight patients, increasingly there is recognition of the important contribution of the sympathetic nervous system (SNS).[5–10] The major regulatory site relevant to obesity lies in the hypothalamus which is known for its role in regulating fluid balance, blood volume and sodium homeostasis but is also an important center for the integration of neural and hormonal signaling. In obesity, the most relevant of the peripheral signaling molecules are the adipokines leptin and adiponectin, the hormone insulin and gut hormones such as ghrelin. These hormones signal metabolic information to the central nervous system (CNS) *via* designated receptors within the arcuate nucleus of the hypothalamus. We know a great deal about the afferent and efferent neural connections of the hypothalamus, as well as the nature of the neurons within its subnuclei that influence hormonal signaling molecules. The hypothalamus is also an important region for the integration of the neural response to acute emotional stress. There may well be overlap between the hypothalamic pathways processing the autonomic responses to stress with those responding to metabolic signals. Thus, to unravel the molecular links between obesity, hypertension and the metabolic cause of CNS dysfunction in obesity is clearly a major health priority. The purpose of this review is to document the progress in our understanding of the processes involved in the development of obesity-related hypertension, metabolic and cardiovascular disease that has been made in recent years and perhaps to indicate the path for future research.

2.2 Consequences of Obesity on Cardiovascular and Metabolic Disease

There is now convincing evidence for a causative association between obesity and metabolic disease, including dyslipidemia, insulin resistance and the development of type 2 diabetes, although the mechanisms are not fully understood. The dyslipidemia accompanying obesity includes higher levels of low density lipoprotein (LDL) particles,[11] reduced levels of high density lipoproteins (HDL) as well as elevated plasma triglycerides.[12] All of these trends are atherogenic.[12] There is a two-fold increase in the risk of having high triglycerides, reduced HDL and other aspects of metabolic syndrome if you are obese and a four-fold increase in the risk of developing type 2 diabetes.[13] The AusDiab population study shows us that the incidence of impaired fasting glucose or glucose tolerance in Australia in 2005 was 14%.[14] Future adult levels of diabetes (from 2010 to 2030) will increase by nearly 70% in developing countries from a somewhat lower base and by 20% in developed countries.[15] The complications associated with obesity and in particular if type 2 diabetes develops, include coronary artery disease, peripheral vascular disease and endothelial dysfunction. While the metabolic syndrome comprises a number of conditions, obesity is independently associated with endothelial dysfunction.[16] Mechanisms may involve elevation in specific adipokines, reduction in adiponectin as well as cytokine-induced oxidative stress and low level inflammation. Vascular indicators of inflammation like fibrinogen, plasminogen activation inhibitor and C-reactive protein are also positively correlated with obesity.[17] Dyslipidemia is also associated with pulmonary embolism, deep venous thrombosis and atherosclerotic arterial disease.[17] Of particular concern is the much higher rate of type 2 diabetes, dyslipidemia and metabolic syndrome in indigenous Australians where a recent study shows that almost one third of those aged over 35 and over half of those over 55 years have type 2 diabetes.[18]

A major cardiovascular consequence of obesity is the risk of myocardial infarction, particularly in men. While a third of cases of elevated triglyceride and reduced HDL are due to obesity in women, this is associated with only 13% elevated risk of myocardial infarction. However, for men less than 20% of cases of dyslipidemia are due to obesity but the risk of myocardial infarction is over 30%.[13] Perhaps the most important consequence of being overweight is hypertension with over 60% of newly diagnosed cases of hypertension being related to obesity.[4] While most patients are unaware of their hypertension, the first indication can often be a stroke, myocardial infarction or renal failure. The Renfrew–Paisley study which followed over 15 000 subjects for two decades found subjects with a BMI greater than 30 had a 1.5–2.5 fold greater risk of coronary heart disease, heart failure, stroke, venous thromboembolism and atrial fibrillation compared with normal subjects with a BMI less than 25.[19] Both systolic and diastolic dysfunction can occur due to the concentric cardiac hypertrophy from elevated blood pressure which is compounded by advanced coronary artery disease.[17] The greater body mass due to both excess fat and also greater lean body mass results in increased oxygen demand, greater blood volume, cardiac output, stroke volume and eccentric cardiac hypertrophy.[20] Prolonged obesity and hypertension may also cause glomerular hypertension, hyper-filtration and loss of renal function as well as increased salt sensitivity.[21]

2.3 Role of the SNS in Obesity-related Hypertension

The mechanism of hypertension has been extensively studied particularly in animal models where central actions of leptin in the hypothalamus result in increased sympathetic activity to peripheral blood vessels, predominantly to the kidney, causing elevated blood pressure.[22,23] Initial breakthrough discoveries from Landsberg and colleagues recognized the robust effect of eating on the activity of the SNS.[24] Bray was the first to propose a model that implied that the effects of obesity included a reduction in activity of the SNS but this study focused on determining the metabolic effects of the SNS on the liver, brown adipose tissue, pancreas and adrenal glands.[25] Since then, it has become apparent that there are variations in SNS activity dependent upon inputs to differing regional beds that result in changes in activity that are not necessarily uniform.[26] To this end, studies measuring norepinephrine spillover in obese human subjects have identified elevations in renal and muscular levels with a reduction in cardiac levels.[27] There are technical constraints to direct recordings of sympathetic activity in humans, limiting measurements to the muscle bed. Notwithstanding this, activity has been reliably shown to be elevated in both obese normotensive and hypertensive subjects and more so in obese hypertensive subjects.[28–32] Lambert and colleagues, using analysis of single unit recordings from peroneal nerves, revealed that greater sympathetic activity in obese patients was due to extra recruited units firing at normal rates.[29] This was in contrast to lean hypertensive patients where the hypertension was characterized by an increased probability of firing and increased incidence of multiple firing from active units as opposed to the recruitment of new active units.[29] Even though an association between sympatho-excitation and long term obesity and the accumulation of visceral fat rather than subcutaneous fat has been established,[30,33,34] it remains to be determined whether this is a causal relationship in human obesity. Dietary restriction has been shown to induce a reduction in total noradrenaline spillover particularly in individuals with elevated insulin levels.[35] Several months after laparoscopic adjustable gastric band surgery in severely obese patients, a 10% weight loss was observed along with a reduction in both blood pressure and muscle sympathetic activity.[36] Moreover, substantial support arises from studies using ganglionic blockade which demonstrates a greater reduction in blood pressure in obese groups compared with lean controls that essentially abolishes the obesity-related hypertension in humans,[37] rats,[38] rabbits,[23] and mice[39,40] (Figure 2.2).

In addition to the metabolic features of obesity, there are also physical impacts such as intermittent hypoxia during sleep when increased fat and soft tissue in the throat obstructs the airway when the patient is recumbent. This is known as obstructive sleep apnea (OSA)[17] and leads to an elevation of blood pressure in nearly half of those with OSA. Similarly 50% of those

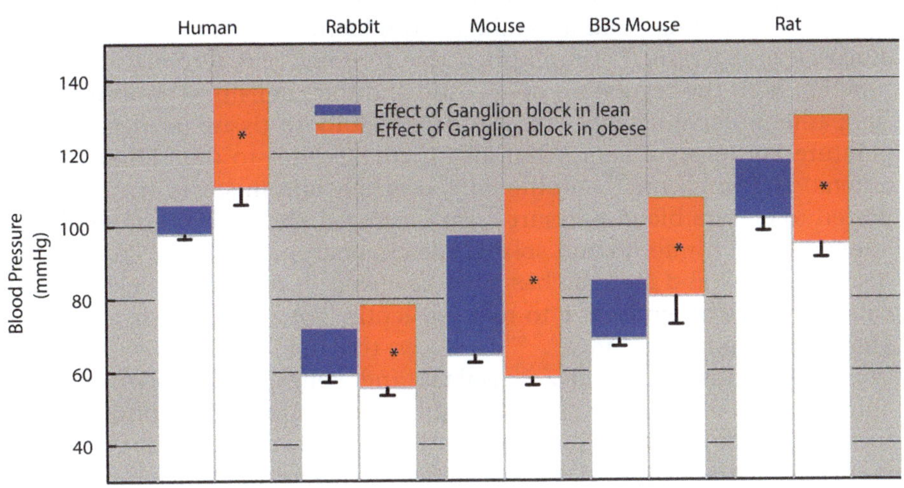

Figure 2.2 Hypotensive responses to ganglionic blockade in obese (red) *versus* lean (blue) controls from humans,[37] rat,[38] rabbit[23] and mouse[39,40] models set according to the basal levels of blood pressure. * Indicates $P<0.05$ between the effect of ganglionic blockade in lean compared to obese.

individuals with morbid obesity have OSA.[41] Airway obstruction causes activation of chemoreceptor afferents and chemoreflex-induced activation of the sympathetic activity.[42] Animal studies have consistently shown that chronic intermittent hypoxia that alters central respiratory rhythms, leads to hypertension.[43] For obese patients, the consequences of OSA would be expected to increase the risk of stroke and pulmonary hypertension and include daytime sleepiness and increased risk of traffic accidents.[17] Another common physical consequence of obesity is breathlessness upon exertion due to the presence of chest and abdominal fat deposits that can limit the movement of the diaphragm.[17] Fortunately, the treatment of using continuous positive airway pressure during sleep in patients with severe obstructive sleep apnea is very effective at reducing blood pressure as well as the frequency of pre-hypertension and hypertension.[44] The current view is that the treatment reduces chemoreceptor afferent activation of central sympathetic vasomotor drive and hence reduces blood pressure.[45,46] Thus, in humans there are two quite separate mechanisms elevating blood pressure in overweight obese patients; the physical effects on breathing as well as the hormonal effects of excess fat accumulation. Both of these involve activation of the SNS.

2.4 Contribution of Leptin and Insulin to Obesity-related Hypertension

Leptin is a product of the *ob* gene and is primarily synthesized and secreted from white adipose tissue (WAT)[47] with peptide plasma concentrations being in proportion to the level of adiposity.[5] While several physiological functions including wound healing, reproduction, immune response and glucose homeostasis[48,49] implicate a role for leptin, its recognized primary function is in the regulation of appetite and metabolism.[7] Leptin increases sympathetic nerve activity (SNA) to brown adipose tissue to increase temperature and expend energy but also increases SNA to the kidney (RSNA), adrenal and hindlimb.[50] Additionally, centrally administered leptin was reported to elevate blood pressure[51] *via* increased SNS activity to the kidney, adrenal, liver, hindlimb and splanchnic circulation.[52–59] Reduced satiety in obese humans and animals has been observed in response to elevated circulating levels of leptin due to reduced leptin signalling that is commonly known as "leptin resistance".[7,60] However, this does not occur with leptin's sympatho-excitatory actions.[7,60] Halaas and colleagues noted that obese animals were resistant to appetite suppression induced by leptin and postulated that the cause was impaired leptin transport into the brain,[61] but this is not the main reason as the resistance occurs to centrally administered leptin. Rahmouni and colleagues reported that obese agouti mice are resistant to the appetite suppressant actions of leptin mediated by neural pathways but have a normal leptin-induced increase in RSNA.[62] Thus, the concept of "selective central leptin resistance", proposed by Mark and colleagues,[63] delineates the loss of specific central actions of leptin in obesity and has been proposed as the fundamental cause of obesity-induced hypertension.[7,60] While there is general agreement for a key role of leptin in diet-induced obesity in experimental animals, Tumer and colleagues offer a contrary view. Central administration of a leptin antagonist administered for a two week period to rats fed a high fat diet, was shown to have no effect on blood pressure.[64] It is important to note that in human obesity-related hypertension the evidence is less convincing given that it is primarily associative.[7] Studies by Alvarez and colleagues are less supportive of leptin causing SNS activation in humans given that SNA correlated with visceral fat rather than subcutaneous fat and not to leptin levels.[33] However, the study was limited to lean and obese young males with excess subcutaneous fat who had low plasma leptin levels roughly in the range that other studies have observed for young lean subjects.[65] This does not therefore preclude an association in subjects with high plasma leptin levels. Mutations in the human leptin gene, albeit rare, constitute the most striking evidence in support of leptin's involvement in obesity. Predisposition to morbid obesity among members of the same family was found to be associated with a homozygous mutation in the Ob-R gene locus.[66] However, this study lacked data on the blood pressure of family members making it difficult to establish an association between this Ob-R gene mutation and hypertension. In a more recent publication, Rosmond and colleagues identified several human leptin receptor gene mutations which appear to be cardio-protective. Men homozygous for the Arg109 and Arg223 genotypes had low systolic and diastolic blood pressure despite being overweight and hyperleptinemic.[67] These findings seem to imply a normal functioning leptin receptor is required to establish obesity-related hypertension, but not obesity alone.

Several investigations have presented evidence that indicate that rabbits exhibit a typical phenotype of human obesity when fed either a high fat or high caloric diet. *Ad libitum* access to a high fat diet, induces obesity and these animals develop mild hypertension, hyperinsulinemia, tachycardia and cardiac hypertrophy compared with rabbits fed a standard diet.[68] The distinct advantage of this model is the ability to record blood pressure and RSNA over extended periods in a conscious animal with the benefit of calibrating total nerve activity to allow for comparisons between distinct groups.[69] Studies from our laboratory have established that rabbits fed a high fat diet (13% fat) show an increase in arterial blood pressure, heart rate and RSNA along with greater levels of plasma noradrenaline, leptin, insulin, visceral fat and total fat compared with rabbits fed a normal diet (4% fat).[22,23,70,71] Moreover, we ascertained a marked facilitation of the sympatho-excitatory effects of central infusion of leptin in these rabbits.[22] While rodent and dog models of obesity have been commonly used rather than the rabbit fat fed model of obesity, the rabbit model has a remarkable resemblance to human obesity-induced hypertension primarily in terms of a modest degree of hypertension (9–12%), and increased heart rate, plasma noradrenaline, renal sympathetic activity, hyperleptinaemia, hyperinsulinaemia and activation of the renin angiotensin system. Moreover, it is striking that the 13% level of fat in the diet given to these rabbits is much less than the 30% level of fat in the Western diet in the USA.[72] Importantly recordings using radiotelemetry to measure blood pressure and RSNA in conscious rabbits in their home cage enabled us to establish that elevated blood pressure, heart rate and RSNA could be detected on the first day of the consumption of a high fat diet.[23] Thus the consumption of one meal containing an extra 10 g of fat had measurable effects on the rabbit's cardiovascular system over the next 24 hours. Continuing the diet over 3 weeks further increased blood pressure and sympathetic activity as well as elevating blood glucose, insulin and leptin levels.[23] At the end of this 3 week period resuming a normal diet returned glucose, insulin, leptin and heart rate to baseline levels but blood pressure and RSNA remained elevated. Importantly these findings indicate that there is no clear association with increased levels of either circulating leptin or insulin and hypertension if the diets are quickly changed back and forth from high to normal fat. Our earlier studies revealed correlations of greater than 0.8 between leptin and visceral fat, blood pressure and RSNA when the diets are stable.[22] Therefore, we propose that the inconsistency may be the consequence of a delay in the longer term effects of leptin in modulating neurons located in central sympatho-excitatory pathways.

Hyperinsulinemia is inextricably linked to obesity due to insulin resistance and may also be associated with elevated blood pressure and amplified sympathetic activity.[73] Several experimental studies have examined the sympatho-excitatory actions of centrally administered insulin.[74–77] Muntzel and colleagues made intracerebroventricular (ICV) injections of insulin and reported activation of lumbar SNA with no effect on either renal or adrenal SNA in anaesthetized rats.[75] Morgan and colleagues found that central infusions of insulin in normal mice increased SNA to renal, hindlimb and brown adipose tissue.[78] They also observed that insulin similarly elevated lumbar SNA of agouti obese mice but lesser renal and brown adipose tissue SNA responses were observed in this strain.[78] Caution is necessary when comparing SNA between groups in studies undertaken in rats and mice as all measurements are standardized to baseline as 100%. Thus, variation in basal SNS activity would lead to "apparent" diminished or exaggerated responses. It should be noted that of the studies described above, none examined the role of either leptin or insulin in influencing obesity-induced hypertension.

The conditions in human obesity-related hypertension remain less clear than in experimental animal studies. The Tecumseh offspring study reported overactivity of the SNS in young individuals with a strong correlation between levels of insulin and left ventricular hypertrophy,[79] likely a manifestation of long term levels of blood pressure. Recently, Straznicky and colleagues have demonstrated that weight loss *via* dietary restriction diminished indices of norepinephrine turnover and that this outcome was greatest in those individuals with hyperinsulinaemia.[35] Additionally, a similar loss of weight attained *via* lap band surgery also reduced SNS activity, insulin, and ambulatory systolic blood pressure with no effect on diastolic blood pressure.[36] While overall there are a number of studies consistent with a likely contribution of hyperinsulinaemia to obesity-related hypertension in humans, the evidence currently remains somewhat inconclusive.

The essential question is whether raised levels of leptin as well as insulin, in addition to sympatho-excitatory effects, also contribute to obesity-induced hypertension. To that end, we recently examined the effects of leptin and insulin antagonists with doses identified as sufficient to block the effects of exogenous leptin and insulin, respectively.[70] We used rabbits that were fed a high fat diet for a 3 week period. These rabbits had previously been implanted with ICV catheters to assess the influence of these antagonists at week 1 and 3 of the high fat diet, given that there is no discernable difference in plasma levels of insulin or leptin at these times.[23] The central leptin antagonist had no effect following 1 week of the diet but at 3 weeks the antagonist normalized blood pressure to levels comparable with rabbits on a normal diet and markedly attenuated RSNA[70] (Figure 2.3). Administration of the insulin antagonist induced a small hypotensive response at both times (Figure 2.3) but had no effect on RSNA.[70] The insulin antagonist accounted for most of the hypertension at week 1 but the further development of hypertension relied on a central leptin pressor effect (Figure 2.3).

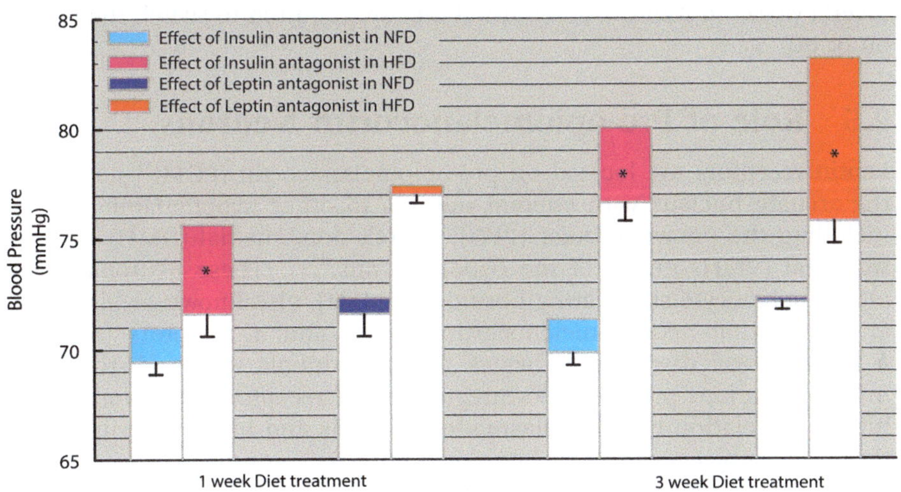

Figure 2.3 Hypotensive responses to administration of insulin antagonist given to normal fed rabbits (light blue) or high fat diet fed rabbits (pink), as well as the leptin antagonist given to normal fed rabbits (dark blue) and high fat diet fed rabbits (red) after 1 and 3 weeks of diet.[70] * Indicates $P < 0.05$ between the effects of the antagonist in obese compare to lean rabbits. Note that the effect of insulin antagonist is greater in fat fed rabbits at both weeks but that the leptin antagonist is only effective at lowering blood pressure after 3 weeks of a high fat diet.

This result is consistent with those findings observed by Muntzel and colleagues where central doses of insulin increased lumbar SNA but had no effect on RSNA.[75] However, our findings are inconsistent with those of Tumer and colleagues described above, where chronic administration of a leptin antagonist to rats on a high fat diet effect had no effect on blood pressure.[64] While it was argued by the authors that the dose in rats was appropriate as it blocked the effect of a virally-induced elevation of leptin, the chronic infusion of the leptin antagonist in the rat also induced a marked increase in food intake and body weight, which would make the interpretation very difficult. Furthermore, calculations based on species differences in brain volume would suggest the rats received a 10 fold lower dose than our studies in rabbits. Thus, we concur with the current view that obesity-related hypertension is due to selective leptin resistance where the sympatho-excitatory actions of leptin are maintained while the anorectic actions develop resistance.[7,80] We have found that throughout sustained fat feeding, rabbits elicit a markedly amplified sympatho-excitatory response that is unchanged following a return to a normal diet for a week[23] or several weeks.[81] By contrast, there is little sympatho-excitatory action of centrally administered leptin in rabbits fed a normal diet. We suggest then that the sympatho-excitation is not simply maintenance of leptin's actions but a marked facilitation of its effect. Our concept of central sympatho-excitation process is that there is a delay in the onset and also in its offset. Thus, several weeks of exposure to a high fat diet or hyperleptinaemia or several weeks to months of a normal diet are required to switch the process on or off.

2.5 Role of Pro-opiomelanocortin Neurons

Leptin receptors are located on pro-opiomelanocortin (POMC) neurons in the arcuate nucleus with efferent outputs to other hypothalamic regions, including the paraventricular (PVN), lateral, dorsomedial (DMH) and ventrolateral (VMH) nuclei of the hypothalamus.[82–84] These neurons release alpha melanocyte stimulating hormone (αMSH), also known as alpha melanotropin that actives melanocortin type 3 and 4 receptors (MC3/4R). A recent novel discovery indicates that individuals with a distinct MC3 missense variant present a downregulation of receptor expression and exhibit a correlation between plasma leptin levels and blood pressure.[85] Furthermore, subjects with a dysfunctional MC4 receptor due to a mutation in the encoding gene have been reported to have lower blood pressure than matched normal subjects.[86] MC4 receptor deficient mice are obese, and demonstrate both hyperinsulinemia and hyperleptinemia but have normal blood pressure.[87] A number of studies across varying species have demonstrated a reduction in blood pressure and sympathetic activity induced by central injections of MC3 and MC4 antagonists. A chronic central infusion of a MC3/4 antagonist attenuated blood pressure in obese rats.[88] Blockade of the MC4 receptor following administration of the antagonist JKC-363 ICV to conscious rabbits was shown to abolish the renal sympatho-excitatory actions of central αMSH.[89] More recently, Li and colleagues reported that both MC3 and MC4 receptor antagonists modulated an increase in sympathetic outflow when applied to anesthetized rats directly into the PVN of the hypothalamus.[90] Moreover, antagonists of both receptors were required to oppose the effects of a MC3/4 agonist.[90] Do Carmo and colleagues demonstrated that chronic central infusion of SHU-9119, which blocks the effects of both MC3 and MC4 receptors, attenuated blood pressure in the Zucker rat model of obesity.[91] These findings support the view that the mechanism responsible for obesity-related hypertension involves signaling in the hypothalamus downstream from the leptin receptor and possibly at second order neurons activated by arcuate POMC neurons. This is supported by our recent studies where we found markedly exaggerated sympatho-excitatory responses to centrally infused αMSH in rabbits fed a high fat diet, which was closely similar to the changes observed with leptin.[92] Thus, the target hypothalamic regions, such as the DMH, VMH and PVN of the hypothalamus, are likely sites of action, as suggested by Enriori and colleagues.[93]

2.6 Contribution of Neuropeptide Y

Neuropeptide Y (NPY) is expressed in arcuate neurons,[94] although high density NPY-containing neurons have been discovered in both the PVN and DMH.[95] NPY is well known as a potent stimulant of appetite[96] and this action is normally suppressed by leptin.[97] Feeding decreases NPY expression in the hypothalamus whilst chronic infusion of NPY into the PVN produces hyperphagia and obesity.[98,99] Co-expressed in NPY containing neurons[100] is the neurotransmitter Agouti Related Peptide (AgRP) that is an endogenous antagonist of the MC3/4 receptor.[101,102] The orexigenic function of AgRP is achieved by inhibiting the melanocortin system. Thus, both melanocortin and NPY/AgRP provide inhibitory regulation of the other. Hypothalamic NPY is known to control VLDL-secretion from the liver *via* the SNS[103] and to reduce sympathetic activity to brown adipose tissue to reduce energy expenditure.[104] Whether NPY regulates autonomic vasomotor function is largely unknown. The intertwined and complex regulation of metabolism, appetite, energy expenditure and autonomic function at the level of hypothalamic neurons and connections has been a considerable barrier to unravelling these processes in obesity. A simple illustration relevant to NPY is a study where NPY was infused ICV for 2 weeks in rats as a method to match the weight gain effects of NPY to evaluate the effect on blood pressure.[105] However, the lack of effects on blood pressure was interpreted as a possible mixture of sympatho-inhibitory action of NPY offsetting the sympatho-excitatory actions of the weight gain. As food intake (appetite) is less affected by cardiovascular effects, the converse experiments are more easily interpreted and thus we know a great deal about the orexigenic actions and mechanisms, but the contribution of these neurons to obesity-related hypertension remains to be elucidated.

2.7 Contribution of Ghrelin

Ghrelin is a 28 amino acid peptide predominantly secreted in the stomach[106] to signal information about nutrients to the brain in order to facilitate meal initiation and meal frequency.[107] Thus, plasma ghrelin concentrations increase before meals and during fasting and decrease after ingestion of food.[107] Ghrelin is the endogenous ligand for growth hormone secretagogue receptor (GHSR) which is highly expressed in the arcuate nucleus.[108] Ghrelin effectively opposes the stimulatory actions of leptin by activating NPY/AgRP neurons and inhibiting POMC neurons.[96] Not surprisingly then, systemically administered ghrelin increases food consumption in humans.[109] Not only are the circulating ghrelin levels generally low among patients with metabolic syndrome and obesity,[110,111] the natural pre-post prandial oscillations in circulating ghrelin concentrations are lost.[112,113] Interestingly, exogenous ghrelin increases hunger and food intake in morbidly obese and lean patients alike suggesting central sensitivity does not change.[113] However, the lack of ghrelin oscillation may contribute to a "grazing" feeding pattern that is observed in obese subjects.[113] Thus, low ghrelin levels in obesity which might appear to be advantageous may in fact be maladaptive.

There have been some limited studies investigating the cardiovascular effects of ghrelin. Matsumura found that ICV ghrelin decreased sympathetic activity in conscious rabbits.[114] However, our studies have shown that short term ghrelin infusion systemically to humans[115] and ICV to rabbits[81] reduces blood pressure and increases sympathetic activity. The latter may be a baroreflex response to the hypotension. This may not have been seen by Matsumura and colleagues as the rabbits were assessed only a few days after surgery and had very high levels of blood pressure and sympathetic activity. Like NPY, the contribution of ghrelin to obesity-induced hypertension is not as yet clear.

2.8 Effects of Obesity Across Generations

Concurrent with the increasing prevalence of obesity across the population is the increasing number of obese females of reproductive age.[116] Obesity is typically associated with insulin resistance and in females this condition leads to increased risk for hypertension, accumulation of central fat deposits and adverse consequences for pregnancy.[117] While infant mortality is a known greater risk should the mother be obese, importantly there is also the long term health impact on cardiovascular outcomes in offspring.[118] Velkoska and colleagues demonstrated that over-nutrition in rats during the suckling period led to a greater body weight in adulthood and hyperleptinaemia compared with normal nutrition.[119] We identified that offspring of rabbits that were fed a high fat diet during pregnancy and lactation had elevated blood pressure, heart rate and RSNA compared with offspring from mothers on a normal diet.[81] Also, in these offspring we noted greater sensitivity to central infusion of leptin and ghrelin, which we postulate is the underlying mechanism contributing to the adverse cardiovascular profile in adulthood.[81] While a modest elevation in visceral fat was observed in the offspring,[81] the elevation of blood pressure, heart rate and RSNA were comparable to adult rabbits that were fed a high fat diet.[22,70] It is difficult to discern the changes associated with developmental programming[120] or those arising from the earlier exposure to high fat fed to their mothers. The potential impact of these findings is enormous should they be translatable to the human population where obese or overweight mothers make up more than 40% of pregnancies.[121]

2.9 The Influence of Aversive Stress on Obesity-related Hypertension

Paul Korner in his epic book on essential hypertension and its causes explored the close link between chronic stress, hypertension and obesity, suggesting that some persons with essential hypertension engage in food consumption as a stress relieving response.[122] Epel and colleagues performed a cross sectional study and identified an association between stress and obesity given that stress-induced cortisol secretion was shown to contribute to central fat accumulation.[123] A similar direct link between stress and obesity was proposed by Bjorntorp, suggesting a commonality of central autonomic systems in that repeated activation of the hypothalamic–pituitary–adrenal axis and the SNS may lead to abdominal obesity and comorbidities.[124] Specifically, it was hypothesized that exposure to chronic stress induces elevated levels of cortisol along with low growth hormone and sex steroids thereby promoting the accumulation of visceral fat. In patients with metabolic disorder, Lambert and colleagues demonstrated a correlation between a pattern of single-unit sympathetic nerve-firing and anxiety levels and depressive states.[125] Furthermore, the pattern of sympathetic activation in hypertensive patients is described as not unlike the defense reaction.[126] Pasquali and colleagues observed that women with increased visceral adiposity evoked greater pressor responses, tachycardia and ACTH release during a stress test compared with controls.[127] Kunyioshi and colleagues reported comparable muscle SNA following mental stress tests in both obese and lean subjects.[128] The findings described above contrast with the conclusion that heart rate reactivity to stress was reduced in individuals with amplified BMI and waist to hip ratio.[129] Furthermore, amplified heart rate reactivity to stress was shown to be accompanied by a diminished likelihood of developing obesity in the following five years.[129]

Notwithstanding the complexity in discerning the interconnection underlying stress and obesity, the neural pathways that play a role in regulating responses to stress are also important in modulating energy homeostasis.[122] Major integrative sites such as the PVN, DMH and VMH are critical for hormonal and autonomic responses to stress,[130] appetite signalling[82] and the sympatho-excitatory actions of leptin.[131] Recently, we detected a steady attenuation of the sympatho-excitatory response to airjet stress over several weeks and a variation in the pattern of the RSNA baroreflex in rabbits fed a high fat diet that paralleled acute stress responses.[23] Therefore, we hypothesize that activation of endogenous stress pathways is responsible for the heightened blood pressure and heart rate induced by a high fat diet and any cardiovascular responses to a subsequent acute stress may be constrained. While our rabbit studies do not confirm the Bjorntorp proposal described above, given the lack of dietary choice, they corroborate the general concensus of an association between psychosocial influences, eating and social behaviours and the underlying causes of hypertension.

2.10 Conclusions

There is overwhelming evidence that the underlying cause of hypertension, a key negative aspect of the obesity phenotype, is principally *via* neurogenic mechanisms. Elevated levels of circulating leptin, insulin and likely other adipokines may reprogram neuronal signalling to produce a state of sympatho-excitation. Evidence suggests a major role for the melanocortin system where changes are dependent upon the type of signal. Insulin can induce a rise in blood pressure almost immediately after beginning a high fat diet, whereas a contribution from leptin may take several weeks to manifest. The contribution of chronic aversive stress to the development of obesity and aberrant eating habits has to date received little attention. Thus, there is a need to understand more widely the contribution and interaction of "stress" and "metabolic" hormonal signalling in the CNS that combine to activate the SNS, leading to hypertension in some obese subjects. Additionally, current studies support the notion of trans-generational cardiovascular effects, in particular hypertension, which is alarming considering the prevalence of obesity in pregnancy in some countries. The profound increase in the proportion of the community that is either overweight or obese must be considered as one of most pressing health issues of our time.

References

1. A. Alwan and World Health Organization, *Global Status Report on Noncommunicable Diseases 2010*, 2010.
2. M. M. Haby and A. Marwick, *Department of Human Services, State Government of Victoria*, 2008.
3. A. J. Cameron, P. Z. Zimmet, J. E. Shaw and K. G. Alberti, *Diabetic Med.*, 2009, **26**, 306.
4. R. J. Garrison, W. B. Kannel, J. Stokes, 3rd and W. P. Castelli, *Prev. Med.*, 1987, **16**, 235.
5. N. Eikelis and M. Esler, *Exp. Physiol.*, 2005, **90**, 673.
6. K. Rahmouni, W. G. Haynes and A. L. Mark, *Curr. Hypertens. Rep.*, 2002, **4**, 119.
7. A. L. Mark, *Am. J. Physiol. Regul. Integr. Comp. Physiol.*, 2013, **305**, R566.

8. J. E. Hall, A. A. da Silva, J. M. do Carmo, J. Dubinion, S. Hamza, S. Munusamy, G. Smith and D. E. Stec, *J. Biol. Chem.*, 2010, **285**, 17271.
9. A. A. da Silva, J. do Carmo, J. Dubinion and J. E. Hall, *Curr. Hypertens. Rep.*, 2009, **11**, 206.
10. J. E. Hall, D. H. Zappe, M. AlonsoGalicia, J. P. Granger, M. W. Brands and S. E. Kassab, *News Physiol. Sci.*, 1996, **11**, 255.
11. P. T. Williams and R. M. Krauss, *Arterioscler., Thromb., Vasc. Biol.*, 1997, **17**, 1082.
12. B. V. Howard, G. Ruotolo and D. C. Robbins, *Endocrinol. Metab. Clin. North Am.*, 2003, **32**, 855.
13. A. J. Cameron, D. W. Dunstan, N. Owen, P. Z. Zimmet, E. L. Barr, A. M. Tonkin, D. J. Magliano, S. G. Murray, T. A. Welborn and J. E. Shaw, *Med. J. Aust.*, 2009, **191**, 202.
14. E. D. Williams, R. J. Tapp, D. J. Magliano, J. E. Shaw, P. Z. Zimmet and B. F. Oldenburg, *Diabetologia*, 2010, **53**, 2538.
15. J. E. Shaw, R. A. Sicree and P. Z. Zimmet, *Diabetes Res. Clin. Pract.*, 2010, **87**, 4.
16. A. Avogaro and S. V. de Kreutzenberg, *Clin. Chim. Acta*, 2005, **360**, 9.
17. V. J. Lawrence and P. G. Kopelman, *Clin. Dermatol.*, 2004, **22**, 296.
18. K. O'Dea, J. Cunningham, L. Maple-Brown, T. Weeramanthri, J. Shaw, T. Dunbar and P. Zimmet, *Diabetes Res. Clin. Pract.*, 2008, **80**, 483.
19. N. F. Murphy, K. MacIntyre, S. Stewart, C. L. Hart, D. Hole and J. J. McMurray, *Eur. Heart J.*, 2006, **27**, 96.
20. J. E. Hall, E. D. Crook, D. W. Jones, M. R. Wofford and P. M. Dubbert, *Am. J. Med. Sci.*, 2002, **324**, 127.
21. J. E. Hall, J. J. Kuo, A. A. da Silva, R. B. de Paula, J. K. Liu and L. Tallam, *Curr. Opin. Nephrol. Hypertens.*, 2003, **12**, 195.
22. L. Prior, N. Eikelis, J. Armitage, P. Davern, S. Burke, J.-P. Montani, B. Barzel and G. Head, *Hypertension*, 2010, **55**, 862.
23. J. A. Armitage, S. L. Burke, L. J. Prior, B. Barzel, N. Eikelis, K. Lim and G. A. Head, *Hypertension*, 2012, **60**, 163.
24. L. Landsberg and J. B. Young, *NEJM*, 1978, **298**, 1295.
25. G. A. Bray, *J. Nutr.*, 1991, **121**, 1146.
26. R. M. McAllen, C. N. May and R. R. Campos, *Clin. Exp. Hypertens.*, 1997, **19**, 607.
27. M. Vaz, G. Jennings, A. Turner, H. Cox, G. Lambert and M. Esler, *Circulation*, 1997, **96**, 3423.
28. N. I. Abate, Y. H. Mansour, M. Tuncel, D. Arbique, B. Chavoshan, A. Kizilbash, T. Howell-Stampley, W. Vongpatanasin and R. G. Victor, *Hypertension*, 2001, **38**, 379.
29. E. Lambert, N. Straznicky, M. Schlaich, M. Esler, T. Dawood, E. Hotchkin and G. Lambert, *Hypertension*, 2007, **50**, 862.
30. G. E. Alvarez, S. D. Beske, T. P. Ballard and K. P. Davy, *Circulation*, 2002, **106**, 2533.
31. G. Grassi, G. Seravalle, R. DellOro, C. Turri, G. B. Bolla and G. Mancia, *Hypertension*, 2000, **36**, 538.

32. E. Lambert, N. Straznicky, N. Eikelis, M. Esler, T. Dawood, K. Masuo, M. Schlaich and G. Lambert, *J. Hypertens.*, 2007, **25**, 1411.
33. G. E. Alvarez, T. P. Ballard, S. D. Beske and K. P. Davy, *Am. J. Physiol. Heart Circ. Physiol.*, 2004, **287**, H414.
34. G. Grassi, R. Dell'Oro, A. Facchini, F. Quarti Trevano, G. B. Bolla and G. Mancia, *J. Hypertens.*, 2004, **22**, 2363.
35. N. E. Straznicky, E. A. Lambert, M. T. Grima, N. Eikelis, K. Richards, P. J. Nestel, T. Dawood, K. Masuo, C. I. Sari, J. B. Dixon, M. D. Esler, E. Paul, M. P. Schlaich and G. W. Lambert, *Obesity*, 2013, **22**, 652.
36. E. A. Lambert, T. Rice, N. Eikelis, N. E. Straznicky, G. W. Lambert, G. A. Head, C. Hensman, M. P. Schlaich and J. B. Dixon, *Am. J. Hypertens.*, 2014, **27**, 1308.
37. C. Shibao, A. Gamboa, A. Diedrich, A. C. Ertl, K. Y. Chen, D. W. Byrne, G. Farley, S. Y. Paranjape, S. N. Davis and I. Biaggioni, *Hypertension*, 2007, **49**, 27.
38. S. H. Carlson, J. Shelton, C. R. White and J. M. Wyss, *Hypertension*, 2000, **35**, 403.
39. K. Rahmouni, M. A. Fath, S. Seo, D. R. Thedens, C. J. Berry, R. Weiss, D. Y. Nishimura and V. C. Sheffield, *J. Clin. Invest.*, 2008, **118**, 1458.
40. N. Rajapakse, F. Karim, G. Head and D. Kaye, *FASEB J.*, 2013, **27**, 905.1.
41. D. S. Silverberg, A. Oksenberg, H. Radwan and A. Iaina, *Isr. J. Med. Sci.*, 1995, **31**, 527. Abstract.
42. K. Narkiewicz and V. K. Somers, *Curr. Hypertens. Rep.*, 1999, **3**, 268.
43. D. B. Zoccal and B. H. Machado, *Curr. Hypertens. Rep.*, 2011, **13**, 229.
44. L. F. Drager, R. P. Pedrosa, P. M. Diniz, L. Diegues-Silva, B. Marcondes, R. B. Couto, D. M. Giorgi, E. M. Krieger and G. Lorenzi-Filho, *Hypertension*, 2011, **57**, 549.
45. K. Narkiewicz, M. Kato, B. G. Phillips, C. A. Pesek, D. E. Davison and V. K. Somers, *Circulation*, 1999, **100**, 2332.
46. N. V. Waravdekar, L. I. Sinoway, C. W. Zwillich and U. A. Leuenberger, *Am. J. Respir. Crit. Care Med.*, 1996, **153**, 1333.
47. Y. Zhang, R. Proenca, M. Maffei, M. Barone, L. Leopold and J. M. Friedman, *Nature*, 1994, **372**, 425.
48. G. Frühbeck, *Nutr. Rev.*, 2002, **60**, S47.
49. M. S. Haque, Y. Minokoshi, M. Hamai, M. Iwai, M. Horiuchi and T. Shimazu, *Diabetes*, 1999, **48**, 1706.
50. W. G. Haynes, D. A. Morgan, S. A. Walsh, A. L. Mark and W. I. Sivitz, *J. Clin. Invest.*, 1997, **100**, 270.
51. E. W. Shek, M. W. Brands and J. E. Hall, *Hypertension*, 1998, **31**, 409.
52. J. C. Dunbar, Y. Hu and H. Lu, *Diabetes*, 1997, **46**, 2040.
53. J. C. Dunbar and H. Lu, *Brain Res. Bull.*, 1999, **50**, 215.
54. W. G. Haynes, D. A. Morgan, A. Djalali, W. I. Sivitz and A. L. Mark, *Hypertension*, 1999, **33**, 542.
55. B. Li, Z. Shi, P. A. Cassaglia and V. L. Brooks, *Hypertension*, 2013, **61**, 812.

56. J. P. Warne, F. Alemi, A. S. Reed, J. M. Varonin, H. Chan, M. L. Piper, M. E. Mullin, M. G. Myers, Jr., C. U. Corvera and A. W. Xu, *Cell Metab.*, 2011, **14**, 791.
57. N. Satoh, Y. Ogawa, G. Katsuura, Y. Numata, T. Tsuji, M. Hayase, K. Ebihara, H. Masuzaki, K. Hosoda, Y. Yoshimasa and K. Nakao, *Diabetes*, 1999, **48**, 1787.
58. M. Tang-Christensen, P. J. Havel, R. R. Jacobs, P. J. Larsen and J. L. Cameron, *J. Clin. Endocrinol. Metab.*, 1999, **84**, 711.
59. K. Matsumura, I. Abe, T. Tsuchihashi and M. Fujishima, *Am. J. Physiol.: Regul., Integr. Comp. Physiol.*, 2000, **278**, R1314.
60. A. L. Mark, M. L. Correia, K. Rahmouni and W. G. Haynes, *J. Hypertens.*, 2002, **20**, 1245.
61. J. L. Halaas, C. Boozer, J. Blair-West, N. Fidahusein, D. A. Denton and J. M. Friedman, *Proc. Natl. Acad. Sci. U. S. A.*, 1997, **94**, 8878.
62. K. Rahmouni, W. G. Haynes, D. A. Morgan and A. L. Mark, *Hypertension*, 2002, **39**, 486.
63. A. L. Mark, M. L. Correia, K. Rahmouni and W. G. Haynes, *Clin. Exp. Hypertens.*, 2004, **26**, 629.
64. N. Tumer, B. Erdos, M. Matheny, I. Cudykier and P. J. Scarpace, *J. Hypertens.*, 2007, **25**, 2471.
65. E. Lambert, C. I. Sari, T. Dawood, J. Nguyen, M. McGrane, N. Eikelis, R. Chopra, C. Wong, K. Chatzivlastou, G. Head, N. Straznicky, M. Esler, M. Schlaich and G. Lambert, *Hypertension*, 2010, **56**, 351.
66. K. Clement, C. Vaisse, N. Lahlou, S. Cabrol, V. Pelloux, D. Cassuto, M. Gourmelen, C. Dina, J. Chambaz, J. M. Lacorte, A. Basdevant, P. Bougneres, Y. Lebouc, P. Froguel and B. Guy-Grand, *Nature*, 1998, **392**, 398.
67. R. Rosmond, Y. C. Chagnon, G. Holm, M. Chagnon, L. Perusse, K. Lindell, B. Carlsson, C. Bouchard and P. Bjorntorp, *J. Clin. Endocrinol. Metab.*, 2000, **85**, 3126.
68. J. F. Carroll, T. M. Dwyer, A. W. Grady, G. A. Reinhart, J. P. Montani, K. Cockrell, E. F. Meydrech and H. L. Mizelle, *Am. J. Physiol. Heart Circ. Physiol.*, 1996, **271**, H373.
69. S. L. Burke and G. A. Head, *J. Neurosci. Methods*, 2003, **127**, 63.
70. K. Lim, S. L. Burke and G. A. Head, *Hypertension*, 2013, **61**, 628.
71. S. L. Burke, L. J. Prior, E. V. Lukoshkova, K. Lim, B. Barzel, P. Davern, J. A. Armitage and G. A. Head, *Chronobiol. Int.*, 2013, **30**, 726.
72. L. Cordain, S. B. Eaton, A. Sebastian, N. Mann, S. Lindeberg, B. A. Watkins, J. H. O'Keefe and J. Brand-Miller, *Am. J. Clin. Nutr.*, 2005, **81**, 341.
73. L. Landsberg, *Yale J. Biol. Med.*, 1989, **62**, 511.
74. S. J. Holt and D. A. York, *Brain Res.*, 1989, **500**, 384.
75. M. Muntzel, D. Morgan, A. Mark and A. Johnson, *Am. J. Physiol. Regul. Integr. Comp. Physiol.*, 1994, **36**, R1350.
76. P. A. Cassaglia, S. M. Hermes, S. A. Aicher and V. L. Brooks, *J. Physiol.*, 2011, **589**, 1643.

77. M. P. Pricher, K. L. Freeman and V. L. Brooks, *Hypertension*, 2008, **51**, 514.
78. D. A. Morgan and K. Rahmouni, *J. Hypertens.*, 2010, **28**, 1913.
79. P. Palatini, S. Majahalme, J. Amerena, S. Nesbitt, O. Vriz, M. Michieletto, L. Krause and S. Julius, *J. Hypertens.*, 2000, **18**, 769.
80. S. E. Simonds and M. A. Cowley, *Trends Neurosci.*, 2013, **36**, 121.
81. L. Prior, P. Davern, S. Burke, K. Lim, J. Armitage and G. Head, *Hypertension*, 2014, **63**, 338.
82. J. K. Elmquist, *Physiol. Behav.*, 2001, **74**, 703.
83. J. K. Elmquist, R. S. Ahima, C. F. Elias, J. S. Flier and C. B. Saper, *Proc. Natl. Acad. Sci. U. S. A.*, 1998, **95**, 741.
84. J. K. Elmquist, R. S. Ahima, E. Maratos-Flier, J. S. Flier and C. B. Saper, *Endocrinology*, 1997, **138**, 839.
85. O. Alsmadi, M. Melhem, P. Hebbar, G. Thareja, S. E. John, F. Alkayal, K. Behbehani and T. A. Thanaraj, *Am. J. Hypertens.*, 2014, **27**, 1429.
86. J. R. Greenfield, J. W. Miller, J. M. Keogh, E. Henning, J. H. Satterwhite, G. S. Cameron, B. Astruc, J. P. Mayer, S. Brage, T. C. See, D. J. Lomas, S. O'Rahilly and I. S. Farooqi, *N. Engl. J. Med.*, 2009, **360**, 44.
87. L. S. Tallam, D. E. Stec, M. A. Willis, A. A. da Silva and J. E. Hall, *Hypertension*, 2005, **46**, 326.
88. J. H. Dubinion, A. A. da Silva and J. E. Hall, *J. Hypertens.*, 2010, **28**, 1466.
89. K. Matsumura, T. Tsuchihashi, I. Abe and M. Iida, *Brain Res.*, 2002, **948**, 145.
90. P. Li, B. P. Cui, L. L. Zhang, H. J. Sun, T. Y. Liu and G. Q. Zhu, *Exp. Physiol.*, 2013, **98**, 435.
91. J. M. do Carmo, A. A. da Silva, J. S. Rushing and J. E. Hall, *Am. J. Physiol.: Regul., Integr. Comp. Physiol.*, 2012, **302**, R561.
92. B. Barzel, S. Burke, J. Armitage and G. Head, *FASEB J.*, 2013, **27**, 955.21. Abstract.
93. P. J. Enriori, P. Sinnayah, S. E. Simonds, C. Garcia Rudaz and M. A. Cowley, *J. Neurosci.*, 2011, **31**, 12189.
94. H.-R. Berthoud and C. Morrison, *Annu. Rev. Psychol.*, 2008, **59**, 55.
95. B. M. Chronwall, D. A. DiMaggio, V. J. Massari, V. M. Pickel, D. A. Ruggiero and T. L. O'Donohue, *Neuroscience*, 1985, **15**, 1159.
96. N. Neary, A. Goldstone and S. Bloom, *Clin. Endocrinol.*, 2004, **60**, 153.
97. J. K. Elmquist, C. F. Elias and C. B. Saper, *Neuron*, 1999, **22**, 221.
98. A. Sahu, P. S. Kalra and S. P. Kalra, *Peptides*, 1988, **9**, 83.
99. B. G. Stanley, K. C. Anderson, M. H. Grayson and S. F. Leibowitz, *Physiol. Behav.*, 1989, **46**, 173.
100. J. R. Shutter, M. Graham, A. C. Kinsey, S. Scully, R. Luthy and K. L. Stark, *Genes Dev.*, 1997, **11**, 593.
101. M. M. Ollmann, B. D. Wilson, Y. K. Yang, J. A. Kerns, Y. Chen, I. Gantz and G. S. Barsh, *Science*, 1997, **278**, 135.
102. T. M. Hahn, J. F. Breininger, D. G. Baskin and M. W. Schwartz, *Nat. Neurosci.*, 1998, **1**, 271.

103. E. Bruinstroop, L. Pei, M. T. Ackermans, E. Foppen, A. J. Borgers, J. Kwakkel, A. Alkemade, E. Fliers and A. Kalsbeek, *Diabetes*, 2012, **61**, 1043.
104. P. T. Chao, L. Yang, S. Aja, T. H. Moran and S. Bi, *Cell Metab.*, 2011, **13**, 573.
105. M. L. Correia, D. A. Morgan, W. I. Sivitz, A. L. Mark and W. G. Haynes, *Am. J. Hypertens.*, 2002, **15**, 137.
106. M. Kojima, H. Hosoda, Y. Date, M. Nakazato, H. Matsuo and K. Kangawa, *Nature*, 1999, **402**, 656.
107. D. E. Cummings and K. E. Foster, *Gastroenterology*, 2003, **124**, 1532.
108. J. M. Zigman, J. E. Jones, C. E. Lee, C. B. Saper and J. K. Elmquist, *J. Comp. Neurol.*, 2006, **494**, 528.
109. A. M. Wren, L. J. Seal, M. A. Cohen, A. E. Brynes, G. S. Frost, K. G. Murphy, W. S. Dhillo, M. A. Ghatei and S. R. Bloom, *J. Clin. Endocrinol. Metab.*, 2001, **86**, 5992.
110. M. Tesauro, F. Schinzari, M. Iantorno, S. Rizza, D. Melina, D. Lauro and C. Cardillo, *Circulation*, 2005, **112**, 2986.
111. M. Tschop, C. Weyer, P. A. Tataranni, V. Devanarayan, E. Ravussin and M. L. Heiman, *Diabetes*, 2001, **50**, 707.
112. B. O. Yildiz, M. A. Suchard, M. L. Wong, S. M. McCann and J. Licinio, *Proc. Natl. Acad. Sci. U. S. A.*, 2004, **101**, 10434.
113. M. S. Huda, T. Dovey, S. P. Wong, P. J. English, J. Halford, P. McCulloch, J. Cleator, B. Martin, J. Cashen, K. Hayden, J. P. Wilding and J. Pinkney, *Int. J. Obes.*, 2009, **33**, 317.
114. K. Matsumura, T. Tsuchihashi, K. Fujii, I. Abe and M. Iida, *Hypertension*, 2002, **40**, 694.
115. E. Lambert, G. Lambert, C. Ika-Sari, T. Dawood, K. Lee, R. Chopra, N. Straznicky, N. Eikelis, S. Drew, A. Tilbrook, J. Dixon, M. Esler and M. P. Schlaich, *Hypertension*, 2011, **58**, 43.
116. P. D. Taylor, A. M. Samuelsson and L. Poston, *Acta Physiol.*, 2014, **210**, 508.
117. P. M. Catalano, *Reproduction*, 2010, **140**, 365.
118. S. Meehan, C. R. Beck, J. Mair-Jenkins, J. Leonardi-Bee and R. Puleston, *Pediatrics*, 2014, **133**, 863.
119. E. Velkoska, T. J. Cole and M. J. Morris, *Am. J. Physiol. Endocrinol. Metab.*, 2005, **288**, E1236.
120. S. L. Henry, B. Barzel, S. L. Burke, G. A. Head and J. A. Armitage, *Clin. Exp. Pharmacol. Physiol.*, 2012, **39**, 799.
121. C. Athukorala, A. R. Rumbold, K. J. Willson and C. A. Crowther, *BMC Pregnancy Childbirth*, 2010, **10**, 56.
122. P. Korner, *Essential Hypertension and its Causes: Neural and Non Neural Mechanisms*, Oxford University Press, New York, 2007.
123. E. S. Epel, B. McEwen, T. Seeman, K. Matthews, G. Castellazzo, K. D. Brownell, J. Bell and J. R. Ickovics, *Psychosom. Med.*, 2000, **62**, 623.
124. P. Bjorntorp, *Obes. Rev.*, 2001, **2**, 73.

125. E. Lambert, T. Dawood, N. Straznicky, C. Sari, M. Schlaich, M. Esler and G. Lambert, *J. Hypertens.*, 2010, **28**, 543.
126. M. S. Rumantir, M. Vaz, G. L. Jennings, G. Collier, D. M. Kaye, D. R. Seals, G. H. Wiesner, H. P. BrunnerLaRocca and M. D. Esler, *J. Hypertens.*, 1999, **17**, 1125.
127. R. Pasquali, B. Anconetani, R. Chattat, M. Biscotti, G. Spinucci, F. Casimirri, V. Vicennati, A. Carcello and A. M. Labate, *Metab., Clin. Exp.*, 1996, **45**, 351.
128. F. H. S. Kuniyoshi, I. C. Trombetta, L. T. Batalha, M. Rondon, M. C. Laterza, M. M. G. Gowdak, A. C. P. Barretto, A. Halpern, S. M. F. Villares, E. G. Lima and C. E. Negrao, *Obesity Res.*, 2003, **11**, 1411.
129. D. Carroll, A. C. Phillips and G. Der, *Psychosom. Med.*, 2008, **70**, 653.
130. R. A. Dampney, J. Horiuchi, S. Killinger, M. J. Sheriff, P. S. Tan and L. M. McDowall, *Clin. Exp. Pharmacol. Physiol.*, 2005, **32**, 419.
131. A. J. Marsh, M. A. Fontes, S. Killinger, D. B. Pawlak, J. W. Polson and R. A. Dampney, *Hypertension*, 2003, **42**, 488.

CHAPTER 3

Discovery of Biomarkers of Obesity

ANNALOUISE O'CONNOR AND ANDREW G. SWICK*

Metagenics, Research and Development, 9770 44th Ave NW Suite 100, Gig Harbor, WA 98332, USA
*Email: andrewswick@metagenics.com

3.1 Introduction

The negative health risks associated with obesity have been well established, and include, coronary heart disease, type 2 diabetes, cancers (endometrial, breast, and colon), hypertension, dyslipidemia, stroke, liver and gallbladder disease, sleep apnea and respiratory problems, osteoarthritis, and gynecological problems (abnormal menses, infertility).[1] These health consequences of obesity highlight the importance of identifying biomarkers that accurately assess all stages of obesity development and risk of down-stream metabolic disease.

Obesity biomarkers serve several functions. Firstly, biomarkers of obesity classification enable us to track changing prevalence of obesity on a population level, permitting insight into groups requiring public health focus. Secondly, at the level of the individual, biomarkers facilitate the identification of obese individuals who require intervention for weight management, or who may require targeted intervention to reduce metabolic risk. Early biomarkers of obesity development may highlight 'at-risk' individuals before the onset of obesity occurs, hence improving the possibility of effective body weight management and metabolic control. Thirdly, the discovery and validation of obesity biomarkers enhances our understanding of

RSC Drug Discovery Series No. 45
Cardiovascular and Metabolic Disease: Scientific Discoveries and New Therapies
Edited by Philip Peplow, James Adams and Tim Young

Published by the Royal Society of Chemistry, www.rsc.org

the processes at play in obese states, shedding light on mechanisms of obesity development and metabolic dysfunction. Fundamentally, this knowledge may lead to improved individualized treatments for obesity and healthy weight maintenance, as well as prevention of further metabolic disease.

Consistent with the varying functions of obesity biomarkers, several biomarker levels exist. At the broadest level, population-appropriate biomarkers allow data critical to the monitoring of population trends, such as the obesity epidemic, to be captured. While these biomarkers lead to some loss of accuracy at the level of the individual, they provide a starting point for assessing risk. Ease of measurement, time and cost are key factors here and need to be weighed against the value of the data. Clinically-appropriate obesity biomarkers should then refine risk assessment and focus intervention strategies if possible. Ease of measurement, cost and time are again important factors to consider, however these represent smaller obstacles than at a population level. At a basic-science level, discovery of novel processes is the central focus, and more detailed characterization that would not be appropriate at the higher levels, is possible here.

These levels of obesity biomarkers form a cycle in which population groups *e.g.* specific age or ethnic groups, are identified to be at increased risk. Next, clinical research or basic science efforts are focused on ascertaining processes driving the increased obesity prevalence. Finally, these biomarkers are tested for predictive power in clinical or population-based cohorts, and may inform guidelines for the management of at-risk individuals, or develop into pharmaceutical or other obesity therapies.

This chapter will focus initially on anthropometric biomarkers of obesity and then move to adipose tissue, biochemical and novel 'omics-level biomarkers characterizing the obese state. The biology underpinning these biomarkers will be briefly reviewed, and lessons learned and existing gaps highlighted where appropriate. The biomarkers discussed will be based on the biological processes which characterize the obese state. Focus will not be given to dietary or lifestyle biomarkers that could potentiate obesity development, or help to alleviate the condition once established.

3.2 Identifying and Classifying Obese Individuals

A critical first step in identifying and discovering novel biomarkers of a disease is the accurate and reproducible measurement of the disease or condition. Several tools relating to the identification and classification of obese individuals have been developed and will be discussed in the following sections, with focus given to the widely adopted World Health Organization (WHO) tools and recommendations. Consensus guidelines for the classification of the obese state are critical when comparing the public health and research data from groups around the world. The WHO recommendations for the assessment of obesity have been broadly accepted, however as will be discussed, are not without criticism and some individual countries have adapted the recommendations and tools slightly so as to be more appropriate for their populations.

3.2.1 Classifying Obesity in Adults

The most common anthropometric tool used to classify obesity is body mass index (BMI). BMI is a weight-for-height index, and is calculated as weight (in kg) divided by height (in metres squared). Thus, for a given height, BMI will be proportional to mass.[2] Although initially developed in the 19th century (then termed the Quetelet index), the use of Body Mass Index (BMI) came into vogue during the 1980s as it was adopted by various expert working groups and large epidemiological studies.[3] A significant contributor to the appeal of BMI lies in the fact that it is independent of age, sex and reference population datasets, which facilitates direct comparison of data sets across countries and studies. BMI involves the simple anthropometric measures of weight and height which are inexpensive and straightforward to collect in a reproducible manner, meaning that BMI has become the biomarker of choice for clinicians and researchers wishing to classify an individual and track the effects of an intervention. Additionally, individuals can simply calculate their own BMI, and understand a weight or weight range that would be considered healthy for them. This can readily be determined using various apps and internet-based calculators.

Similar to any such surrogate measure, there is an ongoing discussion with respect to the suitability of BMI.[4] BMI was initially developed as a tool for use in population-based research. Additionally, as obesity refers to an excess in fat mass (and not lean mass), a biomarker/classification tool for obesity should relate specifically to fat mass. For these reasons, much of the criticism of BMI centers on its failure to account for differences in overall body composition *i.e.* lean *versus* fat mass, which varies from person-to-person, and is dependent on lifestyle variables such as physical activity, genetics, age, and gender.[2] Interestingly, a J-shaped or U-shaped curve has been shown to describe the relationship between BMI and mortality, suggesting an apparent 'obesity paradox'.[5] However, as previously mentioned, BMI cannot differentiate between tissue types, and so this paradox appears to be due at least in part to differences in adiposity and distribution of body fat and lean body mass.[6]

Although not a perfect relationship, an approximately linear relationship between BMI and adiposity exists, with BMI explaining on average ~80% of the variance in body fat percentage.[7] However, for a general population engaging in moderate to low physical activity, increasing body weight correlates well with increasing adiposity, and so BMI is considered a proxy for estimating body fat mass.[7,8] BMI cut-off points have been developed as a way of classifying individuals based on body weight. These cut-offs for overweight and obesity are based on the negative health effects of excess adiposity.[2,7] Although the cut-off points for normal weight, overweight and obesity have shifted slightly since the wide-spread adoption of BMI as a tool

Table 3.1 World Health Organization[2] body mass index (BMI) cut-off points for the classification of body weight status in adults.

Classification	Principal cut-off points (kg m^{-2})
Underweight	<18.50
Severe thinness	<16.00
Moderate thinness	16.00–16.99
Mild thinness	17.00–18.49
Normal range	18.50–24.99
Overweight	≥25.00
Pre-obese	25.00–29.99
Obese	≥30.00
Obese class I	30.00–34.99
Obese class II	35.00–39.99
Obese class III	≥40.00

for classifying obesity, since 2004 these cutoffs have stabilized[2] (Table 3.1). Continued debate surrounding the global appropriateness of BMI cut-offs due to the variation in body proportions and health risk characteristic of certain population groups[9] prompted a later WHO Expert Consultation.[10] However, this group recommended that previously established cut-offs should be retained (as in Table 3.1) to facilitate international comparisons.[10] Recommendations for alternative population-specific cut-off points have been developed for use in Asian countries. These cut-offs represent triggers for health care professionals to intervene.[10]

In summary, although adopted for use in individuals and in the clinical setting, BMI is fundamentally a biomarker best suited to characterize the body weight status of populations,[2] and does not adequately account for the wide variation in body fat distribution present in individuals. For this reason, BMI should be considered the first step for clinicians making a risk assessment, and additional tools have been developed for more direct assessment of adiposity and fat deposition patterns that will be discussed in Section 3.3.

3.2.2 Classifying Obesity in Infants and Children

Obesity during early life is one of the strongest predictors of future body weight status, with 50–70% of childhood obesity tracking into adulthood.[11,12] Furthermore, obesity beginning at an early age particularly increases health risk, and alarmingly, the prevalence of metabolic syndrome risk factors is rising among young children.[13] As the negative consequences of obesity can begin in childhood, the early identification of obesity and propensity to become obese is essential.

With the rapid growth that occurs during infancy and childhood, tracking the potential development and resolution of obesity in a manner that takes growth into consideration is necessary. In 2006, the WHO released growth standards for height, weight and head circumference for children from birth to 20 years of age.[14] These standards were derived from growth data obtained from children participating in the Multi Centre Growth Reference Study (MGRS) living under optimal growth conditions, set by the WHO as: breastfeeding exclusively or predominantly for at least 4 months; timely introduction of appropriate solid foods; no socioeconomic constraints on growth; must have access to health care; and be from non-smoking mothers.[14] These optimal growth conditions have led to slight controversy and a reluctance of some countries to fully adopt the growth standards, as many infants for example are not breastfed, resulting in a growth pattern at odds with that used in the creation of the standard.[15] The use of the WHO growth standards has been shown to result in many more infants and children diagnosed to be on an obese trajectory compared with older growth references.[16,17] Countries adopting the WHO charts suggested that differences in growth would help to identify infants and children with high BMI and help support efforts to reduce future childhood obesity. This creates an onus on public health agencies to develop effective recommendations and support mechanisms for parents with an obese infant or child. Nevertheless, there has been a move by the majority of countries to adopt the WHO growth standards, and follow the recommendation that a weight-for-height of greater than the 98th percentile when using these charts is considered a biomarker of obesity. Some countries, for example the UK[17] and USA,[18] have slightly altered the use of the growth standards to better represent local populations. Similar to BMI, standardized growth charts create a framework through which to classify infants and children based on weight-for-height, enabling the identification of additional biomarkers of disease and the tracking of intervention efficacy. The concept of individuality again comes into play here, as growth potential and health status, among other factors, will determine the interpretation of the BMI percentile for each individual child.

3.3 Adiposity and Adipose Tissue

Intrinsic to the concept of obesity, is the excessive increase in body fat mass. Although increases in lean body mass occur in parallel to increased fat mass, these are prompted by the need to support a larger frame. As the deleterious health effects of obesity result from increased general adiposity, inappropriate adipose tissue deposits, and dysfunctional adipose tissue, the following sections will discuss biomarkers of obesity as they relate to adipose tissue type, location, structure, and function.

3.3.1 White, Brown and Beige Adipose Tissue

Adipose tissue can be divided into three distinct types – white, brown and beige adipose tissue – each with differing functions and effects on metabolism. The quantity of white, brown and beige adipose tissue are themselves biomarkers of obesity, and additionally express and secrete factors which can be considered biomarkers of the metabolic environment characteristic of the obese state.

3.3.1.1 White Adipose Tissue

White adipose tissue is the predominant adipose tissue type in the human body, functioning as an energy reserve with the capacity to expand massively in the face of excessive nutrient intake. Energy stored in adipose tissue can be mobilized during periods of energy shortage, for example the fasted state or during prolonged exercise, through lipolysis and the release of the resultant free fatty acids into the circulation. This energy storage role of white adipose tissue is not its only function, and white adipose tissue is now established as a dynamic endocrine organ with a pivotal role in systemic glucose homeostasis and energy balance.[19] Many of the classical adipokines discussed in later sections are secreted from white adipose tissue, and are responsible for many of the diverse roles the tissue plays in metabolism.

Assessment of body fat mass almost exclusively refers to measurement of white adipose tissue. Cut-off points for obesity based on body fat percentage have been suggested as >25% in men and >30% in women.[20] Direct assessment of body fat mass has historically required considerable methodological expertise. Underwater or hydrostatic weighing, considered an early gold standard, was one of the first methodologies developed for the assessment of fat mass. However, as the subject is fully submerged in water during the measurement, hydrostatic weighing is unsuitable for many children and individuals with any health conditions, therefore limiting its broad applicability. The development of additional measurement techniques such as dual X-ray absorptiometry (DXA), air displacement plethysmography (commonly referred to by commercial device name BOD POD) and bioelectrical impedance (BIA) have permitted the measurement of body fat in larger groups of people and in many different settings, helping to develop individualized risk assessments and more accurately classify the obese state.

3.3.1.2 Brown Adipose Tissue

Brown adipose tissue is considerably less prevalent in humans than white adipose tissue. Although the amount varies between individuals, when present, brown adipose tissue is located in small anatomically discrete sites (cervical, supraclavicular, paravertebral, pericardial).[21] In smaller animals it is well established that brown adipose tissue serves a key thermogenic role. Brown adipocytes differ significantly from white adipocytes in morphology. Instead of a single large lipid droplet, brown adipocytes contain multiple smaller lipid droplets, and contain a greater degree of mitochondria. The mitochondria of brown adipocytes express uncoupling protein-1 (UCP-1), a brown adipocyte biomarker involved in generation of heat through non-shivering thermogenesis.

While originally thought not to be present in human adults, the advent of fluorodeoxyglucose positron emission tomography (FDG-PET)-scanning method development enabled the visualization of metabolically active brown adipose tissue in human adults.[21] FDG-PET has revolutionized the study of brown adipose tissue, and the ability to visualize this tissue in humans has changed the understanding of the potential role brown adipose tissue may play in human obesity development. It has been established however that the amount of brown adipose tissue notably declines with increasing BMI,[22] resulting in the idea that increasing brown adipose tissue may be a useful treatment for obesity. The FDG-PET methodology is available only at specialized research/clinical centers, and at present insufficient data exists to form definitive targets for brown adipose tissue volumes. However, it has been suggested that activation of an additional 50 g of brown adipose tissue in humans could increase energy expenditure (EE) by 20% in a 2500 kcal energy requiring person,[23] an increase that would facilitate weight loss and maintenance of a healthy body weight.

3.3.1.3 *Brite Adipose Tissue*

The presence of brite (brown in white) adipocytes, also known as beige adipocytes, in areas of white adipose tissue has been the focus of considerable attention in recent years. Under basal conditions, brite adipocytes are phenotypically similar to neighboring white adipocytes, however, in response to thermogenic stimuli such as cold and β-3 adrenergic activators, brite adipocytes exhibit a more brown adipocyte-like phenotype, with many small lipid droplets and UCP-1 expression evident.[24] This discovery has spurred considerable research excitement due to the potential to increase energy expenditure through an upregulation of the beiging or 'browning' of white adipose tissue process. Much of this excitement has been driven by the discovery of myokines such as irisin[24] and more recently meteorin-like protein[25] that suggest that beiging of white adipose tissue is an inducible process which could potentially by upregulated therapeutically. Irisin is increased in obese individuals, is positively associated with BMI, adiposity and body weight,[26] and is positively correlated with a subgroup of subjects whose energy expenditure is greater than predicted by lean mass.[27] Irisin in the cerebrospinal fluid is lower in normal weight pregnant women compared with pregnant women who are obese or who have been diagnosed with gestational diabetes.[28] Irisin is positively correlated with insulin,[28] has also been shown to predict insulin resistance with weight regain,[29] and was associated with a lower risk of major adverse coronary events in subjects with established coronary artery disease.[30] At this point it is unclear if irisin is a protective factor secreted in response to increasing adiposity and poor glucose control. However, since it is involved in the beiging of adipose tissue, irisin, and other proteins with similar beiging effects could be biomarkers for weight gain susceptibility.

3.3.2 Abdominal Body Fat Distribution Patterns

Although an increase in overall fat mass is linked with adverse health complications, the location of the adipose tissue and body fat distribution patterns are more closely related to adverse metabolic outcomes than total fat mass alone. The now seminal work on regional adiposity in the 1950's concluded that individuals with an android body composition ('apple') were more likely to exhibit abnormal glucose tolerance, diabetes, hypertension and atherosclerotic coronary artery disease than those with a gynoid ('pear') composition.[31] For example, for every 0.01 unit increase in waist-to-hip ratio (WHR) (indicating a greater degree of abdominal fat deposition), the risk of future CVD events increases 2% and 5%, respectively.[32] It has become clear that this increased risk likely stems from the increased secretion of adipokines, pro-inflammatory mediators and free fatty acids, in a pattern more akin to visceral adipose tissue than other subcutaneous deposits.

Considering the link between abdominal adipose tissue and metabolic health, the measurement of this biomarker is of importance. Although direct assessment of abdominal body fat volumes would be preferable, methodologies such as magnetic resonance imaging (MRI) and computed tomography (CT) scanning are not appropriate for use clinically or on a large scale due to cost and exposure to radiation. To circumvent these issues, several surrogate biomarkers have been developed, of which waist circumference (WC) and waist/hip ratio (WHR) are the most commonly used.

Although WC and BMI are interrelated, WC provides an additional independent predictive value when assessing metabolic risk.[33,34] Specifically, WC provides additional key information for understanding risk of cardiometabolic disease in individuals with BMI up to 34.9 kg m^{-2} (see Table 3.2). The WHO has proposed cut-off points for WC, where women with a

Table 3.2 Risk of metabolic disease associated with body mass index (BMI) and waist circumference (WC) classification cutoffs relative to individuals of normal weight and waist circumference.[34]

	BMI/kg m^{-2}	WC Men < 102 cm (40 in) WC Women < 88 cm (35 in)	WC Men > 102 cm (40 in) WC Women > 88 cm (35 in)
Underweight	18.5		
Normal weight[a]	18.5–24.9		
Overweight	25.0–29.9	Increased	High
Obesity	30.0–34.9 (Obesity Class I)	High	Very High
	35.0–39.9 (Obesity Class II)	Very High	Very High
Extreme Obesity	≥40 (Obesity Class III)	Extremely High	Extremely High

[a]Increased waist circumference can also be a marker for increased risk even in persons of normal weight.

WC above 80 cm and men with WC above 94 cm are considered at increased metabolic risk. A WC above 88 cm and 102 cm in women and men, respectively, is considered a substantially increased risk of metabolic risk.[34]

3.3.3 Subcutaneous *versus* Visceral Adipose Tissue

White adipose tissue is stored throughout the body. The three major anatomical locations are subcutaneous (beneath the skin), visceral (within the body cavities), and ectopic (within organs such as liver, pancreas and muscle).[24] All deposits are seen to increase in an obese state, however the specific location of adipose tissue deposition has significant consequences for metabolic health. Whereas the subcutaneous depot is by far the largest, and contributes to the adverse metabolic environment characteristic of obesity particularly if stored abdominally, even small amounts of visceral or ectopic lipid storage have highly negative effects on metabolic health.

A certain amount of visceral adipose tissue is normal, and is distributed around the organs, providing a protective covering. However, increasing visceral adipose depot size is a significant problem in obese states, and represents an important biomarker of metabolic risk. Visceral adipose tissue exhibits a metabolically adverse phenotype. An increased infiltration of macrophages, an increased production and secretion of pro-inflammatory mediators and free fatty acids, and a decreased production and secretion of anti-inflammatory, insulin-sensitizing factors such as adiponectin is observed in visceral adipose tissue.[35] Endothelial cells isolated from visceral adipose tissue of obese subjects have higher gene expression related to angiogenesis and inflammation than endothelial cells isolated from subcutaneous adipose tissue.[36] Additionally, visceral pockets such as epicardial or perivascular adipose tissue may have a local proinflammatory effect on vasculature, increasing the risk of CVD.[37]

Considering the importance of this depot for overall metabolic risk, research efforts have focused on the measurement of visceral adipose tissue. Currently only magnetic resonance imaging (MRI) and computed tomography (CT) can provide direct cross-sectional area measurements of visceral adipose tissue volume, and are considered the gold standards, providing the highest resolution. However, these methods can be prohibitive for this type of adiposity measurement outside of a small research study due to high cost and justification for radiation exposure.[38] The use of DXA to estimate visceral adipose tissue has recently been reported,[39,40] and may contribute to a broader understanding of the drivers of visceral adipose tissue development within an individual patient. Due to the ability to measure greater numbers of people, this technology may also lead to the development of cut-off points for visceral adipose tissue amounts. Additionally, as a surrogate measure of visceral adipose tissue, WC is considered the most reliable across a wide age range. Although greater amounts of visceral adipose tissue are associated with a specific pattern of adipokines and inflammatory mediators in the circulation, for example decreased adiponectin and increased RBP4, vaspin, chemerin, fetuin-A, omentin and progranulin, currently no biochemical biomarker unique to visceral fat mass exists.[41] The identification of such a factor, potentially through enhanced omics-level screening technologies would be of considerable advantage to the field. Interestingly, it has been reported that plasma concentrations of the endocannabinoid 2-arachidonylglycerol (2-AG), is correlated with visceral but not subcutaneous adipose tissue, and that reductions in waist circumference and visceral fat accumulation lead to decreased 2-AG in circulation.[42] Endocannabinoid system tone, which plays a crucial role in both the central and peripheral regulation of energy and metabolic homeostasis, is known to be deranged in obese states. Although previous attempts to treat obesity through the endocannabinoid system have not been successful due to unanticipated psychiatric side-effects, novel peripherally acting cannabinoid receptor (CB) agonists have great potential both for body weight management and metabolic control.

Ectopic lipid deposition is also a biomarker of metabolic ill health. Ectopic lipid deposition occurs within organs such as the liver, muscle, pancreas and myocardium, and is associated with significant metabolic health consequences such as insulin resistance,[43,44] due to lipotoxic effects. Visceral fat is linked to ectopic deposition. ^{1}H-magnetic resonance spectroscopy (^{1}H-MRS) has developed as a method of measuring ectopic fat deposition, and changes in ectopic lipid accumulation are associated with improved metabolic functionality.[45] Similar to visceral adipose tissue, ectopic fat accumulation varies between individuals and may be a result of the amount of visceral fat present, and/or the influence of genetics.[46] A greater understanding of the extent of ectopic fat deposition and factors underpinning its development in different organs may lead to treatment strategies targeted at resolution of ectopic fat storage. For both visceral and ectopic fat storage, specific therapeutic modalities aimed at decreasing these fat deposits, even in the absence of significant global weight loss, would be metabolically beneficial.

3.3.4 White Adipose Tissue Function and Adipocyte Size

To complement the multi-faceted role adipose tissue plays in the body, adipose tissue has a heterogeneous structure. White adipose tissue is comprised of adipocytes (fat cells) surrounded by a stromal vascular fraction which incorporates preadipocytes, endothelial cells, vascular smooth muscle cells, leukocytes, monocytes and macrophages, all of which appear to be central in the functioning of adipose tissue as a whole. In obese states, adverse changes in the adipose tissue microenvironment occur, compromising the functionality of the tissue. This obesity-induced departure from the normal functioning of the adipose tissue has negative consequences for metabolic health.

Some of the obesity-associated dysfunction stems from lipid storage within the tissue itself. With increasing energy intakes that occur in obesity, excess energy can be stored in adipocytes as fat, leading to adipocyte hypertrophy or hyperplasia. Hyperplasia, referring to the expansion in adipocyte number, occurs early in life, and fat mass expansion in later life is thought to occur predominantly through an increase in adipocyte size.[23]

Enlarged adipocytes, or hypertrophic obesity, is associated with greater risk of further metabolic complications. The reduction in adipocyte size that occurs with weight loss is strongly associated with improved metabolic risk factors and insulin sensitivity. For example, in a recent study in individuals pre- and post-roux-en-Y gastric bypass (RYGB), reduction in adipocyte size was more strongly related to improvements in metabolic health than reduction in overall body weight, abdominal or visceral adipose tissue.[47] As such, adipocyte size could represent a key biomarker for assessing metabolic response to intervention.

Whereas increasing adipocyte size is technically a normal response to the need for increased energy storage, angiogenesis in expanding adipose tissue is often insufficient, resulting in suboptimal oxygenation of the tissue.[23] Interestingly, augmenting angiogenesis in animal models by overexpression of vascular endothelial growth factor (VEGF) in adipose tissue, attenuated insulin resistance and hepatic steatosis associated with diet-induced obesity and induces a 'browning' of white adipose tissue.[48–50] Much of the metabolic complications of adipocyte hypertrophy may be related to a degree of hypoxia within the tissue. Adipose tissue cellular responses to hypoxia include an upregulation of angiogenic, fibrotic, and inflammatory pathways and processes which in turn can negatively impact on insulin sensitivity. Expression of hypoxia inducible factor-1A (HIF-1A) and fibrotic markers within adipose tissue are established biomarkers of adipose tissue hypoxia and dysfunction, and represent enticing treatment targets.[51,52]

Another mechanism suggested to underpin the adipose tissue dysfunction seen in obese states is endoplasmic reticulum (ER) stress. The ER is the site of triglyceride droplet formation and protein synthesis. Due to the need for excessive energy storage in overweight and obese subjects, the pressure to form increased lipid droplets in the adipocyte may induce ER stress. ER stress in turn can trigger an unfolded protein response, an increase in the production of reactive oxygen species (ROS) and upregulate inflammatory responses.[35]

Adipocyte dysfunction can induce apoptosis leading to the accumulation of dead or dying adipocytes. These apoptotic adipocytes become surrounded by macrophages, creating a crown-like shape seen in histological sections of adipose tissue. These crown-like structures within adipose tissue can serve as obesity-driven biomarkers of metabolic risk. Obese individuals lacking these crown-like structures are metabolically healthier,[53] and the number of crown-like structures in adipose tissue is related to systemic low-grade inflammation and insulin resistance in a group of patients with metabolic syndrome (Table 3.3).[53,54]

Table 3.3 Anthropometric/adipocyte biomarkers of obesity may have adverse metabolic consequences.

Body mass index	Increased
Adiposity	Increased
Abdominal adiposity	Increased
Visceral adiposity	Increased
Adipocyte size	Increased
Crown-like structures	Increased
Brown adipose tissue	Decreased
Irisin	Increased
Free fatty acids	Increased
Pro-inflammatory mediators	Increased
Leptin	Increased
Adiponectin	Decreased

3.4 Adipokines

In addition to functioning as an energy storehouse, adipocytes and adipose tissue help to regulate energy balance and metabolism through the production and release of a large array of secreted factors. Adipokines function locally in an autocrine and paracrine manner where they contribute to the regulation of adipogenesis, immune cell migration into adipose tissue, as well as to general adipocyte metabolism and function. They also act as interorgan signaling molecules, eliciting metabolic response from target organs such as the liver, brain and muscle. Adipokines have been shown to play a role in many aspects of energy homeostasis and metabolism including appetite and satiety, fat distribution, energy expenditure, insulin sensitivity and secretion, inflammation, blood pressure, hemostasis and endothelial function.[55] Dysregulation of adipokine production can occur in obese states, contributing to the development and pathogenesis of more severe metabolic health issues such as cardiovascular disease, type 2 diabetes and the metabolic syndrome.[56] Adverse patterns of adipokine secretion occur as a result of adipocyte dysfunction, and are also impacted by fat distribution patterns. Overall, more than 600 adipokines have been identified, and research is ongoing to fully delineate function in different metabolic states. The following sections focus on the best characterized adipokines, in addition to proinflammatory mediators and novel emerging factors that can be considered biomarkers of the obese state.

3.4.1 Leptin

Leptin is a protein mainly synthesized by white adipose tissue. The discovery that leptin was responsible for the hyperphagia and insulin resistance seen in an obese mouse model (*ob*/*ob*) (which can be rescued by exogenous leptin administration), changed how adipose tissue was viewed and established adipokines as functional proteins with diverse roles in metabolism.[57]

Leptin is considered an adipostat, and is closely tied to metabolic and feeding cues. The release of leptin from the adipocyte is stimulated by insulin, with glucose playing a permissive role, and leptin release is inhibited by free fatty acids, which are typically higher in the fasted state. This regulation leads to adipose and plasma leptin concentrations that fall during the fasted state, and rise again during feeding.[58] Being so fundamentally linked with energy storage, leptin can exert negative feedback on energy intake. Leptin acts centrally, where it binds the leptin receptor (LEPR) and signals through the JAK/STAT pathway in specific populations of neurons in the hypothalamus, midbrain and brainstem neurons to regulate food intake, appetite and energy expenditure.[59]

Leptin is intimately connected to metabolic health. This is clearly evident in individuals with relative leptin deficiency (such as those with generalized loss of body fat). In the absence of leptin, these individuals have profound metabolic dysfunction including insulin resistance, type 2 diabetes, non-alcoholic fatty liver disease and high plasma triglyceride concentrations. Recombinant leptin administration to these individuals reduced hepatic steatosis, and leptin is currently approved in certain jurisdictions for the treatment of lipodystrophy.[60]

Paradoxically, despite its role as a regulator of energy balance, leptin in the circulation rises with increasing fat mass and is high in obese subjects. Leptin appears to lose its regulatory potential in overweight and obese states, and even at high circulating concentrations, fails to adequately reduce/control food intake or regulate body weight. This situation has prompted the idea of 'leptin resistance' in obesity, which has been attributed to defective transport of leptin across the blood–brain barrier secondary to increased plasma triglycerides and C-reactive protein concentrations, SOCS3 inhibition of STAT3 phosphorylation (and therefore activation), genetic variation in the leptin receptor, and endoplasmic reticulum stress, all of which are characteristic of the obese state.[59]

This concept of leptin resistance may explain why despite highly influential studies demonstrating the efficacy of leptin in reducing body weight in leptin deficient (*ob*/*ob*) humans,[61] exogenous leptin as a monotherapy or in combination with a reduced caloric diet, fails to induce weight reduction in people with diet-induced obesity.[62] Currently, despite initial promise, leptin treatments are available only to individuals with genetic or relative leptin deficiency. More recent studies have investigated the potential for co-treatment with a leptin sensitizer to allow the weight loss effect of leptin to be felt. Although co-administration of leptin and amylin (pramlintide) was an effective synergistic treatment for obesity, trials of this combined therapy were stopped prematurely due to potentially neutralizing drug antibody formation.[62]

The concept of leptin resistance in obesity has relevance not only for the use of exogenous leptin as a weight loss treatment modality, but also for metabolic health in obese states. Leptin can contribute to adverse metabolic health through promotion of platelet aggregation, angiogenesis, proliferation and migration of vascular smooth muscle cells, vascular inflammation, oxidative stress, and hypertension.[63] Indeed, higher leptin was associated with an increased risk of congestive heart failure and cardiovascular disease in a cohort of 818 elderly subjects,[64] and leptin was reported as an independent predictor of diabetes particularly in men.[65]

In the approximately 30 years since the initial discovery of leptin, much has been understood about its biology as well as the regulatory networks surrounding food intake, energy expenditure and hence body weight control. Although a promising target for the pharmaceutical industry, leptin trials for the treatment of diet-induced obesity have not been successful, highlighting the complexity and the impaired functionality within an obese system.

3.4.2 Adiponectin

Adiponectin is secreted from the adipocyte, and circulates in the blood in relatively high concentrations (3–30 μg mL^{-1}). There are three oligomeric adiponectin complexes, a trimer (3), hexamer (6) and high molecular weight multimer (12–18).[66] Adiponectin is among a very small number of adipokines, including the recently discovered SFRP5,[67,68] considered to have a protective influence on metabolic health.

Similar to leptin, adiponectin has central effects, binding to its receptors in the hypothalamus to help regulate energy homeostasis.[69] However, many of the most important metabolic effects of adiponectin are exerted peripherally. Adiponectin binds to two distinct receptors (AdipoR1 and AdipoR2), which in turn activate key AMPK- and PPAR-α dependent signaling pathways in target tissues such as liver and muscle.[56] Adiponectin has been shown to enhance glucose utilization and fatty acid oxidation, and the net effect of adiponectin is insulin sensitization.[70] Adiponectin is also considered to have anti-inflammatory and anti-atherogenic effects, may improve endothelial function through activation of nitric oxide (NO) production and reduces foam cell formation.[41,56,66]

Adiponectin levels are markedly reduced in obese individuals, and plasma adiponectin concentrations have been shown to be inversely proportional to the amount of visceral adipose tissue. Adiponectin can be downregulated by proinflammatory cytokines, in addition to catecholamines and insulin and glucocorticoids.[56] Considering its link with metabolic health, low circulating adiponectin concentrations have been associated with an increased risk of diabetes, coronary artery disease, hypertension, left ventricular hypertrophy, myocardial infarction, and impaired endothelial-dependent vasodilation.[66]

Since its discovery in 1995, adiponectin has emerged as a drug candidate. Replenishment of adiponectin reduced atherosclerotic lesion size in *ApoE*$^{-/-}$ mice.[71] Exogenous adiponectin in animal models induces insulin secretion[72] and a small molecule AdipoR agonist was recently demonstrated to ameliorate insulin resistance and glucose intolerance in mouse models of diabetes.[73] However, data in humans is currently lacking.[55]

The discovery of adiponectin heralded an early view of protective factors secreted from adipocytes which can be negatively regulated in obese states. Since its initial discovery, proposed protective mechanisms have been developed, and low levels have been associated with increased metabolic risk in large human cohorts. However, although animal studies have shown promising results, this biomarker of obesity has not yet resulted in a human therapeutic platform.

3.5 Inflammation

With extensive studies highlighting the increase of proinflammatory mediators in obese compared with lean individuals, obesity is now viewed as an inflammatory state. Indeed, this chronic low grade inflammatory environment appears to be at the core of the increased metabolic risk associated with obesity.

Inflammation in the obese states can arise from several sources. As a result of adipose tissue dysfunction, the stromal vascular cells within adipose tissue as well as the adipocytes themselves produce cytokines and chemokines involved in upregulating the immune response. Among the more than 600 or so putative adipokines, the majority have been discovered to play a role in mediating proinflammatory processes, for example, interleukin (IL)-1, -6, -8, tumor necrosis factor (TNF)-α, transforming growth factor (TGF)-β, interferon (IFN)-γ, and plasminogen activator inhibitor (PAI-1).[41] Additionally, free fatty acids that are increased in the obese state can activate key proinflammatory signal transducers such as NF-κB on immune cells and induce oxidative stress, a situation which further augments the circulating levels of pro-inflammatory mediators.[74] These proinflammatory mediators activate atherogenic processes and decrease insulin sensitivity, therefore increasing the risk of CVD and diabetes.[75]

The following sections will discuss several notable proinflammatory mediators characteristic of the obese state that have been shown to predict further metabolic disease.

3.5.1 TNF-α

Tumor necrosis factor alpha (TNF-α) is a proinflammatory cytokine, predominantly produced by macrophages but which can also be induced in adipose tissue in obesity.[76,77] As well as the increase in adipose tissue, plasma TNF-α concentrations are elevated in obesity,[78] and expression is decreased with weight loss.[77] TNF-α is closely linked with insulin resistance,[76,78] inhibiting insulin receptor tyrosine kinase activity in muscle and fat.[79] Signaling through tumor-necrosis factor receptor-1 (TNFR1) is thought to induce proinflammatory and pro-apoptotic actions, and may contribute to the adverse metabolic consequences which occur as a function of obesity.[56]

3.5.2 IL-6

Interleukin-6 (IL-6) is increased in both plasma and adipose tissue of obese individuals, and is reduced by weight loss.[80] IL-6 induces the secretion of additional proinflammatory, insulin de-sensitizing factors such as C-reactive protein (CRP).[81] IL-6 has been demonstrated to be a predictor of cardiovascular disease.[82,83]

3.5.3 CRP

C-reactive protein (CRP) is an acute phase protein produced by the liver in response to infection or trauma. Mildly elevated levels (in comparison with levels during infection) are seen with obesity.[35] Due to its effect of attenuating insulin signaling, CRP contributes to insulin resistance.[84] Additionally, CRP is seen to promote proatherosclerotic processes such as decreasing endothelial nitric oxide synthase (eNOS), activation of the complement cascade and promoting foam cell formation.[81] Indeed, many studies have demonstrated a predictive role for CRP in the development of insulin resistance, diabetes and the metabolic syndrome.[81] CRP is a low cost, non-invasive (plasma) assessment with broad availability, and is frequently measured as a risk marker in clinical settings.

3.6 Omics/Novel Biomarkers of Obesity and Risk of Further Metabolic Disease

In addition to adipokine and cytokine proteins, emerging research is pointing to microRNAs as novel biomarkers of obesity and metabolic risk. MicroRNAs are highly conserved noncoding RNA molecules, typically approximately 22 nucleotides in length. MicroRNAs are expressed in a – cell type-specific manner, and are dysregulated in tissues and in the circulation in many disease states including obesity.[85] Mechanistically, emerging reports implicate microRNAs in the fine-tuning of lipid metabolic pathways and adipogenesis.[85] MicroRNA patterns have been shown to vary between human subcutaneous and visceral adipose tissue deposits; however, a visceral specific microRNA has not been identified to date.[86] In adults, plasma miR-130b was increased in obesity, and was seen to predict the presence of metabolic syndrome to a greater extent than plasma triglycerides.[87] Expression of miR-935 and miR-4772 was increased and expression of miR-223, miR-224, and miR-376b decreased in peripheral blood mononuclear cells of responders to a reduced calorie weight loss intervention.[86] Further evidence of the predictive value of microRNAs was seen in a recent study of 7 year-old healthy children with normal body weight. Plasma concentrations of miR-486-5p, miR-486-3p, miR-221 and miR-130b at 7 years were associated with increased BMI in the same children after a 3 year follow up period.[88] At 7 years of age, plasma concentration of miR-142-3p was associated with increased fat in the umbilical region at 10 years in this same group.[88]

Although further studies are needed to replicate these findings before a strong risk biomarker can be established, the predictive potential of circulating microRNAs is intriguing.

3.7 Conclusions

The obesity epidemic continues to put strain on economic and healthcare systems and significantly affects individual quality of life, necessitating the development of effective anti-obesity strategies.

Overweight and obese individuals should always be encouraged to lose weight. More specifically, it is recommended that patients with obesity (BMI > 30 kg m^{-2}) lose at least 10% of their body weight as even this modest reduction has measurable benefits with respect to metabolic health and improvement of obesity associated co-morbidities. Lifestyle modification, such as increasing physical activity, altering dietary macronutrient intake, and cognitive behavioral therapy, are preferred first line treatments for obesity. Although the value of lifestyle interventions has been established, adherence to these changes can be problematic in the longer term. Even relatively small body weight losses are difficult to maintain and it has been estimated that significant weight regain occurs in 80–90% of individuals.[89]

An increased focus on obesity biomarker discovery does not detract from the recommendation that all obese individuals should lose weight. Instead, biomarker discovery helps to refine risk and develop targeted treatments which may facilitate easier weight loss or offset some of the metabolic risk associated with obesity.

As it stands, BMI represents a widely used biomarker for obesity classification most appropriate when used on a population level. For an individual, the metabolic risk associated with BMI can be refined by direct or surrogate measures of adiposity and body fat distribution patterns as well as circulating factors such as adipokines and inflammatory markers. Adipose tissue markers are not currently used for risk assessment outside of the research setting, but provide crucial insight into the underlying processes contributing to the pathogenesis of obesity. As imaging techniques improve, and more sensitive and specific biomarkers are identified in the circulating, screening for these mechanistic tissue markers may cross to the clinical setting.

An overarching goal of biomarker discovery should be to refine obesity/metabolic risk assessment and develop interventions that are as individualized as the complex network of genetic and environmental interactions driving individual obesity susceptibility.

References

1. SurgeonGeneral, *The Surgeon General's Call To Action To Prevent and Decrease Overweight and Obesity*, Office of the Surgeon General (US); Office of Disease Prevention and Health Promotion (US); Centers for

Disease Control and Prevention (US); National Institutes of Health (US). 2001.
2. WHO, *Obesity: Preventing and Managing the Global Epidemic*, The World Health Organization, 2000.
3. R. J. Kuczmarski and K. M. Flegal, *Am. J. Clin. Nutr.*, 2000, **72**, 1074.
4. N. R. Shah and E. R. Braverman, *PLoS One*, 2012, 7, e33308.
5. A. Romero-Corral, V. M. Montori, V. K. Somers, J. Korinek, R. J. Thomas, T. G. Allison, F. Mookadam and F. Lopez-Jimenez, *Lancet*, 2006, **368**, 666.
6. M. Bastien, P. Poirier, I. Lemieux and J. P. Despres, *Prog. Cardiovasc. Dis.*, 2014, **56**, 369.
7. D. Gallagher, S. B. Heymsfield, M. Heo, S. A. Jebb, P. R. Murgatroyd and Y. Sakamoto, *Am. J. Clin. Nutr.*, 2000, **72**, 694.
8. P. Brambilla, G. Bedogni, L. A. Moreno, M. I. Goran, B. Gutin, K. R. Fox, D. M. Peters, P. Barbeau, M. De Simone and A. Pietrobelli, *Int. J. Obes.*, 2006, **30**, 23.
9. M. Deurenberg-Yap and P. Deurenberg, *Nutr. Rev.*, 2003, **61**, S80.
10. WHO Expert Consultation, *Lancet*, 2004, **363**, 157.
11. D. S. Freedman, L. K. Khan, M. K. Serdula, W. H. Dietz, S. R. Srinivasan and G. S. Berenson, *Obes. Res.*, 2005, **13**, 928.
12. P. Deshmukh-Taskar, T. A. Nicklas, M. Morales, S. J. Yang, I. Zakeri and G. S. Berenson, *Eur. J. Clin. Nutr.*, 2006, **60**, 48.
13. J. M. Van Grouw and S. L. Volpe, *Curr. Opin. Endocrinol., Diabetes Obes.*, 2013, **20**, 396.
14. WHO, *WHO child growth standards and the identification of severe acute malnutrition in infants and children: A Joint Statement*, World Health Organization and United Nations Children's Fund, 2009.
15. K. G. Dewey, *J. Hum. Lact.*, 1998, **14**, 89.
16. W. Johnson, J. Wright and N. Cameron, *BMC Pediatr.*, 2012, **12**, 104.
17. Z. Maalouf-Manasseh, E. Metallinos-Katsaras and K. G. Dewey, *J. Nutr.*, 2011, **141**, 1154.
18. L. M. Grummer-Strawn, C. Reinold, N. F. Krebs, Centers for Disease and Prevention, *MMWR. Recommendations and reports : Morbidity and mortality weekly report. Recommendations and reports/Centers for Disease Control*, 2010, vol. **59**, p. 1.
19. E. D. Rosen and B. M. Spiegelman, *Nature*, 2006, **444**, 847.
20. D. O. Okorodudu, M. F. Jumean, V. M. Montori, A. Romero-Corral, V. K. Somers, P. J. Erwin and F. Lopez-Jimenez, *Int. J. Obes.*, 2010, **34**, 791.
21. M. Stephens, M. Ludgate and D. A. Rees, *Clin. Endocrinol.*, 2011, **74**, 661.
22. A. M. Cypess, S. Lehman, G. Williams, I. Tal, D. Rodman, A. B. Goldfine, F. C. Kuo, E. L. Palmer, Y. H. Tseng, A. Doria, G. M. Kolodny and C. R. Kahn, *N. Engl. J. Med.*, 2009, **360**, 1509.
23. B. Feng, T. Zhang and H. Xu, *Ann. N. Y. Acad. Sci.*, 2013, **1281**, 160.
24. A. Park, W. K. Kim and K. H. Bae, *World. J. Stem. Cells.*, 2014, **6**, 33.
25. R. R. Rao, J. Z. Long, J. P. White, K. J. Svensson, J. Lou, I. Lokurkar, M. P. Jedrychowski, J. L. Ruas, C. D. Wrann, J. C. Lo, D. M. Camera,

J. Lachey, S. Gygi, J. Seehra, J. A. Hawley and B. M. Spiegelman, *Cell*, 2014, **157**, 1279.
26. M. Pardo, A. B. Crujeiras, M. Amil, Z. Aguera, S. Jimenez-Murcia, R. Banos, C. Botella, R. de la Torre, X. Estivill, A. B. Fagundo, J. M. Fernandez-Real, J. C. Fernandez-Garcia, G. Fruhbeck, J. Gomez-Ambrosi, R. Rodriguez, F. J. Tinahones, F. Fernandez-Aranda and F. F. Casanueva, *Int. J. Endocrinol.*, 2014, **2014**, 857270.
27. A. G. Swick, S. Orena and A. O'Connor, *Metabolism*, 2013, **62**, 1070.
28. M. K. Piya, A. L. Harte, K. Sivakumar, G. Tripathi, P. D. Voyias, S. James, S. Sabico, N. M. Al-Daghri, P. Saravanan, T. M. Barber, S. Kumar, M. Vatish and P. G. McTernan, *Am. J. Physiol.: Endocrinol. Metab.*, 2014, **306**, E512.
29. A. B. Crujeiras, M. A. Zulet, P. Lopez-Legarrea, R. de la Iglesia, M. Pardo, M. C. Carreira, J. A. Martinez and F. F. Casanueva, *Metabolism*, 2014, **63**, 520.
30. K. N. Aronis, M. Moreno, S. A. Polyzos, J. M. Moreno-Navarrete, W. Ricart, E. Delgado, J. de la Hera, A. Sahin-Efe, J. P. Chamberland, R. Berman, A. Spiro III, P. Vokonas, J. M. Fernandez-Real and C. S. Mantzoros, *Int. J. Obes.*, 2015, **39**, 156.
31. J. Vague, *Am. J. Clin. Nutr.*, 1956, **4**, 20.
32. L. de Koning, A. T. Merchant, J. Pogue and S. S. Anand, *Eur. Heart J.*, 2007, **28**, 850.
33. I. Janssen, P. T. Katzmarzyk and R. Ross, *Am. J. Clin. Nutr.*, 2004, **79**, 379.
34. WHO, *Waist circumference and waist–hip ratio: Report of a WHO expert consultation*, 2011.
35. F. P. de Heredia, S. Gomez-Martinez and A. Marcos, *Proc. Nutr. Soc.*, 2012, **71**, 332.
36. A. Villaret, J. Galitzky, P. Decaunes, D. Esteve, M. A. Marques, C. Sengenes, P. Chiotasso, T. Tchkonia, M. Lafontan, J. L. Kirkland and A. Bouloumie, *Diabetes*, 2010, **59**, 2755.
37. N. Katsiki, D. P. Mikhailidis and A. S. Wierzbicki, *Curr. Opin. Cardiol.*, 2013, **28**, 458.
38. A. Shuster, M. Patlas, J. H. Pinthus and M. Mourtzakis, *Br. J. Radiol.*, 2012, **85**, 1.
39. Y. Xia, D. L. Ergun, W. K. Wacker, X. Wang, C. E. Davis and S. Kaul, *J. Clin. Densitom.*, 2014, **17**, 78.
40. A. D. Karelis, R. Rabasa-Lhoret, R. Pompilus, V. Messier, I. Strychar, M. Brochu and M. Aubertin-Leheudre, *Obesity*, 2012, **20**, 886.
41. M. Bluher, *Diabetes Metab.*, 2012, **36**, 317.
42. C. Silvestri and V. Di Marzo, *Cell Metab.*, 2013, **17**, 475.
43. M. Krssak, K. Falk Petersen, A. Dresner, L. DiPietro, S. M. Vogel, D. L. Rothman, M. Roden and G. I. Shulman, *Diabetologia*, 1999, **42**, 113.
44. K. M. Utzschneider and S. E. Kahn, *J. Clin. Endocrinol. Metab.*, 2006, **91**, 4753.

45. B. Gaborit, I. Abdesselam, F. Kober, A. Jacquier, O. Ronsin, O. Emungania, N. Lesavre, M. C. Alessi, J. C. Martin, M. Bernard and A. Dutour, *Int. J. Obes.*, 2015, **39**, 480.
46. B. Bjorndal, L. Burri, V. Staalesen, J. Skorve and R. K. Berge, *J. Obes.*, 2011, **2011**, 490650.
47. D. P. Andersson, D. Eriksson Hogling, A. Thorell, E. Toft, V. Qvisth, E. Naslund, A. Thorne, M. Wiren, P. Lofgren, J. Hoffstedt, I. Dahlman, N. Mejhert, M. Ryden, E. Arner and P. Arner, *Diabetes Care*, 2014, **37**, 1831.
48. K. Sun, I. Wernstedt Asterholm, C. M. Kusminski, A. C. Bueno, Z. V. Wang, J. W. Pollard, R. A. Brekken and P. E. Scherer, *Proc. Natl. Acad. Sci. U. S. A.*, 2012, **109**, 5874.
49. H. K. Sung, K. O. Doh, J. E. Son, J. G. Park, Y. Bae, S. Choi, S. M. Nelson, R. Cowling, K. Nagy, I. P. Michael, G. Y. Koh, S. L. Adamson, T. Pawson and A. Nagy, *Cell Metab.*, 2013, **17**, 61.
50. I. Elias, S. Franckhauser, T. Ferre, L. Vila, S. Tafuro, S. Munoz, C. Roca, D. Ramos, A. Pujol, E. Riu, J. Ruberte and F. Bosch, *Diabetes*, 2012, **61**, 1801.
51. T. Ichiki and K. Sunagawa, *Trends. Cardiovasc. Med.*, 2014, **24**, 197.
52. K. Sun, J. Tordjman, K. Clement and P. E. Scherer, *Cell Metab.*, 2013, **18**, 470.
53. C. M. Apovian, S. Bigornia, M. Mott, M. R. Meyers, J. Ulloor, M. Gagua, M. McDonnell, D. Hess, L. Joseph and N. Gokce, *Arterioscler., Thromb., Vasc. Biol.*, 2008, **28**, 1654.
54. S. Cinti, G. Mitchell, G. Barbatelli, I. Murano, E. Ceresi, E. Faloia, S. Wang, M. Fortier, A. S. Greenberg and M. S. Obin, *J. Lipid Res.*, 2005, **46**, 2347.
55. M. Bluher, *Mol. Metab.*, 2014, **3**, 230.
56. K. Nakamura, J. J. Fuster and K. Walsh, *J. Cardiol.*, 2014, **63**, 250.
57. Y. Zhang, R. Proenca, M. Maffei, M. Barone, L. Leopold and J. M. Friedman, *Nature*, 1994, **372**, 425.
58. R. S. Ahima, D. Prabakaran, C. Mantzoros, D. Qu, B. Lowell, E. Maratos-Flier and J. S. Flier, *Nature*, 1996, **382**, 250.
59. H. Pan, J. Guo and Z. Su, *Physiol. Behav.*, 2014, **130**, 157.
60. K. Chou and C. M. Perry, *Drugs*, 2013, **73**, 989.
61. I. S. Farooqi, G. Matarese, G. M. Lord, J. M. Keogh, E. Lawrence, C. Agwu, V. Sanna, S. A. Jebb, F. Perna, S. Fontana, R. I. Lechler, A. M. DePaoli and S. O'Rahilly, *J. Clin. Invest.*, 2002, **110**, 1093.
62. A. DePaoli, *J. Endocrinol.*, 2014, **223**, T71.
63. S. Musaad and E. N. Haynes, *Epidemiol.Rev.*, 2007, **29**, 98.
64. W. Lieb, L. M. Sullivan, T. B. Harris, R. Roubenoff, E. J. Benjamin, D. Levy, C. S. Fox, T. J. Wang, P. W. Wilson, W. B. Kannel and R. S. Vasan, *Diabetes Care*, 2009, **32**, 612.
65. S. Soderberg, P. Zimmet, J. Tuomilehto, P. Chitson, H. Gareeboo, K. G. Alberti and J. E. Shaw, *Int. J. Obes.*, 2007, **31**, 1126.
66. S. Lee and H. B. Kwak, *J. Exerc. Rehabil.*, 2014, **10**, 54.

67. N. Ouchi, A. Higuchi, K. Ohashi, Y. Oshima, N. Gokce, R. Shibata, Y. Akasaki, A. Shimono and K. Walsh, *Science*, 2010, **329**, 454.
68. W. Hu, L. Li, M. Yang, X. Luo, W. Ran, D. Liu, Z. Xiong, H. Liu and G. Yang, *J. Clin. Endocrinol. Metab.*, 2013, **98**, 290.
69. I. Klein, M. Sanchez-Alavez, I. Tabarean, J. Schaefer, K. H. Holmberg, J. Klaus, F. Xia, M. C. Marcondes, J. S. Dubins, B. Morrison, V. Zhukov, A. Sanchez-Gonzalez, K. Mitsukawa, J. R. Hadcock, T. Bartfai and B. Conti, *Brain Res.*, 2011, **1423**, 1.
70. J. Fruebis, T. S. Tsao, S. Javorschi, D. Ebbets-Reed, M. R. Erickson, F. T. Yen, B. E. Bihain and H. F. Lodish, *Proc. Natl. Acad. Sci. U. S. A.*, 2001, **98**, 2005.
71. T. Yamauchi, J. Kamon, H. Waki, Y. Imai, N. Shimozawa, K. Hioki, S. Uchida, Y. Ito, K. Takakuwa, J. Matsui, M. Takata, K. Eto, Y. Terauchi, K. Komeda, M. Tsunoda, K. Murakami, Y. Ohnishi, T. Naitoh, K. Yamamura, Y. Ueyama, P. Froguel, S. Kimura, R. Nagai and T. Kadowaki, *J. Biol. Chem.*, 2003, **278**, 2461.
72. M. Okamoto, M. Ohara-Imaizumi, N. Kubota, S. Hashimoto, K. Eto, T. Kanno, T. Kubota, M. Wakui, R. Nagai, M. Noda, S. Nagamatsu and T. Kadowaki, *Diabetologia*, 2008, **51**, 827.
73. M. Okada-Iwabu, T. Yamauchi, M. Iwabu, T. Honma, K. Hamagami, K. Matsuda, M. Yamaguchi, H. Tanabe, T. Kimura-Someya, M. Shirouzu, H. Ogata, K. Tokuyama, K. Ueki, T. Nagano, A. Tanaka, S. Yokoyama and T. Kadowaki, *Nature*, 2013, **503**, 493.
74. G. Boden, *Curr. Opin. Endocrinol., Diabetes Obes.*, 2011, **18**, 139.
75. N. Esser, S. Legrand-Poels, J. Piette, A. J. Scheen and N. Paquot, *Diabetes Res.Clin. Pract.*, 2014, **105**, 141.
76. G. S. Hotamisligil, N. S. Shargill and B. M. Spiegelman, *Science*, 1993, **259**, 87.
77. P. A. Kern, M. Saghizadeh, J. M. Ong, R. J. Bosch, R. Deem and R. B. Simsolo, *J. Clin. Invest.*, 1995, **95**, 2111.
78. P. A. Kern, S. Ranganathan, C. Li, L. Wood and G. Ranganathan, *Am. J. Physiol.: Endocrinol. Metab.*, 2001, **280**, E745.
79. G. S. Hotamisligil, A. Budavari, D. Murray and B. M. Spiegelman, *J. Clin. Invest.*, 1994, **94**, 1543.
80. J. P. Bastard, C. Jardel, E. Bruckert, P. Blondy, J. Capeau, M. Laville, H. Vidal and B. Hainque, *J. Clin. Endocrinol. Metab.*, 2000, **85**, 3338.
81. C. J. Roos, P. H. Quax and J. W. Jukema, *Biomarkers Med.*, 2012, **6**, 35.
82. I. R. G. C. E. R. F. Collaboration, N. Sarwar, A. S. Butterworth, D. F. Freitag, J. Gregson, P. Willeit, D. N. Gorman, P. Gao, D. Saleheen, A. Rendon, C. P. Nelson, P. S. Braund, A. S. Hall, D. I. Chasman, A. Tybjaerg-Hansen, J. C. Chambers, E. J. Benjamin, P. W. Franks, R. Clarke, A. A. Wilde, M. D. Trip, M. Steri, J. C. Witteman, L. Qi, C. E. van der Schoot, U. de Faire, J. Erdmann, H. M. Stringham, W. Koenig, D. J. Rader, D. Melzer, D. Reich, B. M. Psaty, M. E. Kleber, D. B. Panagiotakos, J. Willeit, P. Wennberg, M. Woodward, S. Adamovic, E. B. Rimm, T. W. Meade, R. F. Gillum, J. A. Shaffer, A. Hofman, A. Onat,

J. Sundstrom, S. Wassertheil-Smoller, D. Mellstrom, J. Gallacher, M. Cushman, R. P. Tracy, J. Kauhanen, M. Karlsson, J. T. Salonen, L. Wilhelmsen, P. Amouyel, B. Cantin, L. G. Best, Y. Ben-Shlomo, J. E. Manson, G. Davey-Smith, P. I. de Bakker, C. J. O'Donnell, J. F. Wilson, A. G. Wilson, T. L. Assimes, J. O. Jansson, C. Ohlsson, A. Tivesten, O. Ljunggren, M. P. Reilly, A. Hamsten, E. Ingelsson, F. Cambien, J. Hung, G. N. Thomas, M. Boehnke, H. Schunkert, F. W. Asselbergs, J. J. Kastelein, V. Gudnason, V. Salomaa, T. B. Harris, J. S. Kooner, K. H. Allin, B. G. Nordestgaard, J. C. Hopewell, A. H. Goodall, P. M. Ridker, H. Holm, H. Watkins, W. H. Ouwehand, N. J. Samani, S. Kaptoge, E. Di Angelantonio, O. Harari and J. Danesh, *Lancet*, 2012, **379**, 1205.
83. S. Kaptoge, S. R. Seshasai, P. Gao, D. F. Freitag, A. S. Butterworth, A. Borglykke, E. Di Angelantonio, V. Gudnason, A. Rumley, G. D. Lowe, T. Jorgensen and J. Danesh, *Eur. Heart J.*, 2014, **35**, 578.
84. C. D'Alessandris, R. Lauro, I. Presta and G. Sesti, *Diabetologia*, 2007, **50**, 840.
85. Y. Peng, S. Yu, H. Li, H. Xiang, J. Peng and S. Jiang, *Cell. Signalling*, 2014, **26**, 1888.
86. F. I. Milagro, J. Miranda, M. P. Portillo, A. Fernandez-Quintela, J. Campion and J. A. Martinez, *PLoS One*, 2013, **8**, e54319.
87. Y. C. Wang, Y. Li, X. Y. Wang, D. Zhang, H. Zhang, Q. Wu, Y. Q. He, J. Y. Wang, L. Zhang, H. Xia, J. Yan, X. Li and H. Ying, *Diabetologia*, 2013, **56**, 2275.
88. A. Prats-Puig, F. J. Ortega, J. M. Mercader, J. M. Moreno-Navarrete, M. Moreno, N. Bonet, W. Ricart, A. Lopez-Bermejo and J. M. Fernandez-Real, *J. Clin. Endocrinol. Metab.*, 2013, **98**, E1655.
89. M. Rosenbaum and R. L. Leibel, *Int. J. Obes.*, 2010, **34**(Suppl 1), S47.

CHAPTER 4

Risk Factors for Obesity

JAMES DAVID ADAMS, JR.

Department of Pharmacology and Pharmaceutical Sciences, School of Pharmacy, University of Southern California, Los Angeles, CA 90089, USA
Email: jadams@usc.edu

4.1 Introduction

Starting in the 1980s in the US, it became acceptable to be overweight or obese. Laws were passed to prevent employment discrimination based on body weight. Even discussing a person's body weight issues at work was not and is not allowed. Children were given the option to forgo physical education in school, at the request of their parents. The US public had decided it was politically acceptable to be obese. The Centers for Disease Control website shows the consequences. Obesity increased every year until about 2010. The adult population is now stable at 35% obesity and 68% overweight.[1,2] The incidence of obesity in youth increased yearly in the US and is now stable at 17%.[1]

The food and nutrition industries in the US responded to the public demands for high caloric foods. Flavor enhancers were added to foods in increasing amounts, including salt, monosodium glutamate, high fructose corn syrup and solid fats. Diacetyl, which is butane-2,3-dione, is usually added to solid fats to give a butter flavor. These flavor enhancers are used to help stimulate the consumption of the foods. High fructose corn syrup and solid fats greatly increase the caloric content of foods. Processed foods are common in fast food restaurants and markets. Processed foods are usually high in fat, high in calories and high in flavor enhancers. Fast food establishments in the US serve high fat foods in increasing amounts every year.

RSC Drug Discovery Series No. 45
Cardiovascular and Metabolic Disease: Scientific Discoveries and New Therapies
Edited by Philip Peplow, James Adams and Tim Young

Published by the Royal Society of Chemistry, www.rsc.org

The US public has asked for foods that promote obesity. The food and nutrition industries have responded by offering for sale high caloric foods.

Foods that contain high fructose corn syrup and solid fats can increase body weight. However, diets that contain natural fructose have been found to be superior to low fructose diets for weight loss.[3] Hyperlipidic, in other words high fat, diets are known to cause obesity in rats and humans.[4,5] It should be mentioned that high protein diets, that were promoted for weight loss, are toxic to the human kidney and have killed several people.

4.2 Obesity Can be Caused by Genetic and Other Diseases

There are diseases that can cause obesity: hypothyroidism, polycystic ovarian syndrome, depression, Cushing's syndrome, hypothalamic disorders, Alstrom syndrome, Bardet–Biedl syndrome, adenovirus-36 infection, ciliary protein CEP19 inactivation, pseudohypoparathyroidism, Prader–Willi syndrome, and many others.[6–8] Hypothyroidism may affect about 0.4% of the US population. Polycystic ovarian syndrome affects about 10% of the women of child bearing age in the US. About 10% of US adults are depressed. The other diseases are rare and tend to affect 1 in 10 000 or fewer people. These diseases are major causes of obesity, but do not account for all of the obesity in the US.

4.3 Obesity Can be Caused by Drugs

Several drugs can make people more prone to becoming obese: antidepressants, lithium, antipsychotics, diabetic medicines, antihypertensives, anticonvulsants, some anticancer drugs, migraine headache drugs, steroids, contraceptives, antihistamines, protease inhibitors and others.[2,9,10] Rosiglitazone and pioglitazone are drugs used to treat obesity onset diabetes, also called type 2 diabetes. These drugs can cause patients to become more obese, which increases their diabetes symptoms and can cause congestive heart failure. These drugs are also obesogens, as discussed in Section 4.6. The mechanisms by which all of these drugs induce obesity are not known with certainty. It is likely there are several different mechanisms involved. Smoking cessation can also cause obesity since nicotine is a powerful appetite suppressant.

4.4 The High Fat Food Problem

Eating a high fat diet is the most common cause of obesity. Many people are not aware of the fat content of the foods they eat. For instance, a hamburger can derive 65% or more of its calories from fat. Many people think that a hamburger is a good protein source. It is a better fat source. Cheese can also be very high in fat. Some cheeses, such as brie and camembert, derive 70% or more of their calories from fat. Bacon can derive 80% or more of its calories from fat. Deep fried foods are similarly high in fat. Even deep fried French fries can derive 40% or more of the calories from fat. Ice cream derives about 45% of its calories from fat. Frozen foods can be much higher in fat, due to addition of solid fat as a flavor enhancer. Some frozen foods are so high in fat, that nutritionists do not recommend eating them. Americans, according to the US Department of Agriculture, eat diets that contain about 35% or more dietary fat calories. It is advisable to eat 30% or less dietary fat calories.

4.5 Why is Dietary Fat a Problem?

Eating fat is addictive. This may be because of ancient survival mechanisms in humans. During the fall, people used to put on 8 kg or so in order to survive the winter, the time when food was scarce. Meat and fat were consumed in large amounts in order to put on weight quickly. Fat consumption stimulates opioid peptide release in the brain.[11,12] Release of these opioids decreases pain, causes a sense of satiety and makes people addicted. Dopamine is also released during eating, which causes pleasurable feelings.[13] Eating fat also increases the synthesis and release of endocannabinoids.[14] These endocannabinoids increase the desire to eat, usually about an hour after eating. Endocannabinoids interact with cannabinoid receptors to decrease satiety. They also interact with transient receptor potential cation channels to decrease pain. When people eat fat, they get a sense of satiety and decreased pain. Soon, they want to eat more fat. As the addiction becomes established, they crave fat with every meal.

Of course, fat is available in all foods, even vegetables and fruits. However, as the fat addiction progresses, a tolerance to the opioid peptides released by fat develops. People need to eat ever increasing amounts of fat in order to increase the production of opioid peptides and overcome the tolerance. The endocannabinoids make matters worse by decreasing satiety. There does not appear to be a tolerance to this anti-satiety effect of endocannabinoids.

4.6 Alcohol is the Worst Obesogen

Obesogens are compounds that cause fat deposition in the visceral compartment. Normally, fat is deposited mainly in subdermal areas. Subdermal fat is not as dangerous as visceral fat, since visceral fat produces inflammatory adipokines and increases the production of toxic fats such as ceramide and the endocannabinoids.[15] As humans age, fat deposition shifts from subdermal to visceral fat compartments. However, alcohol and other obesogens greatly increase visceral fat production. Alcohol induces sterol regulatory element binding protein, which alters the deposition of fat, and increases the deposition of visceral fat.[16] This is the basis of beer belly. Alcohol is also addictive. As addiction becomes established, people want more alcohol and develop more visceral fat. They then develop heart disease, type 2 diabetes and arthritis that are caused by the inflammatory adipokines and the toxic fats. Alcohol is considered the number one cause of heart disease in the US, by the American Heart Association. Alcohol is also high in calories. Pure, 100%, alcohol contains 246 calories per serving, 1.5 fluid ounces.

4.7 Fad Foods and Obesity

Advertisements bombard us with misleading information about fad foods, super foods, antioxidant rich foods and other foods that promise to increase our health. Some people add these fad foods in large amounts to their diets in order to greatly increase their health. For instance, pomegranates and acai berries are supposed to be antioxidant rich foods that will improve the health of our hearts and bodies. However, many of these fad foods are flavored with large amounts of sugar or high fructose corn syrup as flavor enhancers. There is nothing wrong with eating pomegranates or acai berries. In fact, eating fruit is advisable since it contains vitamins, minerals, phytonutrients, soluble fiber, protein, carbohydrate, fat and other nutritious compounds. However, people should eat pomegranates and acai berries as part of their normal diet, not in addition to their normal diet. In other words, we are advised to add the fad foods to our diets in addition to what we normally eat. Instead, we should cut out a food and replace it with pomegranates or acai berries. In fact, we would be well advised to cut out half of the meat we eat and replace it with fruit. Most people eat at least twice the amount of protein, especially meat, that they need. It has been known since the 1950s that US vegetarians have much less cancer than meat eaters. Recent research confirms this and shows that eating meat and animal protein can increase the risk of cancer by 4 fold.[17]

4.8 Toxic Fats

Ceramide is a toxic fat that can be made from two fatty acids and serine. As the fat burden in the body increases, triglyceride stores become overloaded. The body starts to make more ceramide. Ceramide causes the dysfunction of nitric oxide synthase, both the inducible and the endothelial forms.[18] This causes endothelial dysfunction and kidney toxicity, leading to hypertension.

During obesity, fatty acids increase in the blood and organs. This increases low density lipoprotein cholesterol (LDL-C) and decreases high density lipoprotein cholesterol (HDL-C). Fatty acid accumulation in organs increases the synthesis and release of endocannabinoids.[18] These endocannabinoids increase the release of inflammatory adipokines from visceral fat.

4.9 Bad Carbohydrates, Empty Calories and High Glycemic Index

Foods made with refined flour, sugar, high fructose corn syrup and other refined carbohydrates, are considered bad carbohydrates and are said to have empty calories. These foods are said to have a high glycemic index, and quickly increase blood sugar levels, which can be bad for diabetics. At this point, there is no evidence that sugar is addictive. However, excessive refined carbohydrate consumption can increase body fat deposition.

4.10 Childhood Obesity

Childhood obesity is a problem. Risk factors for childhood obesity have been identified: parental obesity, very early (by 43 months) high body mass index or adiposity rebound, more than eight hours spent watching television per week by age 3, catch-up growth by age 8 months, excessive weight gain in the first year, birth weight, and short (<10.5 hours) sleep duration by age 3.[19] Parental obesity is important since parents teach their children how to eat high fat diets and how to become obese. The lifestyle issues such as television, sleep duration and weight gain are also controlled by parents.

4.11 Sedentary Lifestyles

When obesity became politically acceptable, exercise became undesirable. Video games, high definition television and other technologies were introduced that quickly reinforced sedentary lifestyles. This was a very rapid change for most people in the US, who had been active and now became sedentary within a few years.

It is clear that *Homo sapiens* has existed for about 200 000 years. During most of that time we have been very active hunters and gatherers. It has only been within the last 10 000 years of so that agriculture was introduced. Of course, early farmers led very active lifestyles. We have had 200 000 years of natural selection for people who lead physically active lifestyles. It has only been since about the 1980s that the US population decided to become sedentary. We are now undergoing a massive natural selection for people who can survive obesity and inactive lifestyles. Since the 1980s, there has been a tremendous increase in heart disease, type 2 diabetes, arthritis and cancer in the US such that healthcare costs amount to 18% of our gross national product.[20]

4.12 Why is Exercise so Important?

We have forgotten the traditional concept of living in balance.[20] This involves keeping ourselves thin and strong, among other things. When the body is in balance, it can heal itself. Most of us are born healthy and can stay healthy if we live in balance. Prevention of disease is the central reason for living in balance.

Exercise keeps the body strong. It is not known why strong muscles are important for health. It is known that exercise stimulates stem cell formation in every organ.[15] Stem cells help restore the body. Exercise helps diminish fat stores in muscle, perivascular sites and other undesirable sites. Fat stores in muscle decrease insulin sensitivity. Perivascular fat secretes inflammatory adipokines and toxic fats that cause atherosclerosis and heart disease. A program of exercise and weight reduction can reverse heart disease, type 2 diabetes and arthritis.[20] It can also help patients survive cancer.[20]

4.13 Present Lessons for the Future

Currently, in the US, the population has decided to be obese. This is part of "living the good life". The food and nutrition industries supply the foods and alcohol that quickly lead to obesity. Keeping these foods and alcohol in the diet maintains obesity. The pharmaceutical industry is now engaged in the major task of keeping these obese people alive with their type 2 diabetes, heart disease, arthritis and cancer. The pharmaceutical industry does not have cures for type 2 diabetes, heart disease or arthritis. Drugs are available to help the body cure itself from cancer. People with heart disease, type 2 diabetes and arthritis must stay on their drugs for the rest of their lives. The healthcare profession has forgotten that drugs do not cure. The body heals itself. Drugs can help the body heal itself. This is a central principle of Pharmacology. The healthcare profession is virtually useless at helping people change their toxic lifestyles and cease being obese. Obesity must be treated as any other addiction, since obesity is usually an addiction to eating fat.

References

1. C. L. Ogden, M. D. Carroll, B. K. Kit and K. M. Flegal, *JAMA*, 2014, **311**, 806.
2. S. M. Wright and L. J. Aronne, *Abdom. Imaging*, 2012, **37**, 730.
3. M. Madero, J. C. Arriaga, D. Jalal, C. Rivard, K. McFann, O. Perez-Mendez, A. Vazquez, A. Ruiz and M. A. Lanaspa, *Metab., Clin. Exp.*, 2011, **60**, 1551.
4. D. Estadella, L. M. Oyama, A. A. Bueno, C. A. Habitante, G. I. Souza, E. B. Ribeiro, C. S. Motoyama and C. M. Oller do Nascimento, *Lipids Health Dis.*, 2011, **10**, 168.
5. A. Astrup, B. Buemann, P. Western, S. Toubro, A. Raben and N. J. Christensen, *Am. J. Clin. Nutr.*, 1994, **59**, 350.
6. J. U. Weaver, *Front. Horm. Res.*, 2008, **36**, 212.
7. R. L. Atkinson, *Int. J. Pediatr. Obes.*, 2011, **6**(Sup 1), 2.
8. A. Shalata, M. C. Ramirez, R. J. Desnick, N. Priedigkeit, C. Buettner, C. Lindtner, M. Mahroum, M. Abdul-Ghani, F. Dong, N. Arar, O. Camacho-Venegas, R. Zhang, S. C. Camacho, Y. Chen, M. Ibdah, R. DeFronzo, V. Gillespie, K. Kelley, B. D. Dynlacht, S. Kim, M. J. Glucksman, Z. U. Borochowitz and J. A. Martignetti, *Am. J. Genet.*, 2013, **93**, 1061.
9. H. Pijl and A. E. MEinders, *Drug Saf.*, 1996, **14**, 329.
10. R. Ness-Abramof and C. M. Apovian, *Timely Top. Med. Cardiovasc. Dis.*, 2005, **9**, E31.

11. G. Q. Chang, O. Karatayev, R. Ahsan, V. Gaysinskaya, Z. Marwil and S. F. Leibowitz, *Am. J. Physiol.: Endocrinol. Metab.*, 2007, **292**, E561.
12. S. Matsumura, A. Eguchi, Y. Okafuji, S. Tatsu, T. Mizushige, S. Tsuzuki, K. Inoue and T. Fushiki, *FEBS Lett.*, 2012, **586**, 1231.
13. N. D. Volkow, G. J. Wang, D. Tomasi and R. E. Baler, *Obes. Rev.*, 2013, **14**, 2.
14. S. Banni and V. Di Marzo, *Mol. Nutr. Food Res.*, 2010, **54**, 82.
15. J. D. Adams, E. J. Lien and K. Parker, in *Extracellular and Intracellular Signaling*, ed. J. D. Adams and K. K. Parker, Royal Society of Chemistry, London, 2011.
16. M. You, M. Fischer, M. A. Deeg and D. W. Crabb, *J. Biol. Chem.*, 2002, **277**, 29342.
17. M. E. Levine, J. A. Suarez, S. Brandhorst, P. Balasubramanian, C. W. Cheng, F. Madia, L. Fontana, M. G. Mirisola, J. Guevara-Aguirre, J. Wan, G. Passarino, B. K. Kennedy, M. Wei, P. Cohen, E. M. Crimmins and V. D. Longo, *Cell Metab.*, 2014, **19**, 407.
18. S. A. Ritchie and J. M. C. Connell, *Nutr., Metab. Cardiovasc. Dis.*, 2007, **17**, 319.
19. J. J. Reilly, J. Armstrong, A. R. Dorosty, P. M. Emmett, A. Ness, I. Rogers, C. Steer and A. Sherriff, *Br. Med. J.*, 2005, **330**, 1357.
20. J. D. Adams, *World J. Pharmacol.*, 2013, **2**, 73.

CHAPTER 5

Gene–Environment Interactions in Obesity

JOSEPH A. CORNICELLI

Charles River Laboratories, 251 Ballardvalle St., Wilmington, MA 01887, USA
Email: joe.cornicelli@crl.com

5.1 Introduction

The world is in the midst of an obesity pandemic. In 2005, the estimated number of adults who were overweight or obese was pegged at 1.1 billion, along with 10% of the world's children.[1] In 2010, the World Health Organization revised that estimate to ~2 billion people, with 700 million defined as obese based upon body mass index (BMI).[2] The rapid increase in incidence is underscored by the experience in the United States. According to the US Center for Disease Control (CDC), in 1991 only 3 states had an obese population of >15%. In 2010, not one state had an incidence of <20%. Thirty six states had an obesity rate of 25%, and in 12 states, 36% of the population could be considered obese on the basis of a BMI ≥ 30.[3] Obesity is not a condition confined to the adult population. Looking again to the US CDC, an estimated 12.5 million US children (ages 2–19 years) are obese.[4] The problem in the US has reached such a point that the American Medical Association issued a policy statement classifying obesity as a disease.[5] The prevalence of obesity is higher in Western and more economically developed countries.[1] However, the acquisition of Western lifestyles in developing nations is allowing those countries to catch up in terms of affected individuals. Since many of those countries have large populations,

RSC Drug Discovery Series No. 45
Cardiovascular and Metabolic Disease: Scientific Discoveries and New Therapies
Edited by Philip Peplow, James Adams and Tim Young

Published by the Royal Society of Chemistry, www.rsc.org

the actual disease burden related to overweight and obesity is larger in the developing world.[6,7] Perhaps for the first time in history, there are more overweight and obese people in the world than there are those who are malnourished.[8]

Examining the demographics of overweight and obesity over the past 50 years leads one to the conclusion that humans have become an obese species. This observation begs the question as to how did this happen. Examined in terms of conservation of energy, the question can be addressed in terms of energy balance; calories taken in *vs.* calories expended. If one side of the equation is greater than the other, body weight will either increase or decrease. The general consensus regarding this paradigm is that obese individuals are taking in more energy than they expend.[9] That statement may seem obvious, but it poses yet another, more difficult question. Why are such a large number of individuals in positive energy balance? This chapter is devoted to an assessment of that question in terms of the genetic and environmental determinants of human obesity and how these factors interact to determine body composition.

5.2 Definition and Methods Used in the Assessment of Obesity

Obesity is defined as the excess accumulation of adipose tissue. This can be measured directly using DEXA imaging (dual energy X-ray absorptiometry), nuclear magnetic resonance (NMR) imaging and analysis, and isotopic dilution methods.[10] These methods are costly, and may involve exposure to X-irradiation or isotope ingestion, and may not be practical to apply to large numbers of individuals and pediatric populations. Fat mass has also been estimated using bioelectrical impedance or skin-fold thickness measurements. These estimates tend to correlate well with adiposity in individuals with normal weight. They are less reliable in the very lean or obese. BMI defined as body weight (in kg) divided by height in m^2 is the most commonly used surrogate for adiposity. It is readily measurable, obtainable from medical records, and can be used with great facility in epidemiological studies. It correlates reasonably well with body composition. For the purposes of this chapter, a BMI of >30 kg m^{-2} is used to define obesity in a clinical sense. This has implications for the treatment and management of the disease. In reality, however, BMI tends to be normally distributed across the population, making a clear cut demarcation between "lean" and obese difficult. This concept has ramifications for the interpretation of population studies. Finally, the distribution of adipose tissue appears to be a significant determinant of metabolic dysfunction in obese individuals, with individuals accumulating abdominal fat at greater risk for subsequent cardiovascular events.[11]

Abdominal obesity is most accurately measured in humans using computer tomography, ultrasound, or magnetic resonance. Again, these methods are costly and may involve irradiation. The measurement of waist to hip ratio (>1.0 for men, and >0.85 for women) has been used to identify individuals with abdominal fat accumulation, as has waist circumference alone. Both correlate well with BMI.[12]

5.3 Genetics and Obesity

Obesity may be one of the most heritable traits that humankind possesses. Heritability defines the per cent variability within a trait that can be accounted for by genetics. One of the earliest records of this concept was published by Davenport in 1923.[13] The author studied families with different body types that were described as very slender, slender, medium, fleshy and very fleshy. BMI within the families was plotted as a function of frequency. The author concluded that there were differences in the "economy of fattening" among individuals and that there were 2 or more genetic factors involved in determining body type, although in some groups, only 1 factor was likely to be involved. This early work not only suggested that genetics plays a role in obesity, but also noted that there may be monogenic and polygenic factors at play.

Undoubtedly, the strongest evidence for the role of genetics in the development of obesity comes from the studies of families, with particular reference to the examination of monozygotic twins. Studies of monozygous twins that were reared apart proved valuable in determining the heritability of obesity. While genetically nearly identical (Bruder[14] and co-workers have demonstrated differences in gene copy numbers in some monozygotic twins), the postnatal environments experienced by these individuals were clearly different, removing that confounding factor. Stunkard *et al.*[15] studied fraternal and identical twins that were either reared together or apart. His sample included nearly 250 pairs of identical twins and over 400 pairs of fraternal twins. Each population was split, more or less equally, into cohorts that were either reared together or apart. The within pair correlation coefficient for BMI for identical twins reared apart was 0.70 for males and 0.66 for females. The authors concluded that sharing the same childhood environment had little to no effect on BMI recorded later in life. Studies of twins from the UK[16] further demonstrated that the heritability of obesity, measured as BMI, did not differ between twins reared apart and twins reared together. That same study also found that the correlation held between twins reared apart with relatives (which tended to be more similar environments) and those raised in less similar environments. Between 1980 and 2006, over 30 published papers have appeared studying the heritability of obesity (BMI), with estimates of heritability ranging between 64 and 84%.[17] While the estimated range is a bit wide, current consensus estimates that about 65% of the variation in obesity is due to genetic factors.[18] Given this level of heritability, the question arises as to which genes are responsible.

5.4 Animal Models and the Genetics of Obesity

5.4.1 Spontaneous Animal Models of Obesity

Much of what we know of human obesity has its roots in animal models. A large number of animal models have been used to study the contribution of genetics and environment to the accumulation of excess adipose tissue. This section reviews some of the more important models that have contributed to our understanding of those mechanisms.

In late 1949, an obese mouse was discovered coincidentally at the Jackson Laboratory in Bar Harbor, Maine. The founder animals were crossed onto a C57BL/6 background and dubbed *ob/ob*, for obese, and characterized as a model of mild diabetes.[19] The animals are hyperphagic, obese, hyperinsulinemic and mildly diabetic. The genetic mutation in these animals is carried on chromosome 6. In 1965, a second obese mouse strain was discovered, but it's mutation was found on chromosome 4.[20] These animals are also obese and hyperphagic, but unlike the *ob/ob* strain, this line developed severe, life threatening diabetes. These strains have subsequently been used as models of obesity driven dysregulation of glycemic control since their discovery. Nearly 50 years would pass before the genetic defect responsible for these phenotypes would be elucidated. In 1994, Zhang and co-workers[21] employed positional cloning to demonstrate that the ob/ob mutation was due to a defect in the leptin gene. Similarly, in 1996[22,23] mutations in the leptin receptor were identified as the determinant of obesity and diabetes in the db/db strain.

Leptin is one of a class of proteins referred to as adipokines that are cytokines secreted from adipocytes. Leptin expression in adipocytes is abundant, and it plays a very important role in appetite control and in thermogenesis. The mutation in the leptin gene in ob/ob mice is due to a single base pair deletion in the coding region of the gene, and is autosomal recessive. This deletion results in a frame shift and generation of a premature stop codon. Leptin deficiency in the ob/ob mice results in uncontrollable appetite, obesity, insulin resistance and hyperinsulinemia, and transient type 2 diabetes. Leptin exerts its actions by binding to its cell surface receptor. A point mutation (G$\rightarrow$T) in the leptin receptor gene results in an abnormal splice variant wherein the receptor is missing the cytoplasmic region, producing a receptor deficient in signaling. The lack of signaling in the hypothalamus in db/db animals homozygous for the mutation leads to hyperphagia in the presence of hyperleptinemia, obesity, insulin resistance and hyperinsulinemia, and frank, persistent diabetes.

Lois and Theodore Zucker observed spontaneously obese 13M strain rats that were crossed to the Sherman strain. They maintained a colony of these animals at the Laboratory for Comparative Pathology in Stowe, MA, and in 1961, reported that the animals had an autosomal recessive mutation on chromosome 5 that they called *fa* for fatty.[24] The rats are hyperphagic and exhibit obesity that is evident by 5 weeks of age due to excessive deposition of subcutaneous adipose tissue. Although the animals are very hyperinsulinemic,[25] they maintain relatively normal blood glucose levels.[26] This is most likely due to the animals' ability to maintain the hyperinsulinemic state. In 1995, it was later determined that the *fa* gene encoded the rat leptin receptor.[27,28] Colonies of ZFR rats have been established at many institutions that were derived from the original, nuclear group of animals.

The relevance of leptin in human obesity came to light in 1997, when Montagu, O'Rahilly, Farooqi and co-workers at the University of Cambridge, had the opportunity to study 2 severely obese children.[29] The patients, aged 8 and 2 years, were Pakistani cousins, weighing 86 kg and 29 kg, respectively. Serum leptin levels in the children were extremely low in the presence of markedly elevated fat mass. Further genetic analysis revealed a single nucleotide deletion (G 133) resulted in a homozygous frame shift mutation in both individuals. These data were the first that provided incontrovertible proof that human obesity can be inherited, and validated the Zucker rat as a model to study the role of leptin in human obesity.

One of the oldest known mouse models of obesity involves a mutation in the *agouti* gene. The activity of this gene is responsible for determining whether an animal's coat color will be solid or banded (agouti). In wild type animals, each hair shaft in the animal's coat is black (or brown or grey, depending upon other alleles at other loci), with only a subapical band of yellow pigment. The yellow agouti mutation (A^y), discovered in 1905, in combination with all of the other agouti locus alleles leads to the production of a uniform yellow coat over the entire body.[30] In addition to determining coat color, A^y has a multitude of effects on other physiologic systems. Early on, investigators found that the A^y mouse developed obesity, increased somatic growth, was more susceptible to tumor formation and had fertility problems. The agouti protein was cloned in 1992 by Bultman.[31] The A^y mutation is the result of a deletion of between 120–170 kb of genomic DNA. That deletion places the expression of the gene product under the control of a ubiquitously expressed promoter, as opposed to the tissue-specific promoter element.[31–33] Transgenic mice, engineered to express *agouti* ubiquitously develop a yellow coat, and are obese, hyperinsulinemic, and hyperglycemic akin to the A^y mouse.[34] Adipose specific expression of *agouti* results in animals with an overgrowth of adipose tissue in the presence of normal food intake, suggesting that the increased fat accumulation is due to changes in energy metabolism.[35] There is reason to believe that *agouti* has relevance to human obesity, as it is expressed in human adipose tissue,[36,37] and is increased in type 2 diabetic adipose tissue.[38] While ubiquitous expression of *agouti* and adipose tissue-specific expression of the protein result in an obese phenotype, expression in the skin does not induce obesity,[39] suggesting that the obesogenic effects of the protein are tissue dependent.

The examples cited above are monogenic models of obesity. The disorder in the vast majority of humans is most likely driven by a number of genetic and environmental factors. The application of polygenic animal models could, therefore, be much more relevant to the study of human obesity.

The New Zealand Obese (NZO) mouse is a polygenic model of obesity. Although the animals exhibit genetically normal levels of leptin and leptin receptors, they are somewhat leptin resistant.[40] This leptin resistance results in hyperphagia and rapid weight gain early in life. Accumulation of adipose tissue in the NZO mouse is quite pronounced with fat depots accounting for more than 40% of total body weight in aged animals.[40] In addition, NZO mice exhibit a decreased propensity to spontaneous exercise when compared to wild type or even ob/ob mice.[41] Decreased physical activity, hyperphagia and reduced energy expenditure in the NZO closely mimics conditions of human obesity.

Selective inbreeding of mice for large body size generated the Kuo Kondo (KK) mouse in Japan.[42] The animals are hyperphagic, hyperinsulinemic, insulin resistant and are moderately obese at 2 months of age.[43] The A^y mutation was bred onto the KK strain to develop the KKA^y mouse which develops frank diabetes.[44]

Otsuka Pharmaceuticals in Tokushima, Japan selectively bred Long Evans rats from Charles River that had developed spontaneous type 2 diabetes, producing, what has come to be known as, the Otsuka Long Evans Tokushima Fatty (OLETF) rat.[45] OLETF rats are hyperphagic, which is manifested several weeks after birth. While many genes appear to be at work in this strain, OLETF rats are a naturally occurring knockout of the cholecystokinin type A receptor (CCKAR).[46] CCK is a potent hormone with multiple activities, including the induction of satiety.[47] The lack of receptor activity in the OLETF rat drives meal size which, in turn, participates in the development of moderate obesity in the strain. Diabetes is evident in all male animals by 25 weeks of age, while less than a third of the females develop the disease even out to the age of 60 weeks.[48]

5.4.2 Genetically Engineered Animal Models of Obesity

All of the genetic models of obesity discussed up to this point are the result of selective breeding of serendipitously discovered mutations. This rather arbitrary approach to discovering loss of function mutations has led to efforts to artificially accelerate the mutation rate using chemicals and radiation.[49,50] This "blunt instrument" approach is relatively efficient at inducing random mutations. However, the cost to phenotype the resultant offspring of the treated animals would be prohibitively expensive, especially in light of the probability that most mutations would not be expected to produce an obese phenotype.

A great deal of progress has been made in the area of transgenic (Tg) and knockout (KO) models. The Human Obesity Gene Map is a collaborative effort to comprehensively monitor published studies demonstrating a role for genes and quantitative trait loci (QTLs) that play a role in the development of obesity. The approach monitors single gene mutations from obesity cases, Mendelian disorders with obesity as a clinical feature, Tg and KO murine models relevant to obesity and energy metabolism, QTLs from animal breeding studies, associations from candidate genes, and linkages from genome wide array scans (GWAS). The 2005 update of the map reported that there are 244 genes that, when mutated or overexpressed as transgenes in mice, result in phenotypic changes to body weight or adiposity.[51]

The initial work to genetically engineer obesity involved overexpression of candidate genes by inserting the full length coding sequence downstream from a promoter that would provide either global or tissue specific expression of the gene product. The approach is relatively inexpensive and straightforward. The level of expression that can be achieved, however, might not always demonstrate a physiologic effect. Global KO models have proven to be more predictable and reproducible. This method silences the gene of interest in all tissues. KO models have, in some instances, resulted in unexpected phenotypes, such as embryonic or neonatal lethality. Additionally, the silencing effects of one gene may be counteracted by compensatory increases in others. This might be best illustrated by studies on neuropeptide Y (NPY). Introduction of NPY directly into the brain produces a robust stimulation of feeding behavior. When animals are fed, the levels decline and when they are starved, NPY increases.[52] Surprisingly, NPY KO mice exhibit no obvious abnormal phenotype.[53–55]

In some instances, insight is gained from unanticipated results using Tg technology. A case in point is that of the *ax1* mouse. *Ax1* is a tyrosine kinase member of the platelet-derived growth factor receptor family. It is overexpressed in many pathologic conditions including some cancers and in the synovium of rheumatoid arthritis patients. The *ax1* Tg was developed to assess the role of this receptor in leukemia, by ectopically expressing the receptor in myeloid cells. The resulting animals did not exhibit hematologic malignancies; they did exhibit a diabetogenic phenotype.[56] The animals were hyperinsulinemic, insulin resistant, and morbidly obese, and yet were not hyperphagic. Additional work demonstrated elevated levels of TNF-α in the serum of the Tg animals that could be further elevated upon stimulation with LPS.[57] TNF-α has been shown to be elevated in both rodent and human obesity.[58,59] These data suggest that expression of *ax1* in myeloid cells modulates TNF-α production, and that this directly contributes to the obesity phenotype.

5.5 Genetic Approaches to Studying Human Obesity

There is general consensus within the scientific community that obesity is one of the most highly heritable traits shared by all humans.[60,61] Given this high degree of heritability, genetic approaches can be useful in understanding the susceptibility to obesity and the mechanisms involved in weight regulation. A number of avenues of attack have been employed to identify human obesity associated genes. These include candidate genes, and association and linkage studies.

The candidate gene approach to the genetics of obesity has been fueled, in large part, by discoveries made in animal models, as described above, and from investigation of familial patterns of severe obesity. These studies have helped to disclose forms of obesity that are inherited in a more or less straightforward manner, whether they are monogenic or syndromic in origin.

Monogenic disorders resulting in obesity are quite rare, very severe and can generally be observed early in life.[62] Less than 200 cases of human obesity caused by a single gene mutation in 11 different genes have been reported.[63] The list of mutations include leptin, the leptin receptor, proopiomelanocortin (POMC), PCSK1/3, BDNF, NTRK2, SIM1, BBS1-16, ALMS1, GNAS1, and the genes encoding the melanocortin 4 receptor (MC4R).[64] Of these, MC4R is the most frequently observed mutation in humans, inherited as an autosomal dominant trait. Bell and co-workers estimate that mutations in MC4R can be observed in 1–6% of obese individuals of various ethnicities.[65]

Syndromic forms of obesity are also quite rare, and can be caused either by discrete genetic defects or abnormalities involving relatively large areas of chromosomes. They can be inherited either in an autosomal fashion or X-linked.[65] Bardet–Biedl syndrome (BBS), an autosomal inherited disorder, is a disease entity characterized by a number of physical and neurologic pathologies, including learning difficulties, polydactyly, retinal dystrophy, renal impairment, hypogonadism, and obesity.[66] Mutations in more than 14 genes have been identified to date that produce this genetically heterogeneous set of symptoms. Prader–Willi syndrome is another example of an inherited condition that involves deletion of a number of genes on chromosome 15q11–13.[67] The genes contained within this region of DNA are imprinted in such a way as that they are only expressed on the chromosome that was paternally inherited.

Several genes have been identified that modify satiety and energy intake. In addition to leptin and its receptor (discussed above), the list includes a number of proteins that interact with receptors in certain hypothalamic areas. Neuropeptide Y (NPY), peptide YY (PYY), corticotrophin releasing hormone (CRH), α-melanocyte-stimulating hormone (αMSH), agouti-related protein, melanin-concentrating hormone (MCH), proopiomelanocortin (POMC), cocaine- and amphetamine-regulated transcript, and the melanocortin 4 receptor (MC4R) have been shown to be important in regulating energy intake.[68] Of all of these factors, MC4R has been most extensively implicated in human obesity. It is one of the most potent known regulators of energy intake, balancing a satiety signal, provided by α-MSH, and an orexigenic signal from the agouti-related protein.[68] Over 50 mutations in the MC4R have been reported in morbidly obese adults or in early onset obesity in children.[69]

The central pathways for energy intake are integrated with pathways for energy expenditure. Adaptive thermogenesis and uncoupling proteins are concerned with mobilization of lipid stores from fat. For that reason, they are important in obesity.[70] Adaptive thermogenesis is under the control of the sympathetic nervous system, and members of the β-adrenoreceptor family (ADRB) have been studied as candidate genes for obesity because of their role in the regulation of energy expenditure.[71] The family consists of 3 proteins; ADRB1, ADRB2, and ADRB3. ADRB2 and 3 are prime candidates as genetic factors predisposing to obesity. ADRB2 is a receptor on adipocytes that regulates lipolysis. Two mutations have been discovered in this receptor and one of these has been linked to obesity in some populations.[72] Agonists of ADRB3 have been shown to have therapeutic effects with respect to diabetes and obesity. A polymorphism in this receptor has been linked to increases in BMI.[73]

A family of carrier proteins on the inner mitochondrial membrane called the uncoupling proteins (UCP) modulates heat-generating uncoupled respiration.[74] The proteins UCP-1, -2, and -3 are differentially expressed in various tissues.[75,76] Brown fat is the primary site of expression of UCP-1, where it participates in thermogenesis. UCP-2 is expressed universally, and UCP-3 can be found in skeletal muscle and brown fat. The UCPs have been extensively studied for their roles in energy metabolism. The UCPs are reported to mediate proton leak in the mitochondria, thereby releasing energy stores as heat. A study of the UCP2–UCP3 gene cluster has identified a haplotype that is significantly linked to the risk of obesity and insulin resistance.[77]

The candidate gene approach required investigators to make educated guesses as to which genes might be involved in the generation of the obese phenotype, and then to examine variants within that gene to determine whether or not any variants could be tied to obesity. The advent of rapid DNA sequencing and hybridization technology has permitted genetic epidemiologists to circumvent the guessing game and to examine the entire genome from obese and lean individuals, in order to disclose small differences in the genetic code that could be linked to, or associated with obesity. These genome wide array scans, or GWAS, studies are predicated on the concept that there are variants in the genome of the obese population that are not present in lean individuals. Investigators specifically search for single nucleotide polymorphisms (SNPs) in defined regions of DNA where variations are known to occur across all chromosomes, from individuals in each population. The pot of gold at the end of this rainbow is the demonstration that obese individuals will show a strong association with specific SNPs that are not seen in the lean samples.

A GWAS study conducted in 2007 revealed that individuals carrying 2 copies of a variant of the *FTO* gene are, on average, 3–4 kg heavier than those individuals with no copies of the variant.[78] The variant appears to be associated with appetite, and predisposes people who carry the variant to consume greater quantities of food. Over 50 variants have been linked to obesity through GWAS studies.[79] It is important to note that GWASs provide an association between a disease and chromosomal regions; not specific genes, and many of the signals map to non-coding regions of the genome. This can lead to difficulties in interpretation of the data. For example, a recent investigation questions the link between *FTO* and obesity;[80] one of the strongest GWAS identified associations to date. Those investigators suggest that the defect might lie in the promoter for a gene called *IRX3*, which is in close proximity to *FTO*. Understanding the physiology that results from the SNP variations has proven to be difficult. Outside of a few notable exceptions, almost all of the polymorphisms that have been discovered are within, or are close to, genes that would not have been suspected of playing an important role in obesity. This makes for very interesting science, and, perhaps, the discovery of novel pathways for therapy. On the other hand, getting to that point might take some time.

Because obesity is such a highly heritable trait, the initial thinking with respect to GWAS studies was that they would uncover a number of common genetic variants that would explain familial patterns. In fact, a large number of variants have been found. However, each variant, in and of itself, appears to have very subtle effects on body weight. So, while estimates of heritability of obesity are in the range of 65%,[18] large human GWAS studies can only explain less than 4% of this.[81] These data suggest that common variants simply cannot account for most of the heritability of obesity. It could be that the genetics of obesity are determined by a very large number (hundreds, perhaps) of variants with small effects, or by low frequency, rare mutations that have strong effects. GWAS studies look for SNPs in areas in which they are known to occur with a reasonable frequency. Most often, those stretches of DNA occur between coding regions. Low frequency, rare alleles with stronger effects, however, might be expected to occur within protein encoding stretches. The technological challenges to test that hypothesis are substantial, requiring the ability to perform high throughput sequencing of the entire exome of large numbers of individuals.

An often unappreciated complication in the study of the genetics of obesity is that the investigations tend to focus either on related individuals (*e.g.* twins), or on unrelated individuals at either end of the BMI spectrum. As was mentioned earlier, obesity is defined as a BMI ≥ 30 kg m^{-2}. That number has implications for categorization and treatment. In reality, adiposity is a continuous variable, and assigning risk to genetic variation in this respect can be quite complicated. Surprisingly, little attention has been focused on the resistance to obesity. Genetic studies of "thinness" suggest that the resistance to developing obesity is also a highly heritable trait (see Bulik[82] for an excellent review). Interestingly, a mutation in PPAR-γ has been identified that is associated with low body mass and increased insulin sensitivity.[83] Searching for genes and variants associated with very thin, healthy individuals might provide significant insight into the genetics of fat accumulation.

5.6 Epigenetics and Obesity

While GWAS studies have provided some insight into genetic determinants of obesity, a large amount of genetic variation remains a mystery. Several investigators have turned to assessing external environmental influences that can have profound influences on phenotype when provided in a specific genetic context. Epigenetic alterations are heritable changes in gene expression that occur without changes in the sequence of the genome.[84] Epigenetic alterations are important determinants of differentiation. The DNA within any one individual is essentially the same in each and every cell, yet liver tissue looks and behaves very differently than kidney tissue, and so on. This level of regulation is achieved by changes in DNA methylation, histone modifications, and chromatin binding, which are major epigenetic regulators in mammalian cells. Additionally, the relatively recent discovery of small, interfering RNAs (siRNA) and micro-RNAs demonstrate another level of control. Further, the epigenetic framework varies across cell types and is dynamic in nature, turning genes on and off at specific times, which is in stark contrast to the relatively static genome.[85] Functionally linking DNA methylation and histone modification in transcription would impart a mechanism for stably propagating gene activity from one generation to the next.[86]

In an effort to understand the role of epigenetics in modifying BMI risk, a group of investigators in the UK assessed genome wide methylation in over 400 individuals in relation to their BMI.[87] The study disclosed that increased BMI in the study population was associated with increased methylation at the locus for hypoxia inducing factor 3-α (HIF3A) in both blood cells and adipocytes. Another study[88] examined the response of obese, post-menopausal women to 6 months of caloric restriction, with respect to epigenetic modifications. The authors found significant differences in DNA methylation in adipose tissue at 35 loci between women who were characterized as either responders (*i.e.* lost significantly more fat mass) *vs.* non-responders. Many of the differences were in genes known to be associated with weight control and insulin sensitivity. Others were found in known genomic imprinted areas.

Environmental stimuli during embryogenesis in humans can have lasting effects on gene expression, and influence the susceptibility to disease in later (adult) life. The concept is known as fetal programming. Events that occurred during the winter of 1944–1945 provided data demonstrating the effect of fetal programming on the development of obesity. The "Hongerwinter" (Hunger Winter) or Dutch Famine of 1944 occurred in occupied Netherlands during the Second World War. Throughout that particularly harsh winter, food stocks in the cities in the western Netherlands dropped precipitously such that by February 1945 adult daily food intake dropped to 580 kCal.[89,90] The male children born during that period of time were the subjects of a remarkable study designed to determine whether or not the *in utero* environment they experienced had any effect on their risk of developing obesity later in life.[91] The study, populated by 300 000 men in the Dutch army, showed that the offspring of women who became pregnant at the end of the famine, and were malnourished only during their first trimester had a significantly higher incidence of obesity at the age of 19, when compared to males born before or after the famine. Interestingly, women who were in their second trimester of pregnancy when the famine began had sons with a lower risk of obesity. The data suggest that nutritional deprivation has effects on adipose development during a critical period of time. The most current evidence now suggests that epigenetic modifications to insulin-like growth factor 2 (*IGF2*) may be responsible for the differences in risk.[92] *IGF2* is an important regulator of growth and development and is maternally imprinted.[93] Further, *IGF2* is well characterized as an epigenetically regulated gene, being differentially methylated, and the methylation patterns are stable through middle age.[94] Investigators studied over 300 members of the Hunger Winter Famine study, and examined methylation patterns from whole blood, comparing the results with those obtained from same-sex siblings, as well as from "unexposed" controls. The results demonstrated that individuals who were periconceptually exposed to the famine had significantly less methylation of *IGF2* when compared to both control groups. Since methylation, in general, tends to be associated with gene silencing, it is unclear what significance *IGF2* hypomethylation has to the risk of obesity. Nonetheless, the study demonstrated an association between *in utero* environmental changes and DNA methylation changes in later life. It is particularly noteworthy that the mean age of the subjects in the DNA methylation study was approximately 60 years. The effects on *IGF2* methylation may not be limited to maternal imprinting. Soubry and colleagues[95] found that children born to obese fathers exhibited diminished methylation in a particular region of the *IGF2* gene than did the children of lean fathers.

Proponents feel that epigenome-wide association studies (EWAS) will elucidate why some individuals carrying genomic variants associated with obesity develop the disease and others do not, or, perhaps more importantly, why some respond better to caloric restriction than others. Critics, on the other hand are quick to point out that, because the epigenome is so dynamic, a host of external and environmental factors can influence the data and complicate interpretation of the data. The difference between cell types is a major factor to consider in assessing EWAS data. Blood cells are easy to collect, but they may not reflect changes in other tissues that are more germane to metabolic disturbances such as liver, muscle and adipose tissue. Further, peripheral blood is comprised of a number of different cell types that can change in the face of infection, chronic disease or even with age. A recent study demonstrated that a series of several epigenetic studies that demonstrated differences in DNA methylation with age were, in fact, due to changes in the proportions of the various cell types that were sampled.[96] Striking differences in epigenetic modifications can be observed among different cell types from the same tissues, and it can be difficult to determine whether any of the modifications are cause or effect. To be fair, the field of epigenomics is in its relative infancy; not yet a decade old. The Wellcome Trust Sanger Institute, Epigenomics AG, and The Center National de Genotypage have formed a consortium to fund the Human Epigenome Project.[97] The project's stated goal is to "identify, catalog, and interpret genome-wide DNA methylation patterns of all human genes in all major tissues". This ambitious project should help in the understanding of the role of epigenetic modifications in a number of disease states, including obesity.

5.7 Environmental Genetic Interactions in Obesity

The arguments outlined above provide strong evidence that genetic factors driving obesity do not operate in a vacuum and can be profoundly influenced by environmental factors. With respect to the development of obesity, the equation, on the surface, is quite simple. It is a matter of calories in *vs.* calories out; or more precisely, energy intake *vs.* energy expended. The debate over which side of the equation is responsible for the fattening of society rages on.

The overriding consensus is that obesity is rampant because caloric intake exceeds energy expenditure.[9] A case has been made against the "usual suspects" including food intake and the advent of calorie dense sweeteners such as high fructose corn syrup. While it is tempting to lay blame on these factors, the data supporting those claims is not clear cut. Some studies have shown that certain populations, especially children, are more greatly exposed to calorie dense, large portion foods.[98–101] However, a 1995 study showed that levels of food intake seemed to be relatively stable.[102] The differences in these may be temporally related, or reflect the populations that were studied. High fructose corn syrup has been the subject of many investigations linking its consumption to the increased incidence of obesity.[103,104] The observations are fairly clear, but the conclusions drawn from them should be considered in the proper context. The 2 major forms of high fructose corn syrup used in foods (HFCS-42 and HFCS-55) are almost exactly 50% glucose and 50% fructose, about the same as one would get from an equivalent amount of sucrose.[105] More likely, massive serving sizes are likely contributing to weight gain and other metabolic disturbances.[106,107] It is not uncommon to find convenience stores offering 64 oz cups of sweetened soft drinks. One 12 oz soft drink contains approximately 39 g of sugars. The USDA recommends a sugar intake of 40 g per day on a 2000 cal diet. One 64 oz soft drink contains nearly a week's worth of sugar at the recommended level.

A case-control study has tested hypotheses attempting to determine whether or not an interaction exists between carbohydrate (CHO) intake and polymorphisms in ADRB2 with respect to obesity.[108] The data collected from that study demonstrated that the Glu polymorphism in ADRB2 alone was not associated with obesity. However, the investigators did observe a significant interaction between CHO intake and the polymorphism. They concluded that energy intake in the form of CHO exceeding 49% of energy intake produced a significant increase in the risk of obesity in the women enrolled in the study. Genetic variation in PPARγ2 also seems to interact with excess CHO intake. Carriers of a Pro12Ala mutation appear to be at increased risk for the development of obesity when CHO intake exceeds 49% of total energy intake.[109]

Physical activity or, rather, the lack thereof, is another environmental factor that is associated with obesity. Over the last 60 or so years, society has witnessed a plethora of changes. Accessibility to automobiles, television, electronics and fundamental modifications to such simple chores as housekeeping and shopping have created a societal movement to a less physically active lifestyle, and a number of studies have demonstrated that this move toward a more sedentary lifestyle is strongly associated with increased obesity, worldwide.[110–112] On the other hand, efforts to increase energy expenditure show that this trend can be reduced, especially in young people.[113]

ADRB genes are involved in lipolysis and might be anticipated to have an effect on energy expenditure during physical activity. Both ADRB2 and ADRB3 polymorphisms have been interrogated for interactions with physical activity and the risk of obesity. Carriers of the Trp64Arg polymorphism who were sedentary were shown to be at increased risk for developing obesity, when adjusted for age and gender, in a case-control study.[114] Another study that examined the effect of Gln27Glu mutation in ADRB2 demonstrated that women with this variant did not benefit as much from physical activity as the non-carriers, and may be more resistant to weight reduction from programs of increased physical activity.[115] It is quite interesting in that this mutation appears to be gender specific. A similar study of the same ADRB2 variant was carried out in Spanish children and adolescents (males and females). The 27Glu mutation was associated with obesity only in females.[115]

Over the past century, sleep duration has fallen.[116] The suggestion has been made that there is a correlation between sleep deprivation and/or abnormal sleep patterns, as might be observed in shift workers, and obesity.[117] While the physiologic basis of the need for sleep remains somewhat elusive, it is well known that rest is essential for survival. Rats that were deprived of sleep for an extensive period of time *via* forced activity became severely debilitated and eventually died.[118] One of the proposed functions of sleep is to preserve energy stores.[119] While there is definite, complex energy utilization during the various phases of sleep, a wakeful period of 24 h results in a loss of 134 kcal.[120] Most of that energy ($\sim$96 kcal) can be recovered by "payback" sleep the following night, but there is still a deficit of nearly 40 kcal that could, in aggregate be physiologically significant. The paradox here is apparent. If sleep deprivation results in energy loss, why would we suspect that decreased duration of sleep would predispose to obesity? The answer may lie in the action of melatonin. Melatonin is secreted by the pineal gland and is a component of the sleep–wake cycle.[121] The hormone is produced at night, and exposure to light in the 460–480 nm range (blue light) through the retina inhibits its production.[122] With respect to energy utilization, melatonin has been reported to increase recruitment and activity of brown adipocytes. The hormone can also suppress insulin secretion from pancreatic β-cells by binding to cell surface melatonin receptor-1 and 2.[123] By suppressing insulin secretion at night, the β-cells would experience a respite, and the periphery would be spared chronic exposure to constant insulin signaling, which would prevent pancreatic exhaustion and insulin resistance. In support of this hypothesis, studies have shown that melatonin type 1 receptor knockout mice are insulin resistant,[124] and improvements in insulin sensitivity and glucose tolerance are achieved in wild type mice fed a high fat diet.[125] Variants in melatonin receptors are also associated with increased risk of insulin resistance and type 2 diabetes in a number of different populations.[126–128] By extending an individual's days, it is possible to achieve a tonic suppression of melatonin synthesis that would lead to decreased activity of brown adipose tissue and promote insulin resistance, while offering the additional opportunity to increase food intake at an aberrant time.

A discussion of environmental influences on obesity would not be complete without addressing the topic of the microbiome. The microbiome refers to the collective flora of microbes that inhabit our skin, mucous membranes, gut and other organs. The population turnover rate in these organisms is quite high, making them prime candidates for alteration in response to environmental changes and stresses. The appearance of advanced genetic technology has permitted the impartial analysis of gut bacteria only recently, and studies showing differences in the microbial flora between obese and lean animals have been published, suggesting that microbiota contribute to energy harvest within the organism.[129] While intriguing, the implications of these kinds of studies are difficult to ascertain. A study of microbiome differences between lean and obese humans failed to demonstrate any differences between the two groups.[130] It is also difficult to determine whether or not these differences are cause or effect, with respect to the phenotype. Relatively recent studies by Le Chatelier *et al.*[131] and Cotillard and co-workers[132] have revealed that obese individuals tended to have a less complex gut microbiome, were more prone to metabolic disturbances, and responded better to dietary intervention. These studies predicated microbial diversity on the relative abundance of 3 enterotypes: *Bacterioides, Prevotella,* and *Ruminococcus*. That the intestinal microbiome complexity can be surveyed through a limited number of enterotypes is certainly debatable.

A study led by Lusis[133] investigated the genetic control of microbiome complexity and obesity in 100 inbred strains of mice from the Hybrid Mouse Diversity Panel fed a high fat, high sucrose diet. They discovered shifts between *Bacterioides* and *Firmicutes* among the various mouse strains after feeding the energy rich diets. Those changes were mapped against changes in adiposity, and other indices of metabolic disturbances. The authors demonstrated genetic control over both obesity and the intestinal microbiome as a consequence of feeding a high fat, high sucrose diet. Using association, they could map biologically significant data to loci for the response to between 1–3 genes, as in human GWAS studies. Several candidate genes have emerged from this work. Whether or not the microbiome proves to be a true risk factor for obesity or a predictor of therapeutic response remains to be determined.

5.8 Summary

That obesity is a worldwide problem with far-reaching public health implications is all too obvious. This begs the question, "How did we all get so fat?" Data from animal studies and investigations of body composition in families has shown that body weight is one of our most heritable traits. On the other hand, we also understand that our environment is also extremely important in the determination of body weight. Epigenetic changes can significantly modify adiposity, and some of those changes are heritable for several generations. A good deal of attention has focused on the contribution of the microbiome to the development of obesity, although the relative importance of these findings is still being determined. The ready availability of high caloric content foods is another factor that plays into the equation.

In the end, obesity develops because the energy that one ingests exceeds energy expended. The key issues relate to defining those factors that regulate these variables. The scientific community studies, at length, obesity. A corollary to this is that there is a good deal to be learned by understanding why some individuals remain thin, despite excessive caloric intake.

The questions appear simple. The answers are complex. The ongoing work in this area will surely evolve to shape how obesity is approached pharmacologically and behaviorally.

References

1. D. W. Haslam and W. P. James, *Lancet*, 2005, **366**, 1197–1209.
2. WHO, Fact Sheets, 2014, vol. 2014.
3. CDC, Adult Obesity Facts, 2014.
4. C. L. Ogden, M. D. Carroll, B. K. Kit and K. M. Flegal, *J. Am. Med. Assoc.*, 2014, **311**, 806–814.
5. American Medical Association, AMA Declares Obesity a Disease, 2013, http://www.medscape.com/viewarticle/806566.
6. D. Gu, K. Reynolds, X. Wu, J. Chen, X. Duan, R. F. Reynolds, P. K. Whelton, J. He and A. C. G. Inter, *Lancet*, 2005, **365**, 1398–1405.
7. K. S. Reddy, D. Prabhakaran, P. Shah and B. Shah, *Obes. Rev.*, 2002, **3**, 197–202.
8. J. R. Speakman and S. O'Rahilly, *Dis. Models & Mech.*, 2012, **5**, 569–573.
9. K. D. Hall, S. B. Heymsfield, J. W. Kemnitz, S. Klein, D. A. Schoeller and J. R. Speakman, *Am. J. Clin. Nutr.*, 2012, **95**, 989–994.
10. B. H. Goodpaster, *Curr. Opin. Clin. Nutr. Metab. Care*, 2002, **5**, 481–487.
11. U. A. White and Y. D. Tchoukalova, *Biochim. Biophys. Acta*, 2014, **1842**, 377–392.
12. L. J. Aronne, *Obes. Res.*, 2002, **10**(Suppl. 2), 105S–115S.
13. C. B. Davenport, *Proc. Natl. Acad. Sci. U. S. A.*, 1923, **9**, 226–230.

14. C. E. Bruder, A. Piotrowski, A. A. Gijsbers, R. Andersson, S. Erickson, T. Diaz de Stahl, U. Menzel, J. Sandgren, D. von Tell, A. Poplawski, M. Crowley, C. Crasto, E. C. Partridge, H. Tiwari, D. B. Allison, J. Komorowski, G. J. van Ommen, D. I. Boomsma, N. L. Pedersen, J. T. den Dunnen, K. Wirdefeldt and J. P. Dumanski, *Am. J. Hum. Genet.*, 2008, **82**, 763–771.
15. A. J. Stunkard, J. R. Harris, N. L. Pedersen and G. E. McClearn, *N. Engl. J. Med.*, 1990, **322**, 1483–1487.
16. R. A. Price and Gottesman, II, *Behav. Genet.*, 1991, **21**, 1–7.
17. H. H. Maes, M. C. Neale and L. J. Eaves, *Behav. Genet.*, 1997, **27**, 325–351.
18. N. L. Segal and D. B. Allison, *Int. J. Obes. Relat. Metab. Disord.*, 2002, **26**, 437–441.
19. A. M. Ingalls, M. M. Dickie and G. D. Snell, *Obes. Res.*, 1996, **4**, 101.
20. K. P. Hummel, M. M. Dickie and D. L. Coleman, *Science*, 1966, **153**, 1127–1128.
21. Y. Zhang, R. Proenca, M. Maffei, M. Barone, L. Leopold and J. M. Friedman, *Nature*, 1994, **372**, 425–432.
22. G. H. Lee, R. Proenca, J. M. Montez, K. M. Carroll, J. G. Darvishzadeh, J. I. Lee and J. M. Friedman, *Nature*, 1996, **379**, 632–635.
23. H. Chen, O. Charlat, L. A. Tartaglia, E. A. Woolf, X. Weng, S. J. Ellis, N. D. Lakey, J. Culpepper, K. J. Moore, R. E. Breitbart, G. M. Duyk, R. I. Tepper and J. P. Morgenstern, *Cell*, 1996, **84**, 491–495.
24. L. M. Zucker and T. F. Zucker, *J. Hered.*, 1961, **52**, 275–278.
25. L. M. Zucker and H. N. Antoniades, *Endocrinology*, 1972, **90**, 1320–1330.
26. G. A. Bray, *Fed. Proc.*, 1977, **36**, 148–153.
27. Y. Ogawa, H. Masuzaki, N. Isse, T. Okazaki, K. Mori, M. Shigemoto, N. Satoh, N. Tamura, K. Hosoda, Y. Yoshimasa *et al.*, *J. Clin. Invest.*, 1995, **96**, 1647–1652.
28. G. E. Truett, N. Bahary, J. M. Friedman and R. L. Leibel, *Proc. Natl. Acad. Sci. U. S. A.*, 1991, **88**, 7806–7809.
29. C. T. Montague, I. S. Farooqi, J. P. Whitehead, M. A. Soos, H. Rau, N. J. Wareham, C. P. Sewter, J. E. Digby, S. N. Mohammed, J. A. Hurst, C. H. Cheetham, A. R. Earley, A. H. Barnett, J. B. Prins and S. O'Rahilly, *Nature*, 1997, **387**, 903–908.
30. T. T. Yen, A. M. Gill, L. G. Frigeri, G. S. Barsh and G. L. Wolff, *FASEB J.*, 1994, **8**, 479–488.
31. S. J. Bultman, E. J. Michaud and R. P. Woychik, *Cell*, 1992, **71**, 1195–1204.
32. E. J. Michaud, S. J. Bultman, M. L. Klebig, M. J. van Vugt, L. J. Stubbs, L. B. Russell and R. P. Woychik, *Proc. Natl. Acad. Sci. U. S. A.*, 1994, **91**, 2562–2566.
33. E. J. Michaud, S. J. Bultman, L. J. Stubbs and R. P. Woychik, *Genes Dev.*, 1993, 7, 1203–1213.
34. M. L. Klebig, J. E. Wilkinson, J. G. Geisler and R. P. Woychik, *Proc. Natl. Acad. Sci. U. S. A.*, 1995, **92**, 4728–4732.

35. R. L. Mynatt, R. J. Miltenberger, M. L. Klebig, M. B. Zemel, J. E. Wilkinson, W. O. Wilkinson and R. P. Woychik, *Proc. Natl. Acad. Sci. U. S. A.*, 1997, **94**, 919–922.
36. H. Y. Kwon, S. J. Bultman, C. Loffler, W. J. Chen, P. J. Furdon, J. G. Powell, A. L. Usala, W. Wilkison, I. Hansmann and R. P. Woychik, *Proc. Natl. Acad. Sci. U. S. A.*, 1994, **91**, 9760–9764.
37. B. D. Wilson, M. M. Ollmann, L. Kang, M. Stoffel, G. I. Bell and G. S. Barsh, *Hum. Mol. Genet.*, 1995, **4**, 223–230.
38. S. R. Smith, B. Gawronska-Kozak, L. Janderova, T. Nguyen, A. Murrell, J. M. Stephens and R. L. Mynatt, *Diabetes*, 2003, **52**, 2914–2922.
39. G. T. Kucera, D. M. Bortner and M. P. Rosenberg, *Dev. Biol.*, 1996, **173**, 162–173.
40. L. Herberg and D. L. Coleman, *Metab., Clin. Exp.*, 1977, **26**, 59–99.
41. H. S. Jurgens, A. Schurmann, R. Kluge, S. Ortmann, S. Klaus, H. G. Joost and M. H. Tschop, *Physiol. Genomics*, 2006, **25**, 234–241.
42. M. Nakamura and K. Yamada, *Diabetologia*, 1967, **3**, 212–221.
43. M. Igel, B. A. Taylor, S. J. Phillips, W. Becker, L. Herberg and H. G. Joost, *J. Mol. Endocrinol.*, 1998, **21**, 337–345.
44. M. Okazaki, Y. Saito, Y. Udaka, M. Maruyama, H. Murakami, S. Ota, T. Kikuchi and K. Oguchi, *Exp. Anim.*, 2002, **51**, 191–196.
45. K. Kawano, T. Hirashima, S. Mori, Y. Saitoh, M. Kurosumi and T. Natori, *Diabetes*, 1992, **41**, 1422–1428.
46. S. Takiguchi, Y. Takata, N. Takahashi, K. Kataoka, T. Hirashima, K. Kawano, K. Miyasaka, A. Funakoshi and A. Kono, *Am. J. Physiol.*, 1998, **274**, E265–270.
47. T. H. Moran, *Physiol. Behav.*, 2008, **94**, 71–78.
48. K. Kawano, T. Hirashima, S. Mori and T. Natori, *Diabetes Res. Clin. Pract.*, 1994, **24**(Suppl), S317–320.
49. V. Gailus-Durner, H. Fuchs, L. Becker, I. Bolle, M. Brielmeier, J. Calzada-Wack, R. Elvert, N. Ehrhardt, C. Dalke, T. J. Franz, E. Grundner-Culemann, S. Hammelbacher, S. M. Holter, G. Holzlwimmer, M. Horsch, A. Javaheri, S. V. Kalaydjiev, M. Klempt, E. Kling, S. Kunder, C. Lengger, T. Lisse, T. Mijalski, B. Naton, V. Pedersen, C. Prehn, G. Przemeck, I. Racz, C. Reinhard, P. Reitmeir, I. Schneider, A. Schrewe, R. Steinkamp, C. Zybill, J. Adamski, J. Beckers, H. Behrendt, J. Favor, J. Graw, G. Heldmaier, H. Hofler, B. Ivandic, H. Katus, P. Kirchhof, M. Klingenspor, T. Klopstock, A. Lengeling, W. Muller, F. Ohl, M. Ollert, L. Quintanilla-Martinez, J. Schmidt, H. Schulz, E. Wolf, W. Wurst, A. Zimmer, D. H. Busch and M. H. de Angelis, *Nat. Methods*, 2005, **2**, 403–404.
50. L. B. Russell and W. L. Russell, *Mutat. Res.*, 1992, **296**, 107–127.
51. T. Rankinen, A. Zuberi, Y. C. Chagnon, S. J. Weisnagel, G. Argyropoulos, B. Walts, L. Perusse and C. Bouchard, *Obesity*, 2006, **14**, 529–644.
52. S. Lin, D. Boey and H. Herzog, *Neuropeptides*, 2004, **38**, 189–200.
53. J. C. Erickson, K. E. Clegg and R. D. Palmiter, *Nature*, 1996, **381**, 415–421.

54. J. C. Erickson, G. Hollopeter and R. D. Palmiter, *Science*, 1996, **274**, 1704–1707.
55. R. D. Palmiter, J. C. Erickson, G. Hollopeter, S. C. Baraban and M. W. Schwartz, *Recent Prog. Horm. Res.*, 1998, **53**, 163–199.
56. K. A. Augustine, R. M. Rossi, G. Van, J. Housman, K. Stark, D. Danilenko, B. Varnum and E. Medlock, *J. Cell. Physiol.*, 1999, **181**, 433–447.
57. K. A. Augustine and R. M. Rossi, *Anat. Rec.*, 1999, **257**, 64–72.
58. A. Katsuki, Y. Sumida, S. Murashima, K. Murata, Y. Takarada, K. Ito, M. Fujii, K. Tsuchihashi, H. Goto, K. Nakatani and Y. Yano, *J. Clin. Endocrinol. Metab.*, 1998, **83**, 859–862.
59. T. Yamakawa, S. Tanaka, Y. Yamakawa, Y. Kiuchi, F. Isoda, S. Kawamoto, K. Okuda and H. Sekihara, *Clin. Immunol. Immunopathol.*, 1995, **75**, 51–56.
60. G. S. Barsh, I. S. Farooqi and S. O'Rahilly, *Nature*, 2000, **404**, 644–651.
61. A. J. Stunkard, T. T. Foch and Z. Hrubec, *J. Am. Med. Assoc.*, 1986, **256**, 51–54.
62. I. S. Farooqi and S. O'Rahilly, *Recent Prog. Horm. Res.*, 2004, **59**, 409–424.
63. W. Yang, T. Kelly and J. He, *Epidemiol. Rev.*, 2007, **29**, 49–61.
64. S. Ramachandrappa and I. S. Farooqi, *J. Clin. Invest.*, 2011, **121**, 2080–2086.
65. C. G. Bell, A. J. Walley and P. Froguel, *Nat. Rev. Genet.*, 2005, **6**, 221–234.
66. J. L. Tobin and P. L. Beales, *Pediatr. Nephrol.*, 2007, **22**, 926–936.
67. T. Sahoo, C. A. Bacino, J. R. German, C. A. Shaw, L. M. Bird, V. Kimonis, I. Anselm, S. Waisbren, A. L. Beaudet and S. U. Peters, *Eur. J. Hum. Genet.*, 2007, **15**, 943–949.
68. M. M. Kamiji and A. Inui, *Endocr. Rev.*, 2007, **28**, 664–684.
69. C. Vaisse, K. Clement, E. Durand, S. Hercberg, B. Guy-Grand and P. Froguel, *J. Clin. Invest.*, 2000, **106**, 253–262.
70. A. G. Dulloo, J. Seydoux and J. Jacquet, *Physiol. Behav.*, 2004, **83**, 587–602.
71. M. C. Ochoa, A. Marti, C. Azcona, M. Chueca, M. Oyarzabal, R. Pelach, A. Patino, M. J. Moreno-Aliaga, M. A. Martinez-Gonzalez, J. A. Martinez and Groupo de Estudio Navarro de Obesidad Infantil, *Int. J. Obes. Relat. Metab. Disord.*, 2004, **28**(Suppl. 3), S37–41.
72. M. S. Corbalan, A. Marti, L. Forga, M. A. Martinez-Gonzalez and J. A. Martinez, *Eur. J. Nutr.*, 2002, **41**, 114–118.
73. T. Fujisawa, H. Ikegami, Y. Kawaguchi and T. Ogihara, *J. Clin. Endocrinol. Metab.*, 1998, **83**, 2441–2444.
74. J. Mozo, Y. Emre, F. Bouillaud, D. Ricquier and F. Criscuolo, *Biosci. Rep.*, 2005, **25**, 227–249.
75. A. Marti, E. Larrarte, F. J. Novo, M. Garcia and J. A. Martinez, *Int. J. Obes. Relat. Metab. Disord.*, 2001, **25**, 68–74.
76. M. S. Corbalan, J. Margareto, J. A. Martinez and A. Marti, *J. Physiol. Biochem.*, 1999, **55**, 67–72.

77. M. C. Ochoa, J. L. Santos, C. Azcona, M. J. Moreno-Aliaga, M. A. Martinez-Gonzalez, J. A. Martinez, A. Marti and G. Members, *Mol. Genet. Metab.*, 2007, **92**, 351–358.
78. T. M. Frayling, N. J. Timpson, M. N. Weedon, E. Zeggini, R. M. Freathy, C. M. Lindgren, J. R. Perry, K. S. Elliott, H. Lango, N. W. Rayner, B. Shields, L. W. Harries, J. C. Barrett, S. Ellard, C. J. Groves, B. Knight, A. M. Patch, A. R. Ness, S. Ebrahim, D. A. Lawlor, S. M. Ring, Y. Ben-Shlomo, M. R. Jarvelin, U. Sovio, A. J. Bennett, D. Melzer, L. Ferrucci, R. J. Loos, I. Barroso, N. J. Wareham, F. Karpe, K. R. Owen, L. R. Cardon, M. Walker, G. A. Hitman, C. N. Palmer, A. S. Doney, A. D. Morris, G. D. Smith, A. T. Hattersley and M. I. McCarthy, *Science*, 2007, **316**, 889–894.
79. R. J. Loos, *Best Pract. Res. Clin. Endocrinol. Metab.*, 2012, **26**, 211–226.
80. S. Smemo, J. J. Tena, K. H. Kim, E. R. Gamazon, N. J. Sakabe, C. Gomez-Marin, I. Aneas, F. L. Credidio, D. R. Sobreira, N. F. Wasserman, J. H. Lee, V. Puviindran, D. Tam, M. Shen, J. E. Son, N. A. Vakili, H. K. Sung, S. Naranjo, R. D. Acemel, M. Manzanares, A. Nagy, N. J. Cox, C. C. Hui, J. L. Gomez-Skarmeta and M. A. Nobrega, *Nature*, 2014, **507**, 371–375.
81. E. K. Speliotes, C. J. Willer, S. I. Berndt, K. L. Monda, G. Thorleifsson, A. U. Jackson, H. Lango Allen, C. M. Lindgren, J. Luan, R. Magi, J. C. Randall, S. Vedantam, T. W. Winkler, L. Qi, T. Workalemahu, I. M. Heid, V. Steinthorsdottir, H. M. Stringham, M. N. Weedon, E. Wheeler, A. R. Wood, T. Ferreira, R. J. Weyant, A. V. Segre, K. Estrada, L. Liang, J. Nemesh, J. H. Park, S. Gustafsson, T. O. Kilpelainen, J. Yang, N. Bouatia-Naji, T. Esko, M. F. Feitosa, Z. Kutalik, M. Mangino, S. Raychaudhuri, A. Scherag, A. V. Smith, R. Welch, J. H. Zhao, K. K. Aben, D. M. Absher, N. Amin, A. L. Dixon, E. Fisher, N. L. Glazer, M. E. Goddard, N. L. Heard-Costa, V. Hoesel, J. J. Hottenga, A. Johansson, T. Johnson, S. Ketkar, C. Lamina, S. Li, M. F. Moffatt, R. H. Myers, N. Narisu, J. R. Perry, M. J. Peters, M. Preuss, S. Ripatti, F. Rivadeneira, C. Sandholt, L. J. Scott, N. J. Timpson, J. P. Tyrer, S. van Wingerden, R. M. Watanabe, C. C. White, F. Wiklund, C. Barlassina, D. I. Chasman, M. N. Cooper, J. O. Jansson, R. W. Lawrence, N. Pellikka, I. Prokopenko, J. Shi, E. Thiering, H. Alavere, M. T. Alibrandi, P. Almgren, A. M. Arnold, T. Aspelund, L. D. Atwood, B. Balkau, A. J. Balmforth, A. J. Bennett, Y. Ben-Shlomo, R. N. Bergman, S. Bergmann, H. Biebermann, A. I. Blakemore, T. Boes, L. L. Bonnycastle, S. R. Bornstein, M. J. Brown, T. A. Buchanan, F. Busonero, H. Campbell, F. P. Cappuccio, C. Cavalcanti-Proenca, Y. D. Chen, C. M. Chen, P. S. Chines, R. Clarke, L. Coin, J. Connell, I. N. Day, M. den Heijer, J. Duan, S. Ebrahim, P. Elliott, R. Elosua, G. Eiriksdottir, M. R. Erdos, J. G. Eriksson, M. F. Facheris, S. B. Felix, P. Fischer-Posovszky, A. R. Folsom, N. Friedrich, N. B. Freimer, M. Fu, S. Gaget, P. V. Gejman, E. J. Geus, C. Gieger, A. P. Gjesing, A. Goel, P. Goyette, H. Grallert, J. Grassler, D. M. Greenawalt, C. J. Groves, V. Gudnason, C. Guiducci, A. L. Hartikainen, N. Hassanali, A. S. Hall, A. S. Havulinna, C. Hayward,

A. C. Heath, C. Hengstenberg, A. A. Hicks, A. Hinney, A. Hofman, G. Homuth, J. Hui, W. Igl, C. Iribarren, B. Isomaa, K. B. Jacobs, I. Jarick, E. Jewell, U. John, T. Jorgensen, P. Jousilahti, A. Jula, M. Kaakinen, E. Kajantie, L. M. Kaplan, S. Kathiresan, J. Kettunen, L. Kinnunen, J. W. Knowles, I. Kolcic, I. R. Konig, S. Koskinen, P. Kovacs, J. Kuusisto, P. Kraft, K. Kvaloy, J. Laitinen, O. Lantieri, C. Lanzani, L. J. Launer, C. Lecoeur, T. Lehtimaki, G. Lettre, J. Liu, M. L. Lokki, M. Lorentzon, R. N. Luben, B. Ludwig, Magic, P. Manunta, D. Marek, M. Marre, N. G. Martin, W. L. McArdle, A. McCarthy, B. McKnight, T. Meitinger, O. Melander, D. Meyre, K. Midthjell, G. W. Montgomery, M. A. Morken, A. P. Morris, R. Mulic, J. S. Ngwa, M. Nelis, M. J. Neville, D. R. Nyholt, C. J. O'Donnell, S. O'Rahilly, K. K. Ong, B. Oostra, G. Pare, A. N. Parker, M. Perola, I. Pichler, K. H. Pietilainen, C. G. Platou, O. Polasek, A. Pouta, S. Rafelt, O. Raitakari, N. W. Rayner, M. Ridderstrale, W. Rief, A. Ruokonen, N. R. Robertson, P. Rzehak, V. Salomaa, A. R. Sanders, M. S. Sandhu, S. Sanna, J. Saramies, M. J. Savolainen, S. Scherag, S. Schipf, S. Schreiber, H. Schunkert, K. Silander, J. Sinisalo, D. S. Siscovick, J. H. Smit, N. Soranzo, U. Sovio, J. Stephens, I. Surakka, A. J. Swift, M. L. Tammesoo, J. C. Tardif, M. Teder-Laving, T. M. Teslovich, J. R. Thompson, B. Thomson, A. Tonjes, T. Tuomi, J. B. van Meurs, G. J. van Ommen, V. Vatin, J. Viikari, S. Visvikis-Siest, V. Vitart, C. I. Vogel, B. F. Voight, L. L. Waite, H. Wallaschofski, G. B. Walters, E. Widen, S. Wiegand, S. H. Wild, G. Willemsen, D. R. Witte, J. C. Witteman, J. Xu, Q. Zhang, L. Zgaga, A. Ziegler, P. Zitting, J. P. Beilby, I. S. Farooqi, J. Hebebrand, H. V. Huikuri, A. L. James, M. Kahonen, D. F. Levinson, F. Macciardi, M. S. Nieminen, C. Ohlsson, L. J. Palmer, P. M. Ridker, M. Stumvoll, J. S. Beckmann, H. Boeing, E. Boerwinkle, D. I. Boomsma, M. J. Caulfield, S. J. Chanock, F. S. Collins, L. A. Cupples, G. D. Smith, J. Erdmann, P. Froguel, H. Gronberg, U. Gyllensten, P. Hall, T. Hansen, T. B. Harris, A. T. Hattersley, R. B. Hayes, J. Heinrich, F. B. Hu, K. Hveem, T. Illig, M. R. Jarvelin, J. Kaprio, F. Karpe, K. T. Khaw, L. A. Kiemeney, H. Krude, M. Laakso, D. A. Lawlor, A. Metspalu, P. B. Munroe, W. H. Ouwehand, O. Pedersen, B. W. Penninx, A. Peters, P. P. Pramstaller, T. Quertermous, T. Reinehr, A. Rissanen, I. Rudan, N. J. Samani, P. E. Schwarz, A. R. Shuldiner, T. D. Spector, J. Tuomilehto, M. Uda, A. Uitterlinden, T. T. Valle, M. Wabitsch, G. Waeber, N. J. Wareham, H. Watkins, C. Procardis, J. F. Wilson, A. F. Wright, M. C. Zillikens, N. Chatterjee, S. A. McCarroll, S. Purcell, E. E. Schadt, P. M. Visscher, T. L. Assimes, I. B. Borecki, P. Deloukas, C. S. Fox, L. C. Groop, T. Haritunians, D. J. Hunter, R. C. Kaplan, K. L. Mohlke, J. R. O'Connell, L. Peltonen, D. Schlessinger, D. P. Strachan, C. M. van Duijn, H. E. Wichmann, T. M. Frayling, U. Thorsteinsdottir, G. R. Abecasis, I. Barroso, M. Boehnke, K. Stefansson, K. E. North, M. I. McCarthy, J. N. Hirschhorn, E. Ingelsson and R. J. Loos, *Nat. Genet.*, 2010, **42**, 937–948.

82. C. M. Bulik and D. B. Allison, *Obes. Rev.*, 2001, **2**, 107–115.

83. S. S. Deeb, L. Fajas, M. Nemoto, J. Pihlajamaki, L. Mykkanen, J. Kuusisto, M. Laakso, W. Fujimoto and J. Auwerx, *Nat. Genet.*, 1998, **20**, 284–287.
84. A. P. Wolffe and D. Guschin, *J. Struct. Biol.*, 2000, **129**, 102–122.
85. Z. A. Kaminsky, T. Tang, S. C. Wang, C. Ptak, G. H. Oh, A. H. Wong, L. A. Feldcamp, C. Virtanen, J. Halfvarson, C. Tysk, A. F. McRae, P. M. Visscher, G. W. Montgomery, Gottesman, II, N. G. Martin and A. Petronis, *Nat. Genet.*, 2009, **41**, 240–245.
86. A. E. Handel, G. C. Ebers and S. V. Ramagopalan, *Trends Mol. Med.*, 2010, **16**, 7–16.
87. K. J. Dick, C. P. Nelson, L. Tsaprouni, J. K. Sandling, D. Aissi, S. Wahl, E. Meduri, P. E. Morange, F. Gagnon, H. Grallert, M. Waldenberger, A. Peters, J. Erdmann, C. Hengstenberg, F. Cambien, A. H. Goodall, W. H. Ouwehand, H. Schunkert, J. R. Thompson, T. D. Spector, C. Gieger, D. A. Tregouet, P. Deloukas and N. J. Samani, *Lancet*, 2014, **383**, 1990–1998.
88. L. Bouchard, R. Rabasa-Lhoret, M. Faraj, M. E. Lavoie, J. Mill, L. Perusse and M. C. Vohl, *Am. J. Clin. Nutr.*, 2010, **91**, 309–320.
89. Z. Stein and M. Susser, *Pediatr. Res.*, 1975, **9**, 76–83.
90. Z. Stein and M. Susser, *Pediatr. Res.*, 1975, **9**, 70–76.
91. G. P. Ravelli, Z. A. Stein and M. W. Susser, *N. Engl. J. Med.*, 1976, **295**, 349–353.
92. B. T. Heijmans, E. W. Tobi, A. D. Stein, H. Putter, G. J. Blauw, E. S. Susser, P. E. Slagboom and L. H. Lumey, *Proc. Natl. Acad. Sci. U. S. A.*, 2008, **105**, 17046–17049.
93. F. M. Smith, A. S. Garfield and A. Ward, *Cytogenet. Genome Res.*, 2006, **113**, 279–291.
94. B. T. Heijmans, D. Kremer, E. W. Tobi, D. I. Boomsma and P. E. Slagboom, *Hum. Mol. Genet.*, 2007, **16**, 547–554.
95. A. Soubry, J. M. Schildkraut, A. Murtha, F. Wang, Z. Huang, A. Bernal, J. Kurtzberg, R. L. Jirtle, S. K. Murphy and C. Hoyo, *BMC Med.*, 2013, **11**, 29.
96. A. E. Jaffe and R. A. Irizarry, *Genome Biol.*, 2014, **15**, R31.
97. H. E. Consortium, 2014.
98. J. O. Fisher and T. V. Kral, *Physiol. Behav.*, 2008, **94**, 39–47.
99. G. D. Foster, A. Karpyn, A. C. Wojtanowski, E. Davis, S. Weiss, C. Brensinger, A. Tierney, W. Guo, J. Brown, C. Spross, D. Leuchten, P. J. Burns and K. Glanz, *Am. J. Clin. Nutr.*, 2014, **99**, 1359–1368.
100. R. E. Lee, K. M. Heinrich, J. Y. Reese-Smith, G. R. Regan and H. J. Adamus-Leach, *Am. J. Health Behav.*, 2014, **38**, 218–224.
101. R. P. Schwartz, *N. C. Med. J.*, 2002, **63**, 305–307.
102. A. M. Prentice and S. A. Jebb, *BMJ*, 1995, **311**, 437–439.
103. G. A. Bray, S. J. Nielsen and B. M. Popkin, *Am. J. Clin. Nutr.*, 2004, **79**, 537–543.
104. S. S. Elliott, N. L. Keim, J. S. Stern, K. Teff and P. J. Havel, *Am. J. Clin. Nutr.*, 2002, **76**, 911–922.

105. M. F. Jacobson, *Am. J. Clin. Nutr.*, 2004, **80**, 1081, author reply 1081–1082.
106. R. D. Johnston, M. C. Stephenson, H. Crossland, S. M. Cordon, E. Palcidi, E. F. Cox, M. A. Taylor, G. P. Aithal and I. A. Macdonald, *Gastroenterology*, 2013, 145 , 1016–1025 e1012.
107. T. H. Moran, *J. Nutr.*, 2009, **139**, 1253S–1256S.
108. J. A. Martinez, M. S. Corbalan, A. Sanchez-Villegas, L. Forga, A. Marti and M. A. Martinez-Gonzalez, *J. Nutr.*, 2003, **133**, 2549–2554.
109. A. Marti, M. S. Corbalan, M. A. Martinez-Gonzalez, L. Forga and J. A. Martinez, *J. Physiol. Biochem.*, 2002, **58**, 219–220.
110. U. Ladabaum, A. Mannalithara, P. A. Myer and G. Singh, *Am. J. Med.*, 2014, **127**(8), 717–727, e712.
111. D. Sigmundova, E. Sigmund, Z. Hamrik and M. Kalman, *Eur. J. Public Health*, 2014, **24**, 210–215.
112. Y. J. Xie, S. M. Stewart, T. H. Lam, K. Viswanath and S. S. Chan, *PloS One*, 2014, **9**, e85440.
113. E. Kern, N. L. Chan, D. W. Fleming, J. W. Krieger, Centers for Disease Control and Prevention, *MMWR. Morbidity and mortality weekly report*, 2014, vol. 63, pp. 155–157.
114. A. Marti, M. S. Corbalan, M. A. Martinez-Gonzalez and J. A. Martinez, *Diabetes, Obes. Metab.*, 2002, **4**, 428–430.
115. M. S. Corbalan, A. Marti, L. Forga, M. A. Martinez-Gonzalez and J. A. Martinez, *Clin. Genet.*, 2002, **61**, 305–307.
116. S. R. Patel and F. B. Hu, *Obesity*, 2008, **16**, 643–653.
117. C. M. Depner, E. R. Stothard and K. P. Wright, Jr., *Curr. Diabetes Rep.*, 2014, **14**, 507.
118. C. A. Everson, *Behav. Brain Res.*, 1995, **69**, 43–54.
119. R. J. Berger and N. H. Phillips, *Behav. Brain Res.*, 1995, **69**, 65–73.
120. C. M. Jung, E. L. Melanson, E. J. Frydendall, L. Perreault, R. H. Eckel and K. P. Wright, *J. Physiol.*, 2011, **589**, 235–244.
121. R. J. Reiter, *Endocr. Rev.*, 1991, **12**, 151–180.
122. G. C. Brainard, J. P. Hanifin, J. M. Greeson, B. Byrne, G. Glickman, E. Gerner and M. D. Rollag, *J. Neurosci.*, 2001, **21**, 6405–6412.
123. D. X. Tan, L. C. Manchester, L. Fuentes-Broto, S. D. Paredes and R. J. Reiter, *Obes. Rev.*, 2011, **12**, 167–188.
124. S. Contreras-Alcantara, K. Baba and G. Tosini, *Obesity*, 2010, **18**, 1861–1863.
125. C. Sartori, P. Dessen, C. Mathieu, A. Monney, J. Bloch, P. Nicod, U. Scherrer and H. Duplain, *Endocrinology*, 2009, **150**, 5311–5317.
126. A. Karamitri, N. Renault, N. Clement, J. L. Guillaume and R. Jockers, *Mol. Endocrinol.*, 2013, **27**, 1217–1233.
127. J. Y. Song, H. J. Wang, J. Ma, Z. Y. Xu, A. Hinney, J. Hebebrand and Y. Wang, *Obes. Facts*, 2011, **4**, 197–203.
128. C. Holzapfel, M. Siegrist, M. Rank, H. Langhof, H. Grallert, J. Baumert, C. Irimie, N. Klopp, B. Wolfarth, T. Illig, H. Hauner and M. Halle, *Eur. J. Endocrinol.*, 2011, **164**, 205–212.

129. P. J. Turnbaugh, R. E. Ley, M. A. Mahowald, V. Magrini, E. R. Mardis and J. I. Gordon, *Nature*, 2006, **444**, 1027–1031.
130. S. H. Duncan, G. E. Lobley, G. Holtrop, J. Ince, A. M. Johnstone, P. Louis and H. J. Flint, *Int. J. Obes. (2005)*, 2008, **32**, 1720–1724.
131. E. Le Chatelier, T. Nielsen, J. Qin, E. Prifti, F. Hildebrand, G. Falony, M. Almeida, M. Arumugam, J. M. Batto, S. Kennedy, P. Leonard, J. Li, K. Burgdorf, N. Grarup, T. Jorgensen, I. Brandslund, H. B. Nielsen, A. S. Juncker, M. Bertalan, F. Levenez, N. Pons, S. Rasmussen, S. Sunagawa, J. Tap, S. Tims, E. G. Zoetendal, S. Brunak, K. Clement, J. Dore, M. Kleerebezem, K. Kristiansen, P. Renault, T. Sicheritz-Ponten, W. M. de Vos, J. D. Zucker, J. Raes, T. Hansen, H. I. T. c. Meta, P. Bork, J. Wang, S. D. Ehrlich and O. Pedersen, *Nature*, 2013, **500**, 541–546.
132. A. Cotillard, S. P. Kennedy, L. C. Kong, E. Prifti, N. Pons, E. Le Chatelier, M. Almeida, B. Quinquis, F. Levenez, N. Galleron, S. Gougis, S. Rizkalla, J. M. Batto, P. Renault, A. N. R. M. consortium, J. Dore, J. D. Zucker, K. Clement and S. D. Ehrlich, *Nature*, 2013, **500**, 585–588.
133. B. W. Parks, E. Nam, E. Org, E. Kostem, F. Norheim, S. T. Hui, C. Pan, M. Civelek, C. D. Rau, B. J. Bennett, M. Mehrabian, L. K. Ursell, A. He, L. W. Castellani, B. Zinker, M. Kirby, T. A. Drake, C. A. Drevon, R. Knight, P. Gargalovic, T. Kirchgessner, E. Eskin and A. J. Lusis, *Cell Metab.*, 2013, **17**, 141–152.

CHAPTER 6

New Drug Design and Therapeutic Development for Weight Reduction: Preclinical Studies and Clinical Trials

STEVEN P. VICKERS,* HELEN C. JACKSON AND SHARON C. CHEETHAM

RenaSci Ltd., BioCity Nottingham, Pennyfoot Street, Nottingham NG1 1GF, UK
*Email: Steve.vickers@renasci.co.uk

6.1 Obesity and its Metabolic Consequences

The present obesity epidemic is no longer restricted to Western cultures but has a worldwide impact with countries such as Mexico, Brazil, China and Russia affected.[1] Levels of obesity are also increasing in Africa with strong links to increasing urbanisation.[2] Obesity is defined where an individual has a body mass index (BMI) of at least 30 kg m^{-2}. In 2008, the World Health Organisation (WHO) estimated that 200 million men and nearly 300 million women were obese worldwide.[3] In addition, childhood increases in obesity are staggering with more than 40 million children under the age of 5 being overweight or obese in 2012.[3] Obesity acts as a driver for life-threatening disorders that include dyslipidaemia, atherogenesis, hypertension and type 2 diabetes, each of which is a risk factor for cardiovascular disease and mortality.[4–8] Also, obesity not only pre-disposes an individual to some

RSC Drug Discovery Series No. 45
Cardiovascular and Metabolic Disease: Scientific Discoveries and New Therapies
Edited by Philip Peplow, James Adams and Tim Young

Published by the Royal Society of Chemistry, www.rsc.org

cancers including those of the colon, kidney, oesophagus and breast but has been implicated in resistance to chemotherapy such that its efficacy is diminished.[9,10] In some cases, obesity may lead to psychological problems such as depression and non life-threatening diseases such as arthritis and sleep apnoea.[11]

Importantly, weight loss can lead to clinically relevant improvements in obesity related comorbidities. Hence, improvements in blood pressure, triglycerides, HDL cholesterol and glucose control are observed in patients with type 2 diabetes that show only a modest weight loss of 5–10%.[12] Not surprisingly, the primary approach to the management of obesity and its comorbidities is focused on increasing levels of physical exercise and reducing caloric intake. However, this approach is not only associated with poor efficacy but many patients do not maintain their weight loss such that long term success is limited.[13,14] Whilst current prevention policies (*e.g.* the Change4Life programme in England) implemented *via* school or by the mass media to improve diet and increase activity may well have an impact on the future increase in obesity, for many obese individuals such behavioural modification needs to be supplemented by additional approaches such as bariatric surgery or drug intervention.

6.1.1 The Strategic and Commercial Landscape of Drug Intervention

Numerous studies suggest bariatric surgery is efficacious in the treatment of obesity-related comorbidities such as type 2 diabetes as well as improving cardiovascular risk factors, reducing risk of stroke, myocardial infarction and reducing all-cause mortality.[15,16] However, such surgery carries operative risks including organ injuries, anaesthesia events and, albeit rarely, death.[17] Furthermore, recent studies have suggested that such surgery could be associated with bone loss that is not attributable to weight loss *per se*.[18] In addition, due to the obvious requirement for surgical expertise and hospital infrastructure, bariatric surgery could never be performed in anything other than a tiny proportion of the obese population and is restricted to patients with severe or complex obesity. Accordingly, drug intervention is often regarded as being able to bridge the gap between lifestyle and surgical interventions.[19]

Current clinical guidance on the use of anti-obesity drugs states that pharmacotherapy may be used as an adjunct to lifestyle modification to promote weight loss.[20] Historically, a number of drugs have been demonstrated to enhance the weight reduction observed by lifestyle intervention alone,[21–23] and some studies have stressed the importance of the combination of lifestyle and drug intervention in delivering weight reduction.[21] For example, the body weight loss of obese patients that received both lifestyle intervention and the serotonin and noradrenaline re-uptake inhibitor, sibutramine, was greater than groups which received either intervention alone.[21] As discussed previously, such drug-induced reductions in body weight are associated with biologically relevant improvements in comorbidities and cardiovascular risk factors.

The principal regulators that determine whether a drug treatment is approved for treatment of obesity are the U.S. Food and Drug Administration (FDA) and the European Medicines Agency (EMA). Although these agencies have different guidelines in regard to the precise degree of weight loss expected through drug intervention they both require weight reduction to be associated with concomitant improvements in cardiovascular risk factors. Specifically, the FDA requires that in a trial of at least 4500 patients (a minimum 1500 to be placebo controls) drugs should meet at least one of the following criteria: (1) a statistically significant weight loss at least 5% greater than placebo at 1 year; (2) at least 35% of patients should achieve 5% or more weight loss from baseline, and the proportion of drug-treated patients achieving this must be at least twice that of the placebo-treated group and statistically significant.[24] In contrast, the criteria of the EMA are that a drug induces a 10% weight loss from baseline (*i.e.* pre-treatment) at 1 year and which must be 5% greater than observed in patients treated with placebo.[25] Importantly, these criteria are only guidelines and both agencies exercise final decisions in regard to approval (or otherwise) on an individual basis dependent upon the efficacy and safety data available.

As has been discussed at length in other reviews,[26,27] with the withdrawal of two recent drugs due to unacceptable side effect profiles, the strategic and commercial landscape for pharmacotherapy of obesity has changed markedly. Over the past 15–20 years a number of drugs registered for the treatment of obesity have been withdrawn. These include dexfenfluramine (Redux®), sibutramine (Meridia®, Reductil®) and rimonabant (Acomplia®). In the case of rimonabant, the drug reduced body weight in clinical trials and led to improvements in type 2 diabetes and dyslipidaemia that were greater than could be attributable to weight loss alone.[28,29] The drug was approved by the EMA in 2006. However, in the US, concerns relating to psychiatric adverse events associated with rimonabant (suicide ideation, anxiety, depression) led to the withdrawal of the licensing application by Sanofi-Aventis. Subsequently, with increasing reports of such side effects, the EMA suspended rimonabant in 2008. In the case of sibutramine, this drug was suspended by the EMA and withdrawn voluntarily in 2010 by Abbot in the US due to an increased risk of serious non-fatal cardiovascular events such as stroke or heart attack in the Sibutramine Cardiovascular OUTcomes (SCOUT) trial.[30] Of interest is that the outcome trial was mandated by the EMA and performed in a cohort of subjects that was excluded from treatment with the drug in the US Product Label and the European Summary of Product Characteristics. This has led some authors to question the ethics of risking the safety of obese patients with high cardiovascular risk by their inclusion in a clinical trial with sibutramine when the drug was not appropriate to manage their obesity.[26] Not surprisingly, for a disease such as obesity where treatment will be long term and used widely, it is natural that regulators demand a favourable benefit–risk profile for drugs. However, as a result, many pharmaceutical companies viewed the risks of developing anti-obesity agents (especially those acting through the CNS) as being too high and, accordingly, focused on other indications.

The perceived increase in the regulatory hurdles is coupled with commercial issues. Hence, despite the prevalence of obesity globally, and the enormous level of unmet need, sales of anti-obesity drugs are modest. Whilst one reason is that medical insurers in the US are often reluctant to provide reimbursement for anti-obesity drugs, the major cause is more likely to be due to the fact that the 5–8% weight loss typically delivered by such drugs does not meet the often unrealistic expectations of patients (who themselves may have to pay for the drugs). To be a commercial success it is likely that 10–15% (or greater) reductions in body weight are going to be required.[26] Despite such an apparently bleak outlook for the pharmacotherapy of obesity, two drugs (lorcaserin and Qsymia®) have recently been approved in the US (in addition to the lipase inhibitor, orlistat, which has been available in the US since 1999); see Section 6.3. On the back of this apparent success, there may be a renewed interest in the discovery of drugs for the treatment of obesity with a large number of preclinical targets being evaluated and a number of late stage clinical trials ongoing. Indeed, if recently approved drugs also prove to be a commercial success this is likely to stimulate such research even further.

6.2 Preclinical Studies

Like humans, animals gain weight when calorie intake is in excess of energy expenditure over a period of time. Not surprisingly, drug intervention has typically focused on either reducing food intake (including inhibition of the absorption of fat from the gut) or increasing energy expenditure. In general, animal models of food intake, obesity and energy expenditure are well validated, have good predictive validity and are widely used in the design of novel drugs for body weight reduction.[31] The following section reviews promising strategies (both central and peripheral mechanisms) that are currently undergoing pre-clinical evaluation. Examples of drugs and their targets that are being researched at the time of writing are detailed in Table 6.1.

6.2.1 CNS Approaches

Typically, centrally acting anti-obesity drugs have focused on monoaminergic or hypothalamic neuropeptide targets. As can be seen in Table 6.1 such research strategies are still evident (*e.g.* 5-HT_{2C} receptor agonists, dopamine re-uptake inhibitors, melanin concentrating hormone (MCH_1) receptor antagonists). As previously discussed, a number of centrally acting obesity treatments have been withdrawn or suspended due to the incidence of unacceptable side effects. In light of this there may be the perception of a perceived risk in regard to centrally acting targets. For example, although lorcaserin has been approved for the treatment of obesity in the US, few

Table 6.1 Pre-clinical approaches for the treatment of obesity.

Target/mode of action	Examples	Company	Comments
Peripherally-restricted CB_1 antagonists/inverse agonists	JD-5006 and JD-5037	Jenrin Discovery	CB1 receptor antagonists devoid of psychiatric side-effects. Status unknown.
5-HT_{2C} agonists	ATH-91385	Athersys	Potent and selective 5-HT_{2C} agonists.
5-HT_{2C} agonists		University of Florida	Combined 5-HT_{2C} agonists/5-$HT_{2A/2B}$ inverse agonists/antagonists – available for licensing.
5-HT_6 receptor antagonists		Alla Chem LLC	Number of 5-HT6 antagonists. May be developing them for obesity.
5-HT_6 receptor antagonists	SUVN-504	Suven Life Sciences Ltd	Being developed for obesity (2013/2014).
Dopamine reuptake inhibitors		NIH	Selective dopamine reuptake inhibitors without abuse potential. SAR suggests they bind at a different domain to cocaine at the dopamine transporter. Reduce food intake in rats. Available for licensing.
Alpha-3-beta-4 nicotinic receptor antagonist	18-methyoxycoronaridine	Savant HWP Inc	Active in animal models of obesity. Modulates excessive dopamine fluctuations in the mesolimbic system.
Leptin (long-acting)	ARX328	Ambrx	Pegylated form of leptin to extend half-life.
Leptin (long-acting)	PAS-leptin	XL-protein GmbH	Pasylated form of leptin to extend half-life.
MCH_1 receptor antagonists	ALB-127158	Albany Molecular Research Inc	ALB-127158 got as far as Phase 1. AMRI are seeking a partner to progress its MCH1 antagonist program.
Pharmacological chaperones which bind to and correct the function of melanocortin 4 receptors		Pfizer	To treat morbidly obesity caused by mutations in the MC4 receptor. Program licensed from Institute for Research in Immunology and Cancer, Commercialisation of Research, Université de Montréal.
Melanocortin polymorphic receptor ligands		University of Florida	Correct functional defects of melanocortin polymorphic receptors. Available for licensing.
Neuropeptide Y2 agonist	PD4048	P2D Bioscience	
Neuropeptide Y4 agonist	PD4047	P2D Bioscience	
Ghrelin antagonists and inverse agonists		Helsinn	

Ghrelin antagonists		University of Tulane	Available for licensing.
Ghrelin O-acyltransferase (GOAT) inhibitor	CF801	Carleton University	Inhibition of ghrelin acylation. Produces weight loss in mice.
Ghrelin neutralizing antibodies		Sorrento Therapeutics/The Scripps Research Institute	
High-affinity anti acyl ghrelin specific antibodies	33A	Amgen	Block stimulatory effects of ghrelin on food intake when given acutely but no effects on food intake and body weight when given chronically to DIO mice. Compensatory mechanisms? Status unknown.
Growth hormone secretagogue receptor (GHS-R) antagonist	G-51	Janusys Corporation	Inhibits responses to ghrelin.
Bitter taste receptor (T2R) antagonists			Decrease ghrelin secretion.
Anti-somatostatin vaccines	JH-17 and JH-18	Braasch Biotech LLC	
Oxytocin analogues		Albert Einstein College of Medicine of Yeshiva University	Available for licensing. Effective at producing weight loss in animals. Pilot studies suggest may also produce weight loss in man (intranasal oxytocin spray).
Dual GLP-1 and glucagon agonist	MOD-6030/MOD-6031	OPKO Health	Conjugated forms of oxyntomodulin (using reversible Pegylated technology).
G protein coupled receptor (GPR) GPR119 agonist	GSK2041706 (GSK706A)	GSK	
GPR119 agonist	HM-47000	Hanmi	
GPR119 agonist	HD-0471042	Hyundai Pharmaceuticals	
GPR40 agonist	P11187	Piramal	
Lipated neuromedin U analogues		Gubra	Series of lipated neuromedin U analogues created to extend the half-life of neuromedin U. Potent reduction in food intake when given acutely to mice. Gubra discovery programme.

Table 6.1 (*Continued*)

Target/mode of action	Examples	Company	Comments
PEGylated neuromedin U analogues		Merck/Peptipharma	Inhibit food intake in rodents.
NUCB2/nesfatin-1		Semmelweis University	Nesfatin-1 novel modulator of food intake and body weight. Levels increased after bariatric surgery.
Fibroblast growth factor 21 (FGF21) analogues		Eli Lilly	Weight loss in rodents and monkeys.
Fibroblast growth factor 21 (FGF21) analogues		Pfizer	
Novel peptides		Zealand/Eli Lilly	Agreement to develop novel peptides to treat obesity and diabetes. Mode of action not disclosed.
Enteroendocrine cell programme		MedImmune/NGM Biopharmaceuticals	Agreement to identify and develop novel therapeutics (peptides and antibodies) from NGMs enteroendocrine cell programme.
Enteropeptidase inhibitor	OBE-2001	ObeTherapy	Inhibits triglycerides and protein absorption.
Brown fat stem cells	ThermoStem™	BioRestorative Therapies	Agreement with Pfizer to study human BAT tissue and cell lines.
BMP6 (bone morphogenetic protein).		Pfizer	Promotion of brown adipocyte differentiation.
Developing drugs acting at a variety of targets related to brown fat biology		Ember Therapeutics	Augmentation and activation of brown fat to increase energy expenditure.
Proprietary compound that converts WAT to BAT	RZL-12	Raziel Therapeutics	
Inosine monophosphate dehydrogenase inhibitors		Verva	Reduces adipose tissue.

Drugs to reduce or inhibit TRIP-Br2		Joslin Diabetes Center	TRIP-Br2 is a transcription coregulator of metabolic genes involved in fat storage and energy metabolism. Suppression of TRIP-Br2 increases expression of hormone-sensitive lipase and β3 adrenergic receptors.
REV-ERB α/βagonists	SR9011	Scripps Research Institute	REV-ERBα and REV-ERBβ are nuclear receptors that play a role in control of circadian rhythm. SR9011 reduces fat mass and increases energy expenditure in DIO mice.
Protein phosphatase 5 (PP5) inhibitor		University of Toledo	PP5 is a pro-lipogenic phosphatase
Stearoyl-CoA Desaturase (SCD1) inhibitor	SAR224	Sanofi	Status unknown.
Stearoyl-CoA Desaturase (SCD1) inhibitor		Johnson & Johnson	Status unknown.
Stearoyl-CoA Desaturase (SCD1) inhibitor		Viking Therapeutics	
Dual PPAR α/γ agonists	CG301269	Crystal genomics	Enhances glucose and lipid metabolism.
Acetyl-coenzyme A carboxylase inhibitor	CP-640186	Pfizer	Status unknown.
Acetyl-coenzyme A carboxylase inhibitor		Takeda	Status unknown.
Acetyl-coenzyme A carboxylase (ACC) 1/2 allosteric inhibitor	ND-630	Nimbus Discovery	
11βHSD (hydroxysteroid dehydrogenase) type 1 inhibitor	DSP-0011	Sunito Dainippon Pharma Co Ltd	Reduces accumulation of visceral fat mass.
Molecules that block activation of SREBPs (sterol regulatory element-binding proteins)	FGH-10019	FGH	Acts by blocking lipid synthesis and promoting lipid oxidation.

Table 6.1 (*Continued*)

Target/mode of action	Examples	Company	Comments
Diacylglycerol acyltransferase (DGAT) inhibitor 1		Amgen	
DGAT 1 inhibitor			
DGAT 1 inhibitor	P7435	Piramal Enterprises	Active in animal models of obesity. Currently recruiting for Phase 1 studies.
CaSR agonists (Calcium Sensing Receptor)	GSK3004774	GSK	CaSR is part of the GI chemosensory system. GSK3004774 is a luminally-retained CaSR agonist. It reduced body weight in mice on high-fat diet.
Adenosine A2a receptor agonist combined with sodium-coupled glucose transporter 1 (SGLT1) inhibitor or apical sodium-dependent bile transporter (ASBT) inhibitor	GR163819 + GSK2594298 (SGLT1 inhibitor) or GSK2331583 (ASBT inhibitor)	GSK	Weight loss in DIO mice.
Arylhydrocarbon receptors (AhR)	ITE	AhR Pharmaceuticals Inc	Physiological ligand for AhR.
Peripherally-acting TNFα-inhibitors	PD-2024	P2D Bioscience	

companies appear currently to be exploiting this mechanism in a drug research programme. Reasons are likely to include firstly the difficulty of designing a ligand that is truly selective for the 5-HT_{2C} receptor over closely related receptors that have agonism-related side effect liability and secondly, the modest body weight efficacy of the drug in the clinic (see Section 6.3.1).

Despite a number of centrally acting drugs and drug combinations progressing through late stage development (see Section 6.3), there appears to be a limited focus in regard to mechanisms that directly target the brain for the future treatment of obesity. However, other targets are also likely to impinge on central mechanisms of appetite control, even if indirectly. Specifically, there are complex interactions between the gut and brain (gut–brain axis) with peptides such as glucagon like peptide (GLP-1), oxyntomodulin, and ghrelin antagonists all likely to control appetite by acting at the hypothalamus.[32]

6.2.2 Peripheral Approaches

A wide variety of mechanisms are detailed in Table 6.1. Whilst too numerous to discuss all in detail, the following section highlights approaches that may be of special relevance.

6.2.2.1 Ghrelin: Antagonists, Inhibitors, Antibodies

Ghrelin is the only known hormone that stimulates food intake in animals and man.[33] The active, acylated, form of ghrelin is converted from non-acylated ghrelin by the enzyme ghrelin *O*-acyltransferase (GOAT). It is a peptide secreted principally from cells in the fundus of the stomach and a number of studies have shown circadian variation in the plasma levels of ghrelin such that increases are observed before food intake with rapid reductions afterwards.[34] For example, in freely feeding rats, plasma ghrelin is increased prior to presentation of an anticipated chocolate meal.[35] Furthermore, knockout mice lacking either ghrelin or its receptor are reported to be resistant to diet-induced obesity and, moreover, double knockout mice (both ghrelin and receptor) exhibit a lean phenotype characterised by increased energy expenditure.[36] In diabetic animals such as the *ob*/*ob* mouse, deletion of ghrelin increased insulin secretion and improved hyperglycaemia although the obese phenotype (driven by a lack of functional leptin) was not altered.[37] Of note also is the role of the brain in mediating the actions of ghrelin since intracerebroventricular administration leads to increased food intake in rats.[38]

The role of ghrelin in the control of appetite is discussed extensively in numerous reviews.[32] However, based on its role in stimulating food intake, several potential drug targets for the treatment of metabolic disorders are available. One obvious approach is the use of selective antagonists (or inverse agonists since the receptor has high constitutive activity) at the ghrelin receptor to prevent the action of the endogenous hormone.[39] Indeed, in preclinical studies, such compounds (*e.g.* JMV4208) decrease food intake, body weight and fat mass in diet-induced obese (DIO) mice.[40] Clearly this approach has been of interest to the pharmaceutical industry with patent activity appearing to peak in 2005–2007 and compounds disclosed by Abbott, Bayer, Aeterna, Merck, Elixir, Sanofi, AstraZeneca and others.[39] It remains to be seen whether selective orally available compounds will demonstrate efficacy in the clinic. Some commentators have noted that from a medicinal chemistry perspective most novel compounds appear to be lipophilic with high molecular weight which may lead to issues of non-selectivity and limited absorption.[39]

Other means by which novel pharmaceuticals may modulate ghrelin signalling and so reduce body weight include the inhibition of GOAT which blocks the formation of active ghrelin. Since ghrelin is the only known substrate of GOAT, such an approach appears attractive. A number of classes of GOAT inhibitors (typically analogues of GOAT) have been discovered and these are reviewed elsewhere.[41] Preclinical studies demonstrate that GOAT inhibitors can reduce body weight in animal models but this approach remains at an early stage. For example, daily intraperitoneal administration of a GOAT inhibitor for one month in C57BL/6 mice fed a medium chain triglyceride diet led to a reduction in weight gain compared to controls.[41] This effect was associated with a reduction in fat mass and improvements in glucose homeostasis.[41] Whether potent, small molecular weight inhibitors of GOAT can be synthesised remains to be seen.

Antibody targeting has been an immunological approach used for a number of years in the study of ghrelin. In such studies monoclonal antibodies (mAb) specific to active ghrelin are generated such that the activity of the hormone is neutralised. This approach appears to have been pursued by Amgen since the high affinity, anti-acyl-ghrelin-specific antibody, 33A, blocks the stimulatory effects of ghrelin on food intake when given acutely but has no effects on food intake or body weight when administered in a 4 week study in mice with diet-induced obesity.[42] The authors suggest that antagonising the peripheral ghrelin pathway alone may not be a viable strategy for the effective treatment of obesity. More recently administration of a "triplet cocktail" of monoclonal antibodies against ghrelin has been demonstrated to increase energy expenditure and reduce food intake in mice allowed to feed after a fast.[43] This research team argues that an oligoclonal (multiple mAbs) approach is required. Clearly determination of the effects of such oligoclonal antibody targeting of ghrelin on the body weight of obese animals would be of interest. If efficacious, the requirement for multiple mAbs is likely to increase cost and complexity of the development/approval process.

6.2.2.2 Brown Fat-based Strategies

Until 2009, the existence of brown adipose tissue (BAT) in adult humans was unconfirmed.[44] However, demonstration of active human BAT in PET studies has led to renewed interest in this tissue for pharmacological intervention after the past disappointment of agonists targeted at the β_3 adrenergic receptor. Brown fat is of especial interest since unlike white fat which is an energy store, it utilises energy to generate heat, a classic example being its role in core temperature maintenance in infants. Accordingly, drugs that can activate this process may be of use in the treatment of obesity by increasing energy expenditure. BAT can be activated by a number of mechanisms including by adrenergic stimulation (β_3 adrenoceptor agonists), TGR5 receptor (G protein coupled bile acid receptor) agonists and thyroid hormone analogues.[45] A number of companies appear to be focused on BAT biology for the treatment of metabolic disease. Hence, biologics and small molecules are being developed that increase BAT levels (including the conversion of white fat to brown) or activate pathways within BAT (*e.g.* Ember Therapeutics, Raziel Therapeutics). Such strategies appear to be in their infancy with limited preclinical data but may include the disruption of WAT-selective aldehyde dehydrogenases which appears to lead to a "browning" of white adipocytes.[46]

As mentioned, β_3 adrenergic receptor agonists have been developed previously for the treatment of obesity but although efficacy was observed in rodent models, studies in humans were complicated by lack of efficacy or unwanted side effects.[47] Accordingly, it will be of interest to see whether current mechanisms under investigation deliver efficacy and safety in both preclinical and clinical settings. In addition, it needs to be established whether compensatory mechanisms occur during chronic activation of BAT.[45]

6.2.2.3 *Acetyl-coenzyme A Carboxylase-2 Inhibitors (ACC-2)*

ACC-2 is involved in the biosynthesis of fatty acids and interest in the inhibition of the enzyme was a result of work in knockout mice. Specifically, ACC-2$^{-/-}$ mutant mice exhibit continuous fatty acid oxidation, reduced body weight and lower amounts of body fat.[48] A programme appears to have been at least initiated by Pfizer with the novel ACC inhibitor, CP-640186, inhibiting fatty acid synthesis in rats and mice (including *ob/ob* mice).[49] However, in a poster presentation to the American Diabetes Association, the compound was reported to reduce body weight but worsen glucose homeostasis in the *ob/ob* mouse.[50] Accordingly, the potential of this approach is unclear although the compound is not selective for ACC-2 over ACC-1.

6.2.2.4 *Other Targets*

Additional targets that merit brief discussion include inhibitors of stearoyl-CoA desaturase-1 (SCD-1). SCD-1 catalyses the synthesis of monounsaturated fatty acids. Such fatty acids are the substrates for the synthesis of triglycerides and cholesteryl esters and SCD-1 inhibitors reduce liver levels of triacylglycerol (TAG) and cholesterol. SCD-1 inhibitors have been reported to improve insulin sensitivity, confer resistance to diet-induced obesity and have been associated with improvements in non-alcoholic steatohepatitis.[51,52]

Another perhaps more established approach that has been investigated previously is inhibitors of diacylglycerol acyltransferase-1 (DGAT-1).[52] DGAT-1 has a role in the biosynthesis of triacylglycerols and inhibitors appear to reduce body weight in preclinical studies.[52] The results of a Phase 1 trial with the DGAT-1 inhibitor AZD7687 have recently been reported.[53] The trial was successful in that the compound reduced postprandial levels of TAG in healthy male subjects but the authors reported nausea, vomiting and diarrhoea with increasing dose which limited escalation and further development. It remains to be seen whether this is a class effect or specific to AZD7687.

Finally, in light of the marked body weight efficacy seen with the cannabinoid $(CB)_1$ receptor antagonist, rimonabant, there has been reported interest in the development of CB_1 receptor antagonists that do not cross the blood–brain barrier and enter the CNS. Such peripheral CB_1 receptor antagonists (*e.g.* JD-5037) may possess the same ability to reduce body weight and improve HbA1c but without the centrally-mediated side effects that led to withdrawal of rimonabant in Europe.

6.3 Clinical Studies

The clinical development of drugs for weight reduction begins with Phase 1 trials that are typically conducted in small numbers of healthy volunteers. The primary aim is to assess the pharmacokinetics and initial safety although effects on hunger may also be assessed by rating scales or similar scoring systems. In Phase 2 trials the drug is tested in a patient population of several hundred participants for approximately 3 months. The principal dependent variable is % weight loss though waist circumference, and some surrogate markers (*e.g.* plasma insulin, blood pressure) are often determined. The design of the Phase 3 trial is largely a result of the regulatory requirements (previously discussed). They run for at least 1 year, contain several thousand patients with the intention being to generate sufficient data on body weight efficacy, safety, surrogate disease markers and effects in comorbid conditions.

The only drug currently marketed for the treatment of human obesity in both the USA and EU is the lipase inhibitor orlistat (Xenical®; Alli®), which reduces body weight through blocking the gastrointestinal absorption of fats. The lipase inhibitor, cetilistat (Oblean®), is marketed in Japan. Efficacy is modest, with 3% weight loss typical after 1–2 years of treatment.[54] Treatment is also coupled with gastrointestinal side effects (faecal urgency, incontinence) and, accordingly, estimates are that fewer than 10% of patients prescribed the drug continue for more than a year.[19] As discussed previously, the pursuit of clinically effective anti-obesity drugs has had a chequered history with a number of compounds that produce significant weight loss subsequently being withdrawn or refused authorisation by regulators (*e.g.* sibutramine, rimonabant). Current anti-obesity drugs (many of which are marketed only in individual territories) and late stage compounds undergoing clinical evaluation are summarised in Table 6.2.

Table 6.2 Current anti-obesity drugs and late stage candidates.

Drug	Trade name(s)	Company	Mode of action	Status
Orlistat	Xenical Alli	Roche GSK	Lipase inhibitor	Marketed USA; marketed EU
Cetilistat	Oblean	Norgine/Takeda	Lipase inhibitor	Marketed Japan
Phentermine	Various	Generic	NA and DA-releasing agent	Marketed USA; withdrawn EU (2001)
Methamphetamine	Desoxyn	Generic	NA and DA-releasing agent	Marketed USA (refractory obesity); withdrawn EU (2000)
Phendimetrazine Diethylpropion Benzphetamine	Various	Generic	Sympathomimetics	Marketed USA; withdrawn EU (2000)
Sibutramine	Reductil, Meridia	Abbott	NA and 5-HT reuptake inhibitor	Withdrawn USA (2010); withdrawn EU (2010)
Rimonabant	Accomplia	Sanofi-Aventis	Cannabinoid CB_1 antagonist	Not approved USA; withdrawn EU (2008)
Phentermine and topiramate (extended release)	Qsymia Qsiva	Vivus	NA and DA releasing agent/ unknown	Marketed USA; marketing authorisation refused EU (February 2013)
Lorcaserin	Belviq	Arena/Eisai	5-HT_{2C} agonist	Marketed USA (approved June 2012); application to EMA withdrawn (May 2013)
Bupropion/naltrexone (fixed dose combination of bupropion sustained release and naltrexone SR)	Contrave (NB32)	Orexigen Therapeutics Inc Takeda (USA)	DA uptake inhibitor/opioid antagonist	Application for marketing approval submitted to FDA (review date September 2014); application for marketing approval submitted to EMA
Liraglutide	Victoza (brand name for diabetes), Saxenda	Novo Nordisk	GLP-1 agonist	Application for marketing approval submitted to FDA (Advisory committee date September 2014); application for marketing approval submitted to EMA

Table 6.2 (*Continued*)

Drug	Trade name(s)	Company	Mode of action	Status
Beloranib	—	Zafgen	Methionine aminopeptidase 2(MetAP2) inhibitor	Phase 3
Bupropion/zonisamide (fixed dose combination of zonisamide SR and naltrexone SR)	Empatic™	Orexigen Therapeutics	DA uptake inhibitor/ Zonisamide increases monoamines	Phase 2 (completed). In discussion with FDA regarding Phase 3 clinical development plans
PRX-00933		Upsher-Smith	5-HT_{2C} agonist	Phase 2
HPP-404		High Point Pharmaceuticals	Histamine H_3 antagonist	Phase 2
RM-493		Rhythm	Melanocortin 4 receptor agonist	Phase 2
S-237648		Shionogi	Neuropeptide Y5 receptor antagonist	Phase 1
Y-242		Imperial College, London	Long-acting analogue of Peptide YY	Phase 1
PP 1420		Imperial College London	Pancreatic polypeptide analogue (peptidase resistant)	Phase 1
LY-2944876/TT-401		Eli Lilly/Transition Therapeutics	Dual GLP-1 and glucagon agonist (OXM analogue)	Phase 2
HM11260C		Hanmi Pharmaceuticals	LAPS-CA-Exendin-4. Long-acting GLP-1 agonist	Phase 2

Exenatide	GSK	GLP1 agonist (continuous infusion)	Phase 1
Exenatide	Intarcia	Continuous administration *via* sub-dermally inserted osmotic minipumps	Phase 2 for diabetes (weight loss reported)
GSK-2374697	GSK	GLP1-agonist (extendin-4 fused to a half-life improving domain antibody (AlbudAb™)	Phase 1
ZYD-1	Zydus Cadila	GLP-1 receptor agonist	Phase 1
HM12525A	Hanmi Pharmaceuticals	LAPS-GLP-1/GCG. Long-acting dual GLP-1 and glucagon agonist	Phase 1
ZP-2929	Zealand Pharma	Dual GLP-1 and glucagon agonist	Phase 1
TK S1225/OAP-189	Thiakis (acquired by Pfizer)	Dual GLP-1 and glucagon agonist (oxyntomodulin analogue)	Phase 1 or 2?
PSA-oxyntomodulin	Xenetic Biosciences Inc/OJSC Pharmasynthez	Dual GLP-1 and glucagon agonist. Conjugated form of OXM using PolyXen technology	Phase 1
TT-402	Transition Therapeutics/Lilly	Dual GLP-1 and glucagon receptor agonist	Phase 1

Table 6.2 (*Continued*)

Drug	Trade name(s)	Company	Mode of action	Status
KD-026		Kadman Corporation	Inhibitor of enterocytic triglyceride transfer protein (MTP inhibitor)	Phase 2
SYN-010		Synthetic Biologics Inc	Management of methane in the gastrointestinal tract	Phase 1
ISIS-FGFR4Rx		Isis Pharmaceuticals	Antisense that specifically reduces FGFR4 (fibroblast growth factor receptor 4) in liver and fat tissue.	Phase 1
MB-12066 (beta lapachone)		KT&G Life Sciences	Sirtuin activator	Phase 1
Adipotide (prohibitin-targeting peptide 1)		Arrowhead Research Corporation	Targeted induction of apoptosis in the vasculature of adipose tissue	Phase 1
MetaBrake		MetaBrake LLC	Oral mimetic of Roux-en-Y by-pass surgery	Phase 1

6.3.1 Lorcaserin

In 2012, the centrally-acting 5-HT_{2C} receptor agonist, lorcaserin, was approved in the US for the treatment of obesity after demonstration of a 3–3.7% placebo-subtracted weight loss in clinical trials and 0.5% reduction in HbA1c, indicative of improved glucose control in patients with type 2 diabetes.[55,56] There is a large scientific literature on the role of serotonin (5-HT) in the control of food intake and the effect of serotonergic compounds in reducing food intake in preclinical and clinical studies is well established.[57,58] Indeed, the clinically effective weight loss agents sibutramine and *d*-fenfluramine non-selectively increase serotonergic activity either by inhibiting reuptake or promoting its release, respectively. Although both the aforementioned drugs were removed from the market due to the incidence of serious side effects (valvular heart disease in the case of fenfluramine and serious, non-fatal cardiovascular events in the case of sibutramine), selective 5-HT_{2C} receptor activation by lorcaserin is believed to have a specific effect to enhance satiety and has low affinity for receptors such as the 5-HT_{2B} receptor which have been linked with the causation of valvulopathy.[59] The development of lorcaserin illustrates the length of time that it takes to move from preclinical research to a marketed drug. Hence, although lorcaserin was available on the US market in 2013, numerous research manuscripts and reviews advocating the use of 5-HT_{2C} receptor agonists for the treatment of obesity date back to the 1990s.[57,60,61] At the time of writing, lorcaserin is not available outside the US: application to the European regulators was withdrawn by Arena/Eisai in 2013 since they were unable to resolve the objections of the Committee for Medicinal Products for Human Use (CHMP) in regard to the potential risk of tumours (based on preclinical studies) and clinical safety concerns (in regard to the increased risk of psychiatric disorders (*e.g.* depression) and valvulopathy) within the appropriate timeframe.

6.3.2 Phentermine

Although withdrawn in Europe in 2001 due to an unfavourable benefit–risk profile, phentermine (approved for short term use in 1959 by the FDA) is the most widely prescribed weight loss drug in the USA[62] although, in part, this may be related to its low cost. The primary pharmacological action of phentermine is the stimulation of noradrenaline release (also serotonin and dopamine) and, although it is chemically-related to amphetamine (both are phenethylamines), phentermine does not appear to exhibit the same potential for abuse. For example, when phentermine therapy was stopped in obese subjects undergoing treatment with the drug, no craving or withdrawal signs were evident.[63] This is also reflected in the lower US scheduling of phentermine (Schedule IV) compared to amphetamine (Schedule II). In a recent, albeit small, 12 week trial of 74 obese patients, a diffuse-controlled release formulation of phentermine was demonstrated to reduce body weight from baseline by 5% in $\geq$75% of patients and 10% in $\geq$59% of patients.[64] Such effects will need to be confirmed in larger and longer trials but this phentermine formulation appears to compare well to recent drug candidates in terms of body weight efficacy.

Historically, phentermine has been used in combination with other clinically effective agents to manage weight loss.[65] Indeed, its combination with fenfluramine and *d*-fenfluramine was widespread until the withdrawal of these fenfluramines due to association with cardiac valve damage. Such an approach is attractive for the treatment of obesity since additive (or even synergistic) weight loss may be seen through combining different pharmacological mechanisms. More recently a fixed dose combination of phentermine and the anticonvulsant agent, topiramate, has been investigated (Qsymia® in the US; Qsiva® in Europe).

6.3.3 Phentermine + Topiramate

Although approved for use as an anticonvulsant and for the prevention of migraines, anecdotal reports of weight loss in topiramate-treated epileptic patients led to the drug being evaluated in clinical trials for the treatment of obesity.[66] Initial trials were halted due to an unacceptable side effect profile. The mechanisms underlying the action of topiramate to reduce body weight remain unknown and have been the subject of other reviews.[67] However, some authors have suggested that principal mechanisms include the action of the drug to inhibit adipocyte mitochondrial carbonic anhydrase V resulting in reduced lipogenesis.[68]

In light of the weight reducing action of topiramate, low doses of the drug were coupled with low doses of phentermine in the hope of realising a combination (Qsymia®) that had not only an improved safety profile but also improved efficacy compared to the individual agents. A number of Phase 3 clinical trials have now been completed for Qsymia® and these have been discussed in numerous reviews.[19,26,27,69,70] In brief, weight loss from baseline is dose-dependent and between 8.9–10% at the highest dose tested (15/92 mg; Phen/TOP). Qsymia was approved in the US in 2012 as an adjunct to diet and exercise. Despite this, the drug is not approved for use in Europe and in 2013 the EMA highlighted concerns in regard to the long term effect of phentermine on heart rate (as discussed previously, phentermine was withdrawn from use in 2001 in Europe). In addition, issues in regard to potential birth deficits with topiramate and long-term psychiatric and cognitive effects of the drug were also raised.[71]

6.3.4 Naltrexone + Bupropion (NB32, Contrave)

A fixed-dose combination of the opioid receptor antagonist, naltrexone, and the dopamine reuptake inhibitor, bupropion, is being developed for obesity by Orexigen. Naltrexone (Vivitrol®) is used clinically for the treatment of alcohol and opiate dependence. Bupropion is approved for use as an antidepressant (Wellbutrin®) and for smoking cessation (Zyban®). Both naltrexone and bupropion tend to inhibit food intake and body weight in preclinical studies in rodents when dosed alone and in combination.[72] Such effects are evident not only in lean animals but also models of diet-induced obesity.[72] Interestingly, naltrexone has little effect on body weight in man[73] and the body weight loss with bupropion monotherapy is modest, reaching a maximum 6 months after treatment.[74] In regard to reducing body weight, it has been proposed that the key neural substrates are pro-opiomelanocortin (POMC) neurones in the arcuate nucleus of the hypothalamus. Specifically, the body weight reduction with bupropion is proposed to be a consequence of a combined dopaminergic and noradrenergic action to increase POMC neuronal activation.[72] However, it is reported that this increased firing is autoinhibited by a naltrexone-sensitive mechanism such that the body weight reduction observed with bupropion monotherapy is limited. Accordingly, in the presence of naltrexone, bupropion-induced action on POMC neurones of the arcuate nucleus, and therefore the effect of bupropion on body weight, should be maximised.[72]

Four 56 week trials have been undertaken to determine the efficacy and safety of NB32 in obese patients either with or without type 2 diabetes.[75] In terms of weight loss, commentators regard the efficacy of NB32 to be less than the combination drug Qsymia, greater than lorcaserin and equivalent to sibutramine and rimonabant.[26] Benefits in cardiovascular associated risk factors were generally commensurate with weight loss. However, after effects of placebo were taken into account, NB32 produced small increases in both systolic and diastolic blood pressure. Indeed, the clinical briefing document states that NB32 increased hypertension and palpitations which led to discontinuation in 0.3% of the drug-treated subjects (no such effects evident with placebo).[75] Perhaps unsurprisingly in light of the withdrawal of sibutramine (see earlier), the FDA requested a long term cardiovascular trial prior to approval for the treatment of obesity. In November 2013 interim results of this trial (LIGHT study) did not show a significant increase in major cardiac events with the combination drug. Orexigen has submitted applications for marketing authorization in the United States and Europe, which in the case of the US is to be reviewed in September 2014.

6.3.5 Liraglutide

The injectable drug, Liraglutide (Victoza, Saxenda), is a long acting glucagon-like peptide-1 (GLP-1) agonist that is approved for the treatment of type 2 diabetes in the US, EU and Japan. The compound is a GLP-1 analogue with 97% homology to the human peptide but more resistant to enzymatic degradation by dipeptidyl peptidase 4 than endogenous GLP-1. There is a large preclinical and clinical literature on the effect of liraglutide on glucose control and ingestive behaviour in both clinical and preclinical studies.[76–80] Whilst GLP-1 receptor agonists improve glucose control by enhancing glucose-stimulated insulin secretion, they have a widely documented effect to suppress appetite. A recent meta-analysis of clinical trials reported that GLP-1 receptor agonists significantly reduced body weight in overweight or obese patients with or without type 2 diabetes.[81] In light of such data, liraglutide is currently undergoing clinical evaluation in adult patients with obesity. Such trials have not only assessed doses used for the treatment of type 2 diabetes (*i.e.* 1.2 mg daily) but have also investigated higher doses (up to 3.0 mg).[79] In a 20 week Phase 2 trial that compared liraglutide with orlistat and placebo, obese (non-diabetic) patients on a calorie-controlled diet and an exercise programme exhibited a dose-dependent reduction in body weight with the GLP-1 receptor agonist.[79] At the 3.0 mg liraglutide dose, patients lost 7.2 kg compared to the 4.1 kg of orlistat-treated patients and the 2.8 kg of those on placebo. More than 5% weight loss was achieved by 76% of liraglutide-participants in contrast to 30% achieving this loss on placebo and 41% with orlistat treatment.[79] In addition, liraglutide reduced blood pressure and reduced the frequency of prediabetes. Currently Novo Nordisk is running a series of trials (the SCALE clinical programme) to evaluate the safety and efficacy of liraglutide for obesity and comorbidities.

In one such randomised Phase 3 trial (SCALE Maintenance) the effect of liraglutide (3.0 mg daily) to maintain weight loss induced by a low calorie diet was investigated in obese/overweight patients.[80] Briefly, patients who lost 5% of body weight during a run-in period on a restricted calorie diet were assigned to placebo or liragutide. Liraglutide was reported to not only maintain the weight loss achieved compared to placebo (84% *cf.* 49%) but further reduced weight loss over a 56 week compared to placebo by 6.1%.[80] Liraglutide is not without side effects. Hence, commonly reported issues are nausea and gastrointestinal disorders (*e.g.* diarrhoea). However, reports are that such events are transient and mild and rarely lead to the discontinuation of treatment.[79,80,82] GLP-1 receptor analogues have been linked with increased risk of pancreatitis[83,84] and the EMA summary of product characteristics states that patients should be warned of the signs of pancreatitis (severe, persistent abdominal pain) and the drug discontinued.[85] However, currently this increased risk of pancreatitis is not believed to be high. As detailed in Table 6.2, applications for marketing approval have been submitted to the FDA and the EMA. In the case of the FDA an advisory committee date has been set for September 2014.

6.3.6 SGLT2 Inhibitors

Sodium-glucose linked cotransporter-2 (SGLT2) inhibitors block the reabsorption of glucose in the proximal tubule of the kidney. These drugs were developed and approved as treatments for diabetes since they promote the disposal of excess glucose *via* the urine.[86,87] A number of SGLT2 inhibitors (*e.g.* dapagliflozin, empagliflozin *etc.*) have also been found to reduce body weight (fat mass specifically) in both preclinical and clinical studies presumably due to the loss of calories (glucose) *via* the urine.[88–90]. Indeed, dapagliflozin not only augments the reduction in HbA1c induced by pioglitazone but it also prevents manifestation of weight gain, a well characterised side effect of PPARγ agonists.[91] Furthermore, in preclinical studies in dietary-induced obese rats, empagliflozin augmented the effect of the anti-obesity drugs sibutramine and orlistat on body weight and glucose control when given in combination.[88] Such a drug class provides a potential strategy for the treatment of obese patients with prediabetes or type 2 diabetes.[88] However, it should be borne in mind that the weight loss observed with such drugs is small (>5% weight loss was seen in only 26% of dapagliflozin-treated patients[90]) and such drugs are not, to the best of our knowledge, being developed for the treatment of obesity *per se*.

6.3.7 Additional Drugs in Early Stages of Development

This section has detailed a number of the most promising and late-stage approaches for the treatment of obesity. A number of additional strategies and their status are detailed in Table 6.2. It can be seen that the mode of action of these compounds is diverse although many are focused, at least in part, on existing strategies. For example, although there appear to be companies focused on 5-HT_{2C} and GLP-1 receptor agonists (no doubt driven by the progress of lorcaserin in the USA and liraglutide in the US/EU), there are also a number of compounds with dual agonist activity at both GLP-1 and glucagon receptors (see Tables 6.1 and 6.2). Importantly, such an approach is consistent in concept with the combination strategy of Qsymia® and Contrave® and may indicate the future of drug development for obesity. Furthermore, like Qsymia® and NB32, Empatic® (bupropion and the anticonvulsant, zonisamide) is also a combination of CNS active drugs and is currently in Phase 3 development.

Additional strategies of note include RM-493, which is a first-in-class melanocortin 4 receptor agonist that, in apparent contrast to previous compounds, is not associated with increased blood pressure. Interestingly, the treatment approach is highly focused. Two Phase 2a studies appear to be planned for 2014 in order to evaluate the compound in patients with a genetic mutation of the MC4 receptor gene or Prader–Willi syndrome. Other strategies of interest include the methionine aminopeptidase 2 (MetAP2) inhibitor, beloranib. The compound is proposed to inhibit angiogenesis in adipose tissue and has been reported to reduce body weight in early stage clinical trials.[92] The approach was initially developed as an anti-cancer therapy since cancer tissue is highly vascularised and inhibiting the creation of new blood vessels from existing ones may impact on tumour development. In the case of obesity, the drugs arrest the growth of angiogenic vessels so preventing the development of adipose tissue/obesity. Interestingly this is a similar approach to that of an earlier-stage compound, adipotide, which induces apoptosis in white adipose tissue endothelium. In preclinical studies such activity is associated with improvements in adiposity, glucose control and lipid metabolism[93] although, interestingly, Phase 1 trials planned for this injectable drug appear to be in obese patients with prostate cancer.

6.4 Conclusions and Future Developments

Clearly there is a current and enormous unmet need for the treatment of obesity and associated metabolic disorders. This is compounded by setbacks in terms of the recent withdrawal or suspension of approved drugs for obesity (especially in the case of sibutramine which led to increased cardiovascular risk in a group of patients whose obesity was not appropriate for the drug and who, ordinarily, would not have been prescribed it). However, it is clear that there is an emerging pipeline of novel targets and drugs both at the preclinical (*e.g.* brown fat-based strategies) and the clinical stage (*e.g.* liraglutide, Contrave®, *etc.*). Indeed, the fixed dose phentermine and topiramate combination, Qsymia®, appears to be more efficacious than orlistat, sibutramine and rimonabant. That said, teratogenic risk, elevations in heart rate and psychiatric and cognitive side effects are all of concern in regard to the combination and safety considerations rightly remain the priority. Cardiovascular outcome trials for Qsymia will be of special interest. Many authors agree that it is unlikely a single drug will ever be designed that resolves obesity in all people. The future of pharmacological intervention of obesity is likely to involve the use of drug combinations, personalised to an individual's needs. With global obesity showing no sign of decline perhaps mechanisms being targeted today will prove to be part of an effective treatment for tomorrow.

References

1. M. Cecchini, F. Sassi, J. A. Lauer, Y. Y. Lee, V. Guajardo-Barron and D. Chisholm, *Lancet*, 2010, **376**, 1775.
2. B. Adeboye, G. Bermano and C. Rolland, *Cardiovasc. J. Afr.*, 2012, **23**, 512.
3. World Health Organisation Factsheet 311. 2014. http://www.who.int/mediacentre/factsheets/fs311/en/.
4. F. X. Pi-Sunyer, *Obes. Res.*, 2002, **10**(Suppl 2), 97S.
5. D. Lorber, *Diabetes Metab. Syndr. Obes.*, 2014, 7, 169.
6. V. G. DeMarco, A. R. Aroor and J. R. Sowers, *Nat. Rev. Endocrinol.*, 2014, **10**, 364.
7. M. J. van Rooy and E. Pretorius, *Curr. Med. Chem.*, 2014, **21**, 2121.
8. J. O. Ebbert and M. D. Jensen, *Nutrients*, 2013, **5**, 498.
9. J. Park, T. S. Morley, M. Kim, D. J. Clegg and P. E. Scherer, *Nat. Rev. Endocrinol.*, 2014, **10**, 455.
10. L. M. Lashinger, E. L. Rossi and S. D. Hursting, *Clin. Pharmacol. Ther.*, 2014, **96**, 458.
11. R. Wolk, A. S. Shamsuzzaman and V. K. Somers, *Hypertension*, 2003, **42**, 1067.
12. R. Wing, W. Lang, T. A. Wadden, M. Safford, W. C. Knowler, A. G. Bertoni, J. O. Hill, F. L. Brancati, A. Peters and L. Wagenknecht, *Diabetes Care*, 2011, **34**, 148.
13. A. J. Scheen, *Expert Opin. Invest. Drugs*, 2008, **17**, 263.

14. T. A. Wadden, J. A. Sternberg, K. A. Letizia, A. J. Stunkard and G. D. Foster, *Int. J. Obes.*, 1989, **13**, 39.
15. C. S. Kwok, A. Pradhan, M. A. Khan, S. G. Anderson, B. D Keavney, P. K. Myint, M. A. Mamas and Y. K. Loke, *Int. J. Cardiol.*, 2014, **173**, 20–28.
16. L. Sjöström, M. Peltonen, P. Jacobson, C. D. Sjöström, K. Karason, H. Wedel, S. Ahlin, Å. Anveden, C. Bengtsson, G. Bergmark, C. Bouchard, B. Carlsson, S. Dahlgren, J. Karlsson, A. K. Lindroos, H. Lönroth, K. Narbro, I. Näslund, T. Olbers, P. A. Svensson and L. M. Carlsson, *JAMA, J. Am. Med. Assoc.*, 2012, **307**, 56.
17. A. J. Greenstein, A. S. Wahed, A. Adeniji, A. P. Courcoulas, G. Dakin, D. R. Flum, V. Harrison, J. E. Mitchell, R. O'Rourke, A. Pomp, J. Pender, R. Ramanathan and B. M. Wolfe, *J. Am. Coll. Surg.*, 2012, **215**, 271.
18. E. M. Stein and S. J. Silverberg, *Lancet Diabetes Endocrinol.*, 2014, **2**, 165.
19. K. Gadde, *Expert Opin. Pharmacother.*, 2014, **15**, 1.
20. M. D. Jensen, D. H. Ryan, C. M. Apovian, C. M. Loria, J. D. Ard, B. E. Millen, A. G. Comuzzie, C. A. Nonas, K. A. Donato, F. X. Pi-Sunyer, F. B. Hu, J. Stevens, V. S. Hubbard, V. J. Stevens, J. M. Jakicic, T. A. Wadden, R. F. Kushner, B. M. Wolfe and S. Z. Yanovski, *J. Am. Coll. Cardiol.*, 2014, **129**, S102.
21. T. A. Wadden, R. I. Berkowitz, L. G. Womble, D. B. Sarwer, S. Phelan, R. K. Cato, L. A. Hesson, S. Y. Osei, R. Kaplan and A. J. Stunkard, *N. Engl. J. Med.*, 2005, **353**, 2111.
22. T. A. Wadden, J. P. Foreyt, G. D. Foster, J. O. Hill, S. Klein, P. M. O'Neil, M. G. Perri, F. X. Pi-Sunyer, C. L. Rock, J. S. Erickson, H. N. Maier, D. D. Kim and E. Dunayevich, *Obesity*, 2011, **19**, 110.
23. J. W. Anderson, F. L. Greenway, K. Fujioka, K. M. Gadde, J. McKenney and P. M. O'Neil, *Obes. Res.*, 2002, **10**, 633.
24. Food and Drug Administration Guidance for industry developing products for weight management, draft guidance, revision 1. http://www.fda.gov/downloads/Drugs/Guidances/ucm071612.pdf.
25. Guideline on clinical investigation of medicinal products used In weight control http://www.ema.europa.eu/docs/en_GB/document_library/Scientific_guideline/2009/09/WC500003213.pdf.
26. D. J. Heal, J. Gosden and S. L. Smith, *Int. J. Obes.*, 2013, **37**, 107.
27. S. Manning, A. Pucci and N. Finer, *Ther. Adv. Chronic Dis.*, 2014, **5**, 135.
28. J. P. Després, A. Golay and L. Sjöström, *N. Engl. J. Med.*, 2005, **353**, 2121.
29. F. X. Pi-Sunyer, L. J. Aronne, H. M. Heshmati, J. Devin and J. Rosenstock, *JAMA, J. Am. Med. Assoc.*, 2006, **295**, 761.
30. W. P. James, I. D. Caterson, W. Coutinho, N. Finer, L. F. Van Gaal, A. P. Maggioni, C. Torp-Pedersen, A. M. Sharma, G. M. Shepherd, R. A. Rode and C. L. Renz, *N. Engl. J. Med.*, 2010, **363**, 905.
31. S. P. Vickers, H. C. Jackson and S. C. Cheetham, *Br. J. Pharmacol.*, 2011, **164**, 1248.
32. R. C. Troke, T. M. Tan and S. R. Bloom, *Ther. Adv. Chronic Dis.*, 2014, **5**, 4.

33. M. R. Druce, A. M. Wren, A. J. Park, J. E. Milton, M. Patterson, G. Frost, M. A. Ghatei, C. Small and S. R. Bloom, *Int. J. Obes.*, 2005, **29**, 1130.
34. J. J. Carlson, A. A. Turpin, G. Wiebke, S. C. Hunt and T. D. Adams, *Nutr. Metab.*, 2009, **6**, 32.
35. M. Merkestein, M. A. Brans, M. C. Luijendijk, J. W. de Jong, E. Egecioglu, S. L. Dickson and R. A. Adan, *Obesity*, 2012, **20**, 963.
36. P. T. Pfluger, H. Kirchner, S. Günnel, B. Schrott, D. Perez-Tilve, S. Fu, S. C. Benoit, T. Horvath, H. G. Joost, K. E. Wortley, M. W. Sleeman and M. H. Tschöp, *Am. J. Physiol.: Gastrointest. Liver Physiol.*, 2008, **294**, G610.
37. Y. Sun, M. Asnicar, P. K. Saha, L. Chan and R. G. Smith, *Cell Metab.*, 2006, **3**, 379.
38. L. F. Faulconbridge, D. E. Cummings, J. M. Kaplan and H. J. Grill, *Diabetes*, 2003, **52**, 2260.
39. K. O. Cameron, S. K. Bhattacharya and A. K. Loomis, *J. Med. Chem.*, 2014, **57**, 8671.
40. M. Holubová, V. Nagelová, Z. Lacinová, M. Haluzík, D. Sýkora, A. Moulin, A. L. Blayo, J. A. Fehrentz, J. Martinez, A. Stofkova, J. Jurčovičová, B. Zelezná and L. Maletínská, *Mol. Cell. Endocrinol.*, 2014, **393**, 120.
41. M. S. Taylor, Y. Hwang, P. Y. Hsiao, J. D. Boeke and P. A. Cole, *Methods Enzymol.*, 2012, **514**, 205.
42. S. C. Lu, J. Xu, N. Chinookoswong, S. Liu, S. Steavenson, C. Gegg, D. Brankow, R. Lindberg, M. Véniant and W. Gu, *Mol. Pharmacol.*, 2009, **75**, 901.
43. J. S. Zakhari, E. P. Zorrilla, B. Zhou, A. V. Mayorov and K. D. Janda, *Mol. Pharm.*, 2012, **9**, 281.
44. A. M. Cypess, S. Lehman, G. Williams, I. Tal, D. Rodman, A. B. Goldfine, F. C. Kuo, E. L. Palmer, Y. H. Tseng, A. Doria, G. M. Kolodny and C. R. Kahn, *N. Engl. J. Med.*, 2009, **360**, 1509.
45. P. W. Butler, D. Mentuccia and F. S. Celi, *Heart Metab.*, 2010, **48**, 19.
46. F. W. Kiefer, C. Vernochet, P. O'Brien, S. Spoerl, J. D. Brown, S. Nallamshetty, M. Zeyda, T. M. Stulnig, D. E. Cohen, C. R. Kahn and J. Plutzky, *Nat. Med.*, 2012, **18**, 918.
47. J. R. Arch and S. Wilson, *Int. J. Obes. Relat. Metab. Disord.*, 1996, **20**, 191.
48. L. Abu-Elheiga, M. M. Matzuk, K. A. Abo-Hashema and S. J. Wakil, *Science*, 2001, **291**, 2613.
49. H. J. Harwood, S. F. Petras, L. D. Shelly, L. M. Zaccaro, D. A. Perry, M. R. Makowski, D. M. Hargrove, K. A. Martin, W. R. Tracey, J. G. Chapman, W. P. Magee, D. K. Dalvie, V. F. Soliman, W. H. Martin, C. J. Mularski and S. A. Eisenbeis, *J. Biol. Chem.*, 2003, **278**, 37099.
50. J. L. Treadway, R. K. McPherson, S. F. Petras, L. D. Shelly, K. S. Frederick, K. Sagawa, D. A. Perry and J. H. Harwood, 2014 ADA Poster (accessed July 2013) http://professional.diabetes.org/Content/Posters/2004/p679-P.pdf.
51. N. Kurikawa, T. Takagi, S. Wakimoto, Y. Uto, H. Terashima, K. Kono, T. Ogata and J. Ohsumi, *Biol. Pharm. Bull.*, 2013, **36**, 259.
52. A. Chatzigeorgiou, E. Kandaraki, A. G. Papavassiliou and M. Koutsilieris, *Obes. Rev.*, 2014, **15**, 487.

53. H. Denison, C. Nilsson, M. Kujacic, L. Löfgren, C. Karlsson, M. Knutsson and J. W. Eriksson, *Diabetes, Obes. Metab.*, 2013, **15**, 136.
54. D. Rucker, R. Padwal, S. K. Li, C. Curioni and D. C. Lau, *BMJ*, 2007, **335**, 1194.
55. S. R. Smith, N. J. Weissman, C. M. Anderson, M. Sanchez, E. Chuang, S. Stubbe, H. Bays and W. R. Shanahan, *N. Engl. J. Med.*, 2010, **363**, 245.
56. P. M. O'Neil, S. R. Smith, N. J. Weissman, M. C. Fidler, M. Sanchez, J. Zhang, B. Raether, C. M. Anderson and W. R. Shanahan, *Obesity*, 2012, **20**, 1426.
57. M. J. Bickerdike, S. P. Vickers and C. T. Dourish, *Diabetes, Obes. Metab.*, 1999, **1**, 207.
58. S. P. Vickers and C. T. Dourish, *Curr. Opin. Invest. Drugs*, 2004, **5**, 377.
59. J. D. Hutcheson, V. Setola, B. L. Roth and W. D. Merryman, *Pharmacol. Ther.*, 2011, **132**, 146.
60. C. T. Dourish, *Obes. Res.*, 1995, **3**, 449S.
61. R. Samanin and S. Garattini, *Therapie*, 1996, **51**, 107.
62. C. Hampp, E. M. Kang and V. Borders-Hemphill, *Pharmacotherapy*, 2013, **33**, 1299.
63. E. J. Hendricks, F. L. Greenway, E. C. Westman and A. K. Gupta, *Obesity*, 2011, **19**, 2351.
64. J. G. Kang, C. Y. Park, J. H. Kang, Y. W. Park and S. W. Park, *Diabetes Obes. Metab.*, 2010, **12**, 876.
65. M. Weintraub, J. D. Hasday, A. I. Mushlin and D. H. Lockwood, *Arch. Intern. Med.*, 1984, **144**, 1143.
66. J. Wilding, L. Van Gaal, A. Rissanen, F. Vercruysse and M. Fitchet, *Int. J. Obes. Relat. Metab. Disord.*, 2004, **28**, 1399.
67. D. Richard, J. Ferland, J. Lalonde, P. Samson and Y. Deshaies, *Nutrition*, 2000, **16**, 961.
68. G. De Simone, A. Di Fiore and C. T. Supuran, *Curr. Pharm. Des.*, 2008, **14**, 655.
69. D. N. Kiortsis, *Hormones*, 2013, **12**, 507.
70. J. Jordan, A. Astrup, S. Engeli, K. Narkiewicz, W. W. Day and N. J. Finer, *Hypertension*, 2014, **32**, 1178.
71. http://www.ema.europa.eu/docs/en_GB/document_library/Other/human/002350/WC500147577.pdf. Accessed July 2014.
72. F. L. Greenway, M. J. Whitehouse, M. Guttadauria, J. W. Anderson, R. L. Atkinson, K. Fujioka, K. M. Gadde, A. K. Gupta, P. O'Neil, D. Schumacher, D. Smith, E. Dunayevich, G. D. Tollefson, E. Weber and M. A. Cowley, *Obesity*, 2009, **17**, 30.
73. M. W. Lee and K. Fujioka, *Expert Opin. Pharmacother.*, 2009, **10**, 1841.
74. J. W. Anderson, F. L. Greenway, K. Fujioka, K. M. Gadde, J. McKenney and P. M. O'Neil, *Obes. Res.*, 2002, **10**, 633.
75. Contrave NDA 2010. http://www.fda.gov/downloads/advisorycommittees/committeesmeetingmaterials/drugs/endocrinologicandmetabolicdrugsadvisorycommittee/ucm235671.pdf. Accessed July 2014.

76. K. Raun, P. von Voss, C. F. Gotfredsen, V. Golozoubova, B. Rolin and L. B. Knudsen, *Diabetes*, 2007, **56**, 8.
77. D. Nagakubo, M. Shirai, Y. Nakamura, N. Kaji, C. Arisato, S. Watanabe, A. Takasugi and F. Asai, *Comp. Med.*, 2014, **64**, 121.
78. H. Harder, L. Nielsen, D. T. Tu and A. Astrup, *Diabetes Care*, 2004, **27**, 1915.
79. A. Astrup, R. Carraro, N. Finer, A. Harper, M. Kunesova, M. E. Lean, L. Niskanen, M. F. Rasmussen, A. Rissanen, S. Rössner, M. J. Savolainen and L. Van Gaal, *Int. J. Obes.*, 2012, **36**, 843.
80. T. A. Wadden, P. Hollander, S. Klein, K. Niswender, V. Woo, P. M. Hale and L. Aronne, *Int. J. Obes.*, 2013, **37**, 1443.
81. T. Vilsbøll, M. Christensen, A. E. Junker, F. K. Knop and L. L. Gluud, *BMJ*, 2012, **10**, 344.
82. M. E. Lean, R. Carraro, N. Finer, H. Hartvig, M. L. Lindegaard, S. Rössner, L. Van Gaal and A. Astrup, *Int. J. Obes.*, 2014, **38**, 689.
83. P. C. Butler, M. Elashoff, R. Elashoff and E. A. Gale, *Diabetes Care*, 2013, **36**, 2118.
84. D. Cohen, *BMJ*, 2013, 7, 346.
85. Victoza Summary of Product Characteristics http://www.ema.europa.eu/docs/en_GB/document_library/EPAR_-_Product_Information/human/001026/WC500050017.pdf. Accessed July 2014.
86. E. Ferrannini and A. Solini, *Nat. Rev. Endocrinol.*, 2012, **8**, 495.
87. A. A. Tahrani, A. H. Barnett and C. J. Bailey, *Lancet Diabetes Endocrinol.*, 2013, **1**, 140.
88. S. P. Vickers, S. C. Cheetham, K. R. Headland, K. Dickinson, R. Grempler, E. Mayoux, M. Mark and T. Klein, *Diabetes, Metab. Syndr. Obes.: Targets Ther.*, 2014, **7**, 265.
89. S. Grandy, M. Hashemi, A. M. Langkilde, S. Parikh and C. D. Sjöström, *Diabetes, Obes. Metab.*, 2014, **16**, 645.
90. J. Bolinder, Ö. Ljunggren, J. Kullberg, L. Johansson, J. Wilding, A. M. Langkilde, J. Sugg and S. Parikh, *J. Clin. Endocrinol. Metab.*, 2012, **97**, 1020.
91. J. Rosenstock, M. Vico, L. Wei, A. Salsali and J. F. List, *Diabetes Care*, 2012, **35**, 1473.
92. T. E. Hughes, D. D. Kim, J. Marjason, J. Proietto, J. P. Whitehead and J. E. Vath, *Obesity*, 2013, **21**, 1782.
93. D. H. Kim, M. A. Sartor, J. R. Bain, D. Sandoval, R. D. Stevens, M. Medvedovic, C. B. Newgard, S. C. Woods and R. J. Seeley, *Diabetes*, 2012, **61**, 2299.

Part 2
Hypertension and Cardiovascular Disease

CHAPTER 7

Biomarkers of Hypertension and Cardiovascular Disease

PAUL WELSH, DAVID PREISS AND RHIAN M. TOUYZ*

Institute of Cardiovascular and Medical Sciences, BHF Glasgow Cardiovascular Research Centre, University of Glasgow, Glasgow, G12 8TA, UK
*Email: Rhian.Touyz@glasgow.ac.uk

7.1 Introduction

Cardiovascular diseases are a major cause of morbidity and mortality globally with hypertension being the most important modifiable risk factor.[1,2] Hypertension is a complex disease due to multiple interacting factors (genetic, environmental, physiological) and involving many organ systems, including the heart, vasculature, kidney, brain and nerves. Fundamental to the pathophysiology of hypertension is the vascular system, which undergoes characteristic functional and structural changes including endothelial dysfunction, vascular hyper-reactivity, structural remodelling, fibrosis and vascular inflammation.[3–5] These processes may precede overt manifestations of hypertension and often become more exaggerated as hypertension develops and progresses. As such being able to detect early vascular changes with biomarkers of vascular injury may provide an index to predict the development and severity of hypertension and associated cardiovascular diseases. Here we will define biomarkers in the context of vascular damage in hypertension and will focus on some circulating factors that have been

RSC Drug Discovery Series No. 45
Cardiovascular and Metabolic Disease: Scientific Discoveries and New Therapies
Edited by Philip Peplow, James Adams and Tim Young

Published by the Royal Society of Chemistry, www.rsc.org

identified as vascular biomarkers. The clinical significance of biomarkers will be addressed and potential implications in therapeutics will be discussed.

7.2 Definition of a Biomarker

The term "biomarker" is a contraction of "biological marker". A biomarker may be a molecular measure, a physiological measurement, or a derived measure from a visual image. An NIH working group defined biomarkers as "A characteristic that is objectively measured and evaluated as an indicator of normal biological processes, pathogenic processes, or pharmacologic responses to a therapeutic intervention".[6] Under this broad definition, a vascular disease biomarker might be a result of a particular exposure of interest (*i.e.* it is distal to the outcome of vascular disease), it might be an early indicator of risk of disease or an outcome of interest (*i.e.* it is proximal to the outcome of vascular disease), or it might be a mediator of disease lying on the causal pathway between another exposure and disease (Figure 7.1). Biomarkers may be useful in the early detection of vascular injury and the risk of cardiovascular disease. Biomarkers may also provide insights to better profile patients at risk, which should lead to more directed and targeted therapy. However, there are challenges that still need to be

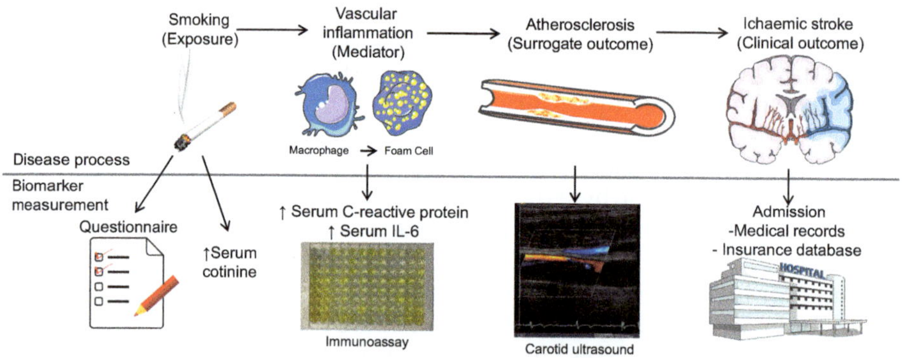

Figure 7.1 Illustration of the potential availability of biomarkers at every stage of disease pathophysiology, from exposures to clinical outcomes, including use of inflammatory markers as either exposures or mediators of disease. As an example, smoking promotes vascular injury and inflammation that leads to atherosclerosis, which in turn causes ischaemic events. Measuring biomarkers at every stage in the progress of the pathological process, such as determination of risk factors by questionnaire, measurement of plasma biomarkers, non-invasive imaging of vascular disease by ultrasound and finally clinical assessment may provide insights into risk of cardiovascular disease, mechanisms of disease and targeted therapies. This figure was generated using adapted images from Servier Medical Art, provided by Servier, under a Creative Commons Attribution 3.0 unported license (available at http://www.servier.com/Powerpointimage-bank).

addressed such as identifying appropriate and validated markers of vascular injury, development of methods to accurately measure biomarkers and establishment of clinically useful algorithms.

7.3 Vascular Damage in Hypertension and Cardiovascular Disease

7.3.1 Endothelial Dysfunction

In hypertension, the healthy endothelium becomes dysfunctional. This is characterised by disruption of the protective endothelial monolayer with reduced endothelium-dependent vasodilation, platelet activation, thrombus formation, increased permeability, leukocyte adhesion and monocyte migration into the vascular wall,[7–9] resulting in a decreased vasodilation and a proinflammatory and prothrombotic phenotype. Molecular mechanisms contributing to this include decreased nitric oxide (NO) production due to reduced activation of endothelial nitric oxide synthase (eNOS), increased bioavailability of reactive oxygen species (ROS) due to increased activation of NADPH oxidase (Nox) and dysregulated redox signaling (oxidative stress).[10,11] In addition there is increased expression of adhesion molecules, increased synthesis of proinflammatory and proatherosclerotic factors, activation of the local renin-angiotensin system (RAS) and increased endothelin-1 (ET-1) secretion.[12–15] Impaired endothelial function may be an important prognosticator of cardiovascular outcomes and is an independent predictor of cardiovascular events in patients with risk factors.[16–18] Improvement in endothelial function in hypertensive patients is associated with reduced target organ damage and more favourable outcomes.

7.3.2 Vascular Contractility

Vascular contraction is regulated primarily by vascular smooth muscle cells (VSMC) in the arterial wall. Vascular tone and contractility are increased in hypertension. The contractile machinery of vascular smooth muscle cells (actin and myosin) is regulated in a Ca^{2+}-dependent and Ca^{2+}-independent manner. The key process triggering vascular smooth muscle contraction is an increase in intracellular free Ca^{2+} concentration ($[Ca^{2+}]_i$) in response to activation of G-protein coupled receptors by vasoconstrictors such as angiotensin II (Ang II), endothelin-1 (ET-1), serotonin and urotensin.[19,20] Physical factors, such as stretch and shear stress also influence vascular contraction.[21] In addition to VSMC-derived factors regulating tone, endothelium-derived factors, such as prostacyclin and thromboxane A_2, act as endothelium-derived contracting factors (EDCF) that counteract the vasodilator effect of endothelium-derived NO.[22] Increased EDCF production has been demonstrated in experimental and human hypertension[23–25] and further contributes to increased vascular tone in hypertension. Smooth muscle myosin can also be phosphorylated in a Ca^{2+}-independent manner by additional proteins like Rho kinases.[24–26]

7.3.3 Structural Remodeling

In hypertension, arteries undergo structural changes (vascular remodelling) characterized by reduced lumen diameter with increased thickness of the vascular wall.[27–30] Whether increased pressure itself or other factors are responsible for the initiation of vascular remodeling remains unclear. Structural remodeling allows arteries to withstand an increased pressure load and under physiological conditions, such as aging and exercise, is adaptive. Under pathological conditions the adaptive process is overwhelmed, resulting in remodeled vessels that are rigid, stiff, and poorly compliant, typically observed in hypertension. At the molecular and cellular levels remodeling involves changes in cytoskeletal organization, cell-to-cell connections, and altered growth/apoptosis, senescence, calcification, inflammation, and rearrangement of VSMCs.[31,32] At the extracellular level there are changes in extracellular matrix protein composition and reorganization of proteoglycans, collagens (type I and III) and fibronectin.[33]

7.3.4 Vascular Inflammation

Vascular inflammation is a complex process that involves neutrophil recruitment to the endothelium with rolling, adhesion and transmigration into the vascular wall, where neutrophils are recruited by resident macrophages to promote an inflammatory response.[34] These inflammatory cells produce ROS, which in the context of chronic low grade vascular inflammation in cardiovascular disease leads to oxidative stress, which further promotes inflammation by activating pro-inflammatory signaling pathways and activation of transcription factors such as NFκβ.[35] Activation of NFκβ in the vascular wall triggers a proinflammatory response characterized by increased levels of TNFα, IL-1β, IL-6, and IL-8 and increased expression of adhesion molecules. These proinflammatory cytokines are also produced in perivascular fat, which further contributes to the vascular phenotype in cardiovascular disease.

In the vascular system the major source of ROS is NADPH oxidases (Nox). Markers of inflammation have been shown to correlate with the levels of oxidative stress in inflammatory cells in hypertension.[36–39] Participation of innate and adaptive immune responses have been suggested to be important in vascular inflammation associated with hypertension, atherosclerosis and cardiac disease, where different subsets of T lymphocytes may be involved in processes leading to inflammation.[40] Effector T-cells including Th1 (interferon-γ-producing) and Th2 (interleukin-4 producing) lymphocytes, as well as Th17 (which produce interleukin-17) and T-suppressor lymphocytes such as T(reg)-cells (regulatory T-cells), participate respectively as pro- and anti-inflammatory cells.[41–43] Involvement of immune mechanisms in cardiac, vascular and renal changes in cardiovascular disease and hypertension has been demonstrated in many experimental models.[44,45] Mechanisms triggering immune responses in hypertension remain unclear but neo-antigens could be generated by elevated blood pressure through damage-associated molecular pattern (DAMP) receptors or other mechanisms.[46] Exactly what these neo-antigens are remains unclear and have yet to be identified.

7.3.5 Vascular Calcification

Arterial calcification is not uncommon in aging, chronic kidney disease, diabetes, atherosclerosis, and hypertension, and is related to cardiovascular morbidity and mortality.[47] Vascular calcification is a tightly controlled process similar to bone formation and involves mineralization of the internal elastic lamina and elastic fibers within the vessel wall. VSMCs play a critical role in this process, through their capacity to undergo phenotypic differentiation.[48] Factors that trigger and promote VSMC osteogenic induction include abnormalities in mineral metabolism, particularly hyperphosphatemia and hypercalcemia. This is driven by upregulation of transcription factors such as cbfa1 (core-binding factor 1α)/Runx2, MSX-2 and bone morphogenetic protein 2 (BMP-2), involved in normal bone development, and which control the expression of osteogenic proteins, including osteocalcin, osteonectin, alkaline phosphatase, collagen-1, and bone sialoprotein.[49,50] Another mechanism contributing to vascular mineralization is loss of calcification inhibitors, such as fetuin-A, matrix Gla protein, pyrophosphate, and osteopontin.[51,52] These factors are all measurable in plasma and have been suggested to be biomarkers of vascular calcification.

7.4 Biomarkers of Vascular Injury in Hypertension and Associated Cardiovascular Disease

7.4.1 Proinflammatory Cytokines

It is clear that there is a relationship between inflammation and vascular disease, and it appears likely that inflammation might play a role in promoting atherosclerosis and fibrosis of blood vessels. This has been dubbed the "inflammatory hypothesis of cardiovascular disease" and has stimulated considerable interest in the possibility that dampening inflammatory responses, through a variety of pathways, might be an efficacious approach to reduce CVD (Table 7.1).

7.4.1.1 C-reactive Protein

The acute phase reactant C-reactive protein (CRP) (Figure 7.2) has been the biomarker of choice for many years when measuring the general inflammatory response in the clinical setting. Early experimental evidence suggested that CRP might be proatherogenic, but subsequent animal models identified that this apparent effect was due to impure CRP preparations (Table 7.1).[53] Subsequently, high quality meta-analyses from Mendelian

Table 7.1 Examples of inflammatory markers associated with vascular disease.

Biomarker(s)	Inflammatory role	Association with vascular disease?	Evidence of causality?
C-reactive protein	Common clinical biomarker of the inflammatory response. A pentraxin produced by hepatocytes during the acute phase response.	Increased levels in the normal range associated with increased CVD, but early studies overestimated association.[54] Controversial use as a predictor of CVD; less utilised in Europe than the USA.	Originally thought to play a role in atherogenesis, but now strong evidence against a causal role.[54]
IL-6	Key proinflammatory cytokine, known for stimulating acute phase response. Strongly correlated with CRP.	Increased levels in the normal range associated with increased CVD. Probably similar CVD prediction potential as CRP,[99] but not normally measured clinically.	Strong evidence that elevated IL-6 causes CVD from genetic studies.[56] RCTs of anti-IL-6 biologics underway.
TNF-α	Key proinflammatory cytokine. Can be produced by multiple non-inflammatory cell types. Multiple complex physiological roles.	Difficult to measure due to very low circulating concentrations in healthy people (<1 pg mL^{-1}). Consequently weaker associations with CVD than CRP and IL-6.[99] Not generally measured clinically.	Some evidence from RA patients that blocking can reduce CVD. Blocking causes harm in heart failure patients (stimulation of apoptosis?).
IL-1α	Key proinflammatory cytokine, classically produced by stimulation of toll-like receptors on innate cells.	No extensive data due to difficulty in detecting circulating levels in healthy people.	No evidence of causality from underpowered genetic studies.[103] RCTs of biologics underway.

IL-18 and IL-33	Proinflammatory members of the IL-1 superfamily.	Weak associations with increased CVD risk.	Causal role presently uncertain.
Chemokines	Attract lymphocytes into the vessel. IL-8 acts as an early acute chemokine.	Low circulating concentrations in healthy people. Some associations with CVD, but strongly confounded by classical risk factors.[100]	Early clinical trials of chemokine receptor blockers under way in inflammatory conditions[104] but efficacy limited thus far.
Soluble integrins and selectins	Membrane bound P- and E-selectin, ICAM-1 and VCAM-1 facilitate extravasation of lymphocytes into vessel wall.	Possible weak association of solubilized molecules with CVD. P-selectin can be artificially elevated by platelet degranulation. sICAM-1 may be more closely associated with diabetes than CVD.[101]	Whether solubilization into circulation is proinflammatory or represents resolution of inflammation unclear
Adipokines	Released from adipose tissue along with classical cytokines (*e.g.* leptin and adiponectin). Leptin produces a sense of satiety in the brain if leptin sensitivity is retained.	Increased circulating adipokines may have weak context dependant associations with CVD, but are clearly more strongly associated with diabetes risk.[102]	Unlikely to play an important direct causal role in vascular disease. Being overweight increases circulating inflammatory markers, and may increase CVD risk in this way.[102]

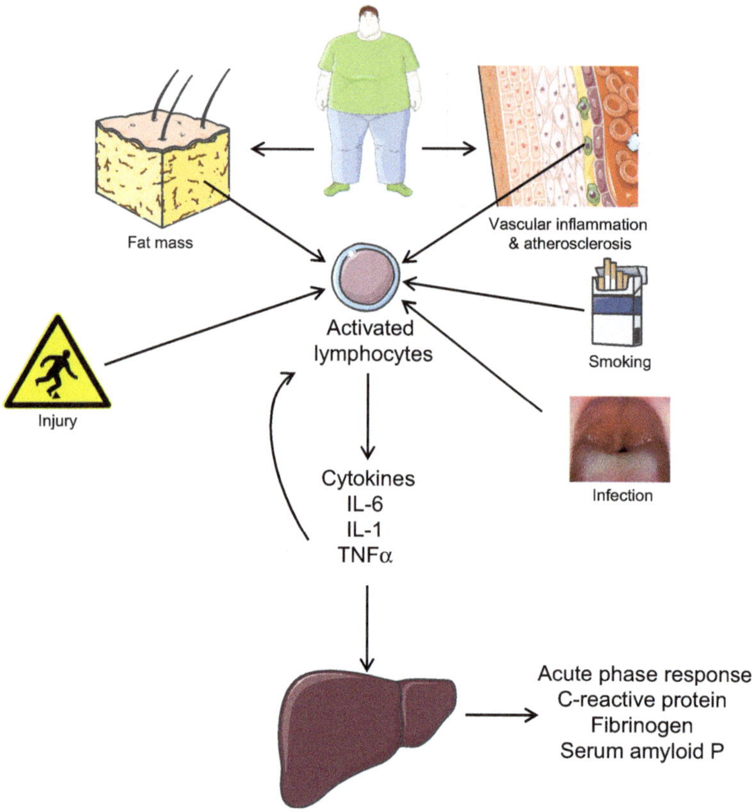

Figure 7.2 Diagram demonstrating the relationship between cardiovascular risk factors, such as obesity, that induces an inflammatory responses and activation of lymphocytes, which lead to production of pro-inflammatory cytokines and mediators that could be measured in plasma as a biomarker of vascular and systemic inflammation.
This figure was generated using adapted images from Servier Medical Art, provided by Servier, under a Creative Commons Attribution 3.0 unported license (available at http://www.servier.com/Powerpointimage-bank).

randomisation genetic studies (which allow causal inference with a greater degree of certainty than traditional observational data) suggested that CRP does not cause vascular disease; the pooled risk ratio for CHD corresponded to 1.00 (95% CI 0.90 to 1.13) per 1 standard deviation increase in CRP caused by genes.[54] This contrasts with a risk ratio of 1.44 (95% CI 1.32–1.57) per standard deviation increase in CRP after adjustment using the traditional observational data,[54] suggesting residual confounding factors may cause traditional observational studies to yield unreliable causal inferences.

7.4.1.2 IL-6 and TNF

IL-6 and TNF-α are proinflammatory cytokines upstream of the acute phase response.[55] Mendelian randomisation studies have been more encouraging than the CRP studies that IL-6 may play a causal role in causing CVD.[56] Anti-IL-6 and anti-TNF-α biologics have been developed for use in patients with inflammatory diseases such as rheumatoid arthritis, and appear highly efficacious at improving disease activity in these patients.[57] It is well known that rheumatoid (RA) patients are at ∼50% increased risk of CVD even after accounting for more adverse classical CVD risk factors compared to the general population.[58] Interestingly there is some evidence that both of these biologics may to some extent ameliorate the increased CVD risk in RA patients. TNF-α blockers appear to increase circulating cholesterol levels in RA patients. Although superficially worrying, this observation simply reflects the fact that acute inflammation lowers circulating cholesterol, which may be sequestered into tissues.[59] Resolution of inflammation allows the circulating cholesterol level to "bounce back" to reflect underlying cholesterol levels. The same appears to be true for IL-6 blockers, which as well as increasing cholesterol, decrease plasma fibrinogen and D-dimer,[60] indicating decreased formation and turnover of subclinical thrombi. Given these improvements in biomarker profile, data from biologics treatment registers of RA patients show patients who respond to anti-TNF biologics have reduced rates of myocardial infarction (3.5 per 1000 person-years) compared to non-responders (9.4 per 1000 person-years) resulting in an adjusted rate ratio of 0.36 (95% CI 0.19–0.69).[60] Similar data have been reported in the US CORRONA database.[61] It should be noted that these findings are based on pharmacoepidemiology, and are therefore subject to potential confounding factors and thus they represent a weaker level of evidence for causality than randomised controlled trials (RCTs). No single sufficiently powered trial to assess the effect of TNF blockers on CVD in RA patients has been possible. Meta-analysis of three small trials did not yield a significant reduction in CVD risk associated with anti-TNF biologics[62] although even this meta-analysis was substantially underpowered.

Other studies investigating anti-inflammatory biologics have also demonstrated disappointing efficacy in CVD. Despite initial excitement at the potential role of TNF-α in heart failure pathophysiology, two separate reports from RCTs showed no clinical benefit of infliximab and etanercept in patients with pronounced clinical heart failure, and some data were consistent with worsening symptoms on treatment.[63] This has led to contraindication of anti-TNF biologics in patients with advanced heart failure. Numerous mechanisms have been suggested to explain these observations; including some biologics cross-linking membrane bound TNF in cardiomyocytes and causing apoptosis. This anecdote serves as a cautionary tale warning against complacency of potential unexpected harms in use of anti-inflammatory interventions, a particularly relevant concern in patients with low grade non-debilitating diseases.

Finally, it should be noted that cytokines may also play a role in peripheral vascular disease. Diabetes is a proinflammatory condition that is associated with increased circulating cytokines, endothelial dysfunction, raised risk of peripheral vascular disease, and risk of microvascular events such as nephropathy, retinopathy, and neuropathy.[64] Recent data from the large ADVANCE clinical trial challenged the inflammatory paradigm of microvascular disease showing that circulating cytokines are not strongly associated with microvascular risk after adjusting for confounding risk factors.[65]

7.4.2 Markers of Oxidative Stress

One of the upstream inflammatory processes thought to play a causal role in the pathogenesis of CVD is production of ROS such as superoxide (O_2^-). Vascular ROS are produced in endothelial cells and VSMCs primarily from NAD(P)H oxidase.[66] In experimental studies increased levels of ROS cause endothelial dysfunction and subsequent vascular inflammation, with associated development of hypertension, although data from human studies are slightly less consistent.[67] Clinical studies measure levels of plasma thiobarbituric acid, 8-isoprostanes and F2-isoprostanes as markers of ROS status. Although small clinical studies demonstrate a relationship between plasma ROS levels and CVD risk, there is a lack of prospective data linking these measures to vascular outcomes in large cohort studies.[68] The theory that ROS may cause CVD has been set back by data from large RCTs which reported that antioxidant vitamin supplements do not prevent CVD, and some may even cause harm.[69] However, there remains optimism that targeted inhibition of ROS pathways, *via* specific NAD(P)H enzymes, may prove a beneficial therapeutic approach.[70,71]

In contrast to ROS, which may promote vascular damage, endothelial-derived NO inhibits coagulation, inflammation and oxidative stress and maintains vascular tone.[72] Plasma nitrite and nitrate are typically measured surrogates of NO bioavailability, although these surrogates are highly confounded by inflammation, smoking status, diet, and have a short half-life, and therefore may be suboptimal biomarkers. Asymmetric dimethylarginine (ADMA) is a competitive inhibitor of L-arginine associated NO production, and has been studied as a predictor of CVD risk, although associations with CVD outcomes are likely to be confounded as well.[73]

7.4.3 Markers of Vascular Calcification

Various biomarkers related to the pathophysiological processes underlying vascular calcification have been identified, including fibroblast growth factor (FGF)-23, BMP-2, osteoprotegerin (OPG), osteopontin, fetuin-A, matrixGla protein (MGP) osteonectin, osteocalcin, pyrophosphate and sclerostin.[74–77] FGF-23 is released by bone. It promotes phosphate excretion and requires the cofactor Klohto for its effects. Increased serum FGF-23 is frequently found in patients with chronic kidney disease and is associated with advanced coronary artery calcification.[77] BMP-2 belongs to a family of TGF-α-like growth factors. It regulates growth, differentiation and bone formation and plays an important role in vascular calcification. OPG is a regulatory factor produced by bone marrow derived stromal cells and is involved in bone turnover. OPG inhibits osteoclast differentiation. High plasma levels have been observed in patients with coronary artery calcification and chronic kidney disease. Osteopontin is a protein expressed in mineralized tissue. It inhibits mineralization by blocking hydroxyapatite formation and activating osteoclast function. OPN is present in calcified vessels and OPN--knockout mice do not develop vascular calcification. Fetuin A and MGP are potent calcification inhibitors and several studies have demonstrated reduced serum levels in patients with vascular calcification. Lower serum fetuin-A concentrations have been associated with increased calcification scores, arterial stiffness, mortality and incidence of cardiovascular disease. It should be highlighted that many factors can influence plasma levels of these biomarkers and as such it still remains to be proven whether they truly reflect calcification status of vessels.

7.4.4 Vitamin D

Vitamin D (1,25(OH)D), the secosteroid derived from circulating calcidiol (25(OH)D) has been shown to have anti-inflammatory effects under experimental conditions; 25(OH)D can be processed by activated lymphocytes into 1,25(OH)D, which acts in an autocrine and paracrine manner to modulate an inflammatory response from Th1 responses to Th2 or T-regulatory responses. As such, vitamin D "deficiency" has been linked to a host of chronic and acute inflammatory conditions, including CVD, with low plasma vitamin D levels suggesting CVD risk. Multiple lines of evidence from observational studies have suggested vitamin D deficiency to be associated with CVD in humans.[78] For instance a recent meta-analysis of observational data suggests that CVD relative risk is 0.66 (95% CI 0.57–0.77).[79] Despite this, randomised controlled trials of vitamin D supplementation to prevent CVD and other inflammatory conditions have generally been negative; the relative risk of CVD in a meta-analysis of supplementation was 0.95 (95% CI 0.86–1.05) among those taking supplements.[80] Further large high quality trials to investigate the role of vitamin D in chronic disease will shed further light on this area.[81]

7.4.5 Microparticles

Microparticles are another emerging potential biomarker that might be interesting in the setting of inflammation and endothelial dysfunction. These are small fragments of cellular membranes (<1 μm) shed from stressed/damaged cells.[82] There is some evidence that microparticles may have distinctive expressions in different patient groups, *e.g.* in patients with unstable (*versus* stable) carotid plaques CD11bCD66b+ and CD15+ leukocyte-derived microparticles are elevated.[83] Levels of circulating endothelial microparticles are elevated in many cardiovascular diseases, including pulmonary hypertension, cardiac failure, sleep apnea, diabetes, chronic kidney disease, preeclampsia and hypertension, and may reflect underlying endothelial dysfunction and vascular injury.[84,85] Originally thought to be inert, there is now considerable interest in the possibility that microparticles may allow information transfer between cells and as such are increasingly being considered as both biomarkers and biovectors of disease.[86–88] Recent clinical and experimental studies demonstrate that increased levels of podocyte-derived microparticles in urine are associated with microalbuminuria and renal injury.[89] Since podocyte microparticles can be detected before overt microalbuminuria, it has been suggested that urinary microparticles may be a potential biomarker of renal dysfunction. The conditions for collecting samples suitable for microparticle phenotyping from plasma and urine are stringent and may be subject to considerable pre-analytical artefact due to post-collection platelet degranulation.[90] This adds difficulty when conducting large clinical studies or applying a biomarker in real life settings.

7.4.6 MicroRNAs

MicroRNAs (miRNAs) are a family of small noncoding RNA fragments found in plasma. Originally thought to be inert fragments of larger RNA molecules, a growing body of evidence indicates that distinct miRNAs perform different regulatory functions and, as such, may represent therapeutic targets.[91] For instance, animal models have demonstrated that inhibition of miRNA33a/b with anti-miRNA oligonucleotide increased hepatic expression of ATP-binding cassette transporter ABCA1 (ABCA1) and induced a ~50% increase in HDL cholesterol and a ~40% decrease in VLDL-cholesterol after 12 weeks.[92] There is emerging evidence that miRNA may be involved in regulating inflammatory responses and also CVD risk.[93] Whether this pathway is directly relevant to CVD is presently unclear; complexity is compounded by the possibility that inflammation itself will impact on miRNA expression. Thus, questions remain regarding under which conditions specific miRNA might cause inflammation, under which conditions specific miRNA may be a consequence of ongoing inflammation and disease, and to what extent miRNAs as a biomarker may be able to improve prediction of CVD.

Overall the strength of evidence linking these biomarkers to CVD in the clinical setting is still quite weak. More work is required to investigate underlying mechanisms and potential pathways for intervention.

7.5 Clinical Significance of Biomarkers: Antihypertensive Treatment and Anti-inflammatory Biomarkers

There is presently limited evidence that antihypertensive drugs have any substantial impact on markers of inflammation. For instance, in the small Comparison of Amlodipine *versus* Enalapril to Limit Occurrences of

Thrombosis (CAMELOT) study, 2 years of treatment with amlodipine or enalapril did not change IL-18 or CRP, and in the Ramipril C-Reactive pRotein Randomized evaluation (4R), 12 week treatment with ramipril in 246 healthy middle-aged people did not lower CRP compared to placebo even though blood pressure was significantly reduced.[94] Hence, blood pressure responses do not seem to be reflected by inflammatory biomarkers, which raises the question as to whether the events underlying the inflammatory markers are even linked to blood pressure elevation or whether inflammation is an independent process from hypertension. It is also possible that the processes targeted by the drugs are unrelated to inflammation. Finally, limitations of the study design may explain the lack of association between blood pressure responses to drug treatment and levels of inflammatory biomarkers, including the relative health of the study participants, the small sample sizes, and the short duration of the intervention. However, it does appear that any relationship between reduction in blood pressure and reduction of inflammation is likely to be moderate.

7.6 Conclusions

Ideally biomarkers should be indicators of early detection of vascular damage and give insights into pathophysiological processes underlying progression of cardiovascular disease. In addition, biomarkers should aid in the identification of patients who may benefit from specific therapeutic interventions, although this stratified medicine approach is still a developing field. Biomarkers are therefore of interest from a research point of view as they may help identify new therapeutic targets for novel interventions, they may improve the prediction of disease, or they may provide mechanisms to target therapies to those who would benefit most. Although numerous biomarkers have been identified in experimental and clinical hypertension and cardiovascular disease, such as acute phase proteins, inflammatory mediators, oxidative stress markers, cell-based fractions, osteogenic factors *etc.*, these still need to be validated clinically in large cohorts.[95–97] Moreover, these biomarkers, which are an indirect index of vascular status, need to be correlated directly with vascular function. Such approaches are being attempted using comprehensive non-invasive imaging of small, medium and large vessels.[98] Although there has been significant progress in the field with many new biomarkers of vascular injury being identified, extensive research is still needed to identify ideal biomarkers. Considering the complexity of hypertension and cardiovascular disease, it is likely that there will not be one biomarker, but rather an array or platform of numerous biomarkers that will define the vascular phenotype in health and disease.

References

1. V. Fuster, *Nat. Rev. Cardiol.*, 2014, **11**, 671.
2. C. M. Lawes, S. Vander Hoorn and A. Rodgers, *Lancet*, 2008, **371**, 1513.

3. A. C. Montezano, A. Nguyen Dinh Cat, F. J. Rios and R. M. Touyz, *Curr. Hypertens. Rep.*, 2014, **16**, 431.
4. M. Ponticos and B. D. Smith, *J. Biomed. Res.*, 2014, **28**, 25.
5. A. Yannoutsos, B. I. Levy, M. E. Safar, G. Slama and J. Blacher, *J. Hypertens.*, 2014, **32**, 216.
6. Biomarkers Definitions Working Group, *Clin. Pharmacol. Ther.*, 2001, **69**, 89.
7. C. Savoia, D. Burger, N. Nishigaki, A. Montezano and R. M. Touyz, *Expert Rev. Mol. Med.*, 2011, **13**, e11.
8. E. L. Schiffrin, *Am. J. Hypertens.*, 2002, **15**, S115.
9. P. Rajendran, T. Rengarajan, J. Thangavel, Y. Nishigaki, D. Sakthisekaran, G. Sethi and I. Nishigaki, *Int. J. Biol. Sci.*, 2013, **9**, 1057.
10. A. C. Montezano and R. M. Touyz, *Basic Clin. Pharmacol. Toxicol.*, 2012, **110**, 87.
11. T. Heitzer, T. Schlinzig, K. Krohn, T. Meinertz and T. Munzel, *Circulation*, 2001, **104**, 2673.
12. P. Libby, M. Aikawa and M. K. Jain, *Handb. Exp. Pharmacol.*, 2006, **176**, 285.
13. S. Xu, *et al.*, *Hypertension*, 2009, **54**, 427.
14. E. R. Regan and W. C. Aird, *Circ. Res.*, 2012, **111**, 1101.
15. L. Xiao, Y. Liu and N. Wang, *Am. J. Physiol. Heart Circ. Physiol.*, 2014, **306**, H317.
16. G. Brevetti, A. Silvestro, V. Schiano and M. Chiariello, *Circulation*, 2003, **108**, 2093.
17. J. P. Halcox, W. H. Schenke and G. Zalos, *Circulation*, 2002, **106**, 653.
18. E. L. Schiffrin, L. Y. Deng and P. Larochelle, *Hypertension*, 1994, **23**, 83.
19. S. de Campos Grifoni and L. M. Bendhack, *Pharmacology*, 2004, **70**, 160.
20. R. M. Touyz, M. El Mabrouk, G. He, X. H. Wu and E. L. Schiffrin, *Circ. Res.*, 1999, **84**, 505.
21. D. Versari, E. Daghini, A. Virdis, L. Ghiadoni and S. Taddei, *Br. J. Pharmacol.*, 2009, **157**, 527.
22. A. Virdis, L. Ghiadoni and S. Taddei, *Pflugers Arch.*, 2010, **459**, 1015.
23. M. Félétou, T. J. Verbeuren and P. M. Vanhoutte, *Br. J. Pharmacol.*, 2009, **156**, 563.
24. G. Loirand and P. Pacaud, *Nat. Rev. Cardiol.*, 2010, 7, 637.
25. M. Uehata, T. Ishizaki, H. Satoh, T. Ono, T. Kawahara, T. Morishita, H. Tamakawa, K. Yamagami, J. Inui, M. Maekawa and S. Narumiya, *Nature*, 1997, **389**, 990.
26. R. Komers, T. T. Oyama, D. R. Beard and S. Anderson, *Br. J. Pharmacol.*, 2011, **162**, 163.
27. E. L. Schiffrin, *J. Am. Soc. Hypertertens.*, 2007, **1**, 56.
28. D. Rizzoni and E. Agabiti-Rosei, *Intern. Emerg. Med.*, 2012, 7, 205.
29. M. Pacurari, R. Kafoury, P. B. Tchounwou and K. Ndebele, *Int. J. Inflammation*, 2014, **2014**, 689360.

30. K. Sonoyama, A. Greenstein, A. Price, K. Khavandi and T. Heagerty, *Ther. Adv. Cardiovasc. Dis.*, 2007, **1**, 129.
31. M. J. Mulvany, G. L. Baumbach, C. Aalkjaer, A. M. Heagerty, N. Korsgaard, E. L. Schiffrin and D. D. Heistad, *Hypertension*, 1996, **28**, 505.
32. E. L. Schiffrin, *Am. J. Hypertens.*, 2004, **17**, 1192.
33. C. A. Lemarié, P. L. Tharaux and S. Lehoux, *J. Mol. Cell. Cardiol.*, 2010, **48**, 433.
34. T. N. Mayadas, X. Cullere and C. A. Lowell, *Annu. Rev. Pathol.*, 2014, **9**, 181.
35. E. Androulakis, D. Tousoulis, N. Papageorgiou, G. Latsios, G. Siasos, C. Tsioufis, A. Giolis and C. Stefanadis, *Curr. Pharm. Des.*, 2011, **17**, 4121.
36. T. Usui, M. Okada, Y. Hara and H. Yamawaki, *Am. J. Physiol. Heart Circ. Physiol.*, 2013, **305**, H756.
37. X. Sun, N. Belkin and M. W. Feinberg, *Curr. Atheroscler. Rep.*, 2013, **15**, 372.
38. Q. N. Dinh, G. R. Drummond, C. G. Sobey and S. Chrissobolis, *Biomed. Res. Int.*, 2014, **2014**, 406960.
39. G. S. Reis, V. S. Augusto, A. P. Silveira, A. A. Jordão Jr, J. Baddini-Martinez, O. Poli Neto, A. J. Rodrigues and P. R. Evora, *Pulm. Circ.*, 2013, **3**, 856.
40. D. G. Harrison, P. J. Marvar and J. M. Titze, *Front. Physiol.*, 2012, **3**, 128.
41. E. L. Schiffrin, *Curr. Opin. Nephrol. Hypertens.*, 2010, **19**, 181.
42. M. Kassan, A. Wecker, P. Kadowitz, M. Trebak and K. Matrougui, *J. Hypertens.*, 2013, **31**, 1939.
43. T. J. Guzik, N. E. Hoch and K. A. Brown, *J. Exp. Med.*, 2007, **204**, 2449.
44. D. L. Mattson, *Am. J. Physiol. Renal. Physiol.*, 2014, **307**, F499.
45. P. J. Marvar, S. R. Thabet, T. J. Guzik, H. E. Lob, L. A. McCann, C. Weyand, F. J. Gordon and D. G. Harrison, *Circ. Res.*, 2010, **107**, 263.
46. C. G. McCarthy, S. Goulopoulou, C. F. Wenceslau, K. Spitler, T. Matsumoto and R. C. Webb, *Am. J. Physiol. Heart Circ. Physiol.*, 2014, **306**, H184.
47. C. V. Serrano, M. Oranges, V. Brunaldi, A. de M. Soeiro, T. A. Torres, J. C. Nicolau and J. A. F. Ramires, *Vasc. Health Risk Manage.*, 2011, **7**, 143.
48. A. C. Montezano, D. Zimmerman, H. Yusuf, A. Z. Chignalia, V. Wadhera and R. M. Touyz, *Hypertension*, 2010, **56**, 453.
49. C. De Ciuceis, F. Amiri, P. Brassard, D. H. Endemann, R. M. Touyz and E. L. Schiffrin, *Arterioscler., Thromb., Vasc. Biol.*, 2005, **25**, 2106.
50. J. M. Davaine, T. Quillard, R. Brion, O. Lapérine, B. Guyomarch, T. Merlini, M. Chatelais, F. Guilbaud, M. Á. Brennan, C. Charrier, D. Heymann, Y. Gouëffic and M. F. Heymann, *PLoS One*, 2014, **9**, e107642.
51. P. Pateinakis, A. Papagianni, S. Douma, G. Efstratiadis and D. Memmos, *BMC Nephrol.*, 2013, **14**, 122.

52. S. Liabeuf, H. Okazaki, L. Desjardins, D. Fliser, D. Goldsmith, A. Covic, A. Wiecek, A. Ortiz, A. Martinez-Castelao, B. Lindholm, G. Suleymanlar, F. Mallamaci, C. Zoccali, G. London and Z. A. Massy, *Nephrol., Dial., Transplant.*, 2014, **29**, 1275.
53. M. B. Pepys, J. R. Gallimore, J. Lloyd, Z. Li, D. Graham, G. W. Taylor, S. Ellmerich, P. P. Mangione, G. A. Tennent, W. L. Hutchinson, D. J. Millar, G. Bennett, J. More, D. Evans, Y. Mistry, S. Poole and P. N. Hawkins, *J. Immunol. Methods*, 2012, **384**, 92.
54. Emerging Risk Factors Collaboration, S. Kaptoge, E. Di Angelantonio, G. Lowe, M. B. Pepys, S. G. Thompson, R. Collins and J. Danesh, *Lancet*, 2010, **375**, 132.
55. J. Wolf, S. Rose-John and C. Garbers, *Cytokine*, 2014, **70**, 11.
56. Interleukin-6 Receptor Mendelian Randomisation Analysis (IL6R MR) Consortium, A. D. Hingorani and J. P. Casas, *Lancet*, 2012, **379**, 1214.
57. C. Gabay, P. Emery, R. van Vollenhoven, A. Dikranian, R. Alten, K. Pavelka, M. Klearman, D. Musselman, S. Agarwal, J. Green, A. Kavanaugh and A. S. Investigators, *Lancet*, 2013, **381**, 1541.
58. L. Lévy, B. Fautrel, T. Barnetche and T. Schaeverbeke, *Clin. Exp. Rheumatol.*, 2008, **26**, 673.
59. G. E. McKellar, D. W. McCarey, N. Sattar and I. B. McInnes, *Nat. Rev. Cardiol.*, 2009, **6**, 410.
60. I. B. McInnes, L. Thompson, J. T. Giles, J. M. Bathon, J. E. Salmon, A. D. Beaulieu, C. E. Codding, T. H. Carlson, C. Delles, J. S. Lee and N. Sattar, *Ann. Rheum. Dis.*, 2013, **2013**, 204345, DOI: 10.1136/annrheumdis.
61. J. D. Greenberg, J. M. Kremer, J. R. Curtis, M. C. Hochberg, G. Reed, P. Tsao, M. E. Farkouh, A. Nasir, S. Setoguchi and D. H. Solomon, *Ann. Rheum. Dis.*, 2011, **70**, 576.
62. C. Barnabe, B.-J. Martin and W. A. Ghali, *Arthritis Care Res.*, 2011, **63**, 522.
63. A. P. Coletta, A. L. Clark, P. Banarjeea and J. G. F. Cleland, *Eur. J. Heart Failure*, 2002, **4**, 559.
64. C. Botti, C. Maione, G. Dogliotti, P. Russo, G. Signoriello, A. M. Molinari, M. M. Corsi, V. Sica and G. Cobellis, *J. Biol. Regul. Homeostatic Agents*, 2012, **26**, 67.
65. G. Lowe, M. Woodward, G. Hillis, A. Rumley, Q. Li, S. Harrap, M. Marre, P. Hamet, A. Patel, N. Poulter and J. Chalmers, *Diabetes*, 2014, **63**, 1115.
66. A. C. Montezano and R. M. Touyz, *Ann. Med.*, 2012, **44**, S2.
67. A. C. Montezano and R. M. Touyz, *Basic Clin. Pharmacol. Toxicol.*, 2012, **110**, 87.
68. S. S. Davies and L. J. Roberts, *Free Radical Biol. Med.*, 2011, **50**, 559.
69. S. P. Fortmann, B. U. Burda, C. A. Senger, J. S. Lin and E. P. Whitlock, *Ann. Intern. Med.*, 2013, **159**, 824.
70. M. Sedeek, A. C. Montezano, R. L. Hebert, S. P. Gray, E. Di Marco, J. C. Jha, M. E. Cooper, K. Jandeleit-Dahm, E. L. Schiffrin, J. L. Wilkinson-Berka and R. M. Touyz, *J. Cardiovasc. Transl. Res.*, 2012, **5**, 509.

71. G. R. Drummond and C. G. Sobey, *Trends Endocrinol. Metab.*, 2014, **25**, 452.
72. A. M. Beyer, M. J. Durand, J. Hockenberry, T. C. Gamblin, S. A. Phillips and D. D. Gutterman, *Am. J. Physiol. Heart Circ. Physiol.*, 2014, **307**, H1587.
73. M. O. Gore, N. Luneburg, E. Schwedhelm, C. R. Ayers, M. Anderssohn, A. Khera, D. Atzler, J. A. de Lemos, P. J. Grant, D. K. McGuire and R. H. Boger, *Arterioscler., Thromb., Vasc. Biol.*, 2013, **33**, 2682.
74. G. Jia, R. M. Stormont, D. M. Gangahar and D. K. Agrawal, *Am. J. Physiol. Heart Circ. Physiol.*, 2012, **303**, H523.
75. S. Evrard, P. Delanaye, S. Kamel, J. P. Cristol, E. Cavalier and On behalf of the SFBC/SN joined working group on vascular calcifications, *Clin. Chim. Acta*, 2014, 00386.
76. A. Ndip, F. L. Wilkinson, E. B. Jude, A. J. Boulton and M. Y. Alexander, *Diabetologia*, 2014, **57**, 2251.
77. C. I. Kiu Weber, G. Duchateau-Nguyen, C. Solier, A. Schell-Steven, R. Hermosilla, E. Nogoceke and G. Block, *Clin.* Kidney J., 2014, 7, 167.
78. J. R. Mora, M. Iwata and U. H. von Andrian, *Nat. Rev. Immunol.*, 2008, **8**, 685.
79. L. Wang, Y. Song, J. E. Manson, S. Pilz, W. März, K. Michaëlsson, A. Lundqvist, S. K. Jassal, E. Barrett-Connor, C. Zhang, C. B. Eaton, H. T. May, J. L. Anderson and H. D. Sesso, *Circ. Cardiovasc. Qual. Outcomes*, 2012, **5**, 819.
80. E. Theodoratou, I. Tzoulaki, L. Zgaga and J. P. A. Ioannidis, *Br. Med. J.*, 2014, **348**, g2035.
81. J. E. Manson, S. S. Bassuk, I.-M. Lee, N. R. Cook, M. A. Albert, D. Gordon, E. Zaharris, J. G. Macfadyen, E. Danielson, J. Lin, S. M. Zhang and J. E. Buring, *Contemp. Clin. Trials*, 2012, **33**, 159.
82. S. F. Mause and C. Weber, *Circ. Res.*, 2010, **107**, 1047.
83. G. Sarlon-Bartoli, Y. Bennis, R. Lacroix, M. D. Piercecchi-Marti, M. A. Bartoli, L. Arnaud, J. Mancini, A. Boudes, E. Sarlon, B. Thevenin, A. S. Leroyer, C. Squarcioni, P. E. Magnan, F. Dignat-George and F. Sabatier, *J. Am. Coll. Cardiol.*, 2013, **62**, 1436.
84. D. Burger, D. G. Kwart, A. C. Montezano, N. C. Read, C. R. Kennedy, C. S. Thompson and R. M. Touyz, *J. Am. Heart Assoc.*, 2010, **1**, e001842.
85. D. Burger, A. C. Montezano, N. Nishigaki, Y. He, A. Carter and R. M. Touyz, *Arterioscler., Thromb., Vasc. Biol.*, 2011, **31**, 1898.
86. N. Narin, E. Yilmaz, O. Pamukcu, A. Baykan, M. Argun, A. Ozyurt, S. Onan, S. Sezer and K. Uzum, *Biomarkers*, 2014, **19**, 319.
87. N. Amabile, S. Cheng, J. M. Renard, M. G. Larson, A. Ghorbani, E. McCabe, G. Griffin, C. Guerin, J. E. Ho, S. Y. Shaw, K. S. Cohen, R. S. Vasan, A. Tedgui, C. M. Boulanger and T. J. Wang, *Eur. Heart J.*, 2014, **35**, 2972.
88. S. D. Crowley, *Antioxid. Redox Signaling*, 2014, **20**, 102.
89. D. Burger, J. F. Thibodeau, C. E. Holterman, K. D. Burns, R. M. Touyz and C. R. Kennedy, *J. Am. Soc. Nephrol.*, 2014, **25**, 1401.

90. D. Burger, S. Schock, C. S. Thompson, A. C. Montezano, A. M. Hakim and R. M. Touyz, *Clin. Sci.*, 2013, **124**, 423.
91. L. He and G. J. Hannon, *Nat. Rev. Genet.*, 2004, **5**, 522.
92. K. J. Rayner, C. C. Esau, F. N. Hussain, A. L. McDaniel, S. M. Marshall, J. M. van Gils, T. D. Ray, F. J. Sheedy, L. Goedeke, X. Liu, O. G. Khatsenko, V. Kaimal, C. J. Lees, C. Fernandez-Hernando, E. A. Fisher, R. E. Temel and K. J. Moore, *Nature*, 2011, **478**, 404.
93. B. Schroen and S. Heymans, *Cardiovasc. Res.*, 2012, **93**, 605.
94. P. Zamani, P. Ganz, P. Libby, S. C. Sutradhar, N. Rifai, S. J. Nicholls, S. E. Nissen and S. Kinlay, *Am. Heart J.*, 2012, **163**, 735.
95. A. C. Montezano and R. M. Touyz, *Antioxid. Redox Signaling*, 2014, **20**, 164.
96. A. Nguyen Dinh Cat, A. C. Montezano, D. Burger and R. M. Touyz, *Antioxid. Redox Signaling*, 2013, **19**, 1110.
97. F. Paneni, S. Costantino and F. Cosentino, *Cardiovasc. Diagn. Ther.*, 2014, **4**, 324.
98. E. Ammirati, F. Moroni, P. Pedrotti, I. Scotti, M. Magnoni, E. P. Bozzolo, O. E. Rimoldi and P. G. Camici, *Front. Immunol.*, 2014, **5**, 399.
99. S. Kaptoge, S. R. Seshasai, P. Gao, D. F. Freitag, A. S. Butterworth, A. Borglykke, E. Di Angelantonio, V. Gudnason, A. Rumley, G. D. Lowe, T. Jorgensen and J. Danesh, *Eur. Heart J.*, 2014, **35**, 578.
100. C. Herder, J. Baumert, B. Thorand, S. Martin, H. Lowel, H. Kolb and W. Koenig, *Arterioscler., Thromb., Vasc. Biol.*, 2006, **26**, 2147.
101. N. Sattar, H. M. Murray, P. Welsh, G. J. Blauw, B. M. Buckley, A. J. de Craen, I. Ford, N. G. Forouhi, D. J. Freeman, J. W. Jukema, P. W. Macfarlane, M. B. Murphy, C. J. Packard, D. J. Stott, R. G. J. Westendorp and J. Shepherd, *Diabetologia*, 2009, **52**, 235.
102. P. Welsh, H. M. Murray, B. M. Buckley, A. J. de Craen, I. Ford, J. W. Jukema, P. W. Macfarlane, C. J. Packard, D. J. Stott, R. G. Westendorp, J. Shepherd and N. Sattar, *Diabetes Care*, 2009, **32**, 308.
103. L. Zhou, J. Cai, G. Liu, Y. Wei and H. Tang, *PLoS One*, 2012, **7**, e45641.
104. R. Horuk, *Nat. Rev. Drug Discov.*, 2009, **8**, 23.

CHAPTER 8

Risk Factors for Hypertension and Cardiovascular Disease

MARY T. BAUSCH-JURKEN AND THEODORE A. KOTCHEN*

Medical College of Wisconsin, Department of Medicine, 8701 Watertown Plank Road, Milwaukee, Wisconsin 53226, USA
*Email: tkotchen@mcw.edu

8.1 Introduction

In 2009, the American Heart Association estimated the total cost, both direct and indirect, of cardiovascular disease (CVD) and stroke in the United States to be $312.6 billion and they projected the cost of cardiovascular care to increase to an estimated $818.1 billion by 2030.[1,2] CVD is the leading cause of death, not only in the US, but globally, as well. The Global Burden of Diseases, Injuries, and Risk Factors Study 2010 (GBD 2010) estimated one in four deaths worldwide in 2010 were attributable to ischemic heart disease and stroke.[3]

Many of the risk factors associated with the development of CVD have been identified through large population studies.[4,5] Non-modifiable factors include age, gender and family history of CVD. Men have a higher risk for CVD than women; and family history of CVD in a first degree relative before 55 years of age for men, and 65 years of age for women imparts an increased risk. Potentially modifiable risk factors are often inter-related and include hypertension, obesity, dyslipidemia, insulin resistance and diabetes, obstructive sleep apnea, tobacco use, poor diet, insufficient physical activity, psychosocial stress, possibly air pollution, and possibly chronic infection and inflammation. Globally, hypertension, high BMI, and high fasting blood

RSC Drug Discovery Series No. 45
Cardiovascular and Metabolic Disease: Scientific Discoveries and New Therapies
Edited by Philip Peplow, James Adams and Tim Young

Published by the Royal Society of Chemistry, www.rsc.org

glucose are among the leading risk factors. The American Heart Association attributes 40.6% of CVD to hypertension, 13.7% to smoking, 13.2% to poor diet, 11.9% to inactivity, and 8.8% to abnormal blood glucose.[1] Moreover, the GBD 2010 study points to hypertension as the leading risk factor for the global disease burden, with tobacco use second, highlighting the shift away from communicable diseases to non-communicable diseases as the leading cause of deaths globally.[3,5]

The purpose of this chapter is to briefly summarize the disease burden of these risk factors and then to describe underlying mechanisms by which they contribute to cardiovascular disease. An understanding of the pathophysiology in terms of inflammation, oxidative stress, and autonomic imbalance may provide impetus for the development of innovative therapies directed at new therapeutic targets.

8.2 Disease Burden of Risk Factors

8.2.1 Hypertension (Table 8.1)[6]

Even a slight elevation in blood pressure confers an increased CVD risk, doubling with each 20/10 mmHg elevation.[6] The Framingham Study concluded the cumulative lifetime risk of developing hypertension for 55 and 65 year olds to be 90% and currently, it is estimated that between 34 and 41% of adults in the US are affected.[1] In the US, 40% of hypertensive individuals are unaware of the disorder, and overall, blood pressure is uncontrolled in 54% of treated hypertensives.[7] 16.5% of all deaths in the US can be attributed to hypertension, including 51% of stroke deaths and 45% of coronary disease deaths.[5] This is not just a problem in the US. In 2010, hypertension was the leading risk factor for the global burden of disease.[3]

8.2.2 Dyslipidemia (Table 8.2)[8]

According to the Global Burden of Disease 2010 Study, high total cholesterol accounts for 29% of ischemic heart disease disability adjusted life-years (DALYs), second only to hypertension. Elevated cholesterol levels are highest in Europe (54%), followed by the Americas (48%), while the lowest levels are found in Africa (22.6%) and South East Asia (29%). There is a positive correlation between the income level of a country and cholesterol levels.[5]

Table 8.1 US Department of Health and Human Services criteria for the diagnosis of pre-hypertension and hypertension.

Category	Systolic BP mmHg	And/or	Diastolic BP mmHg
Normal	<120	And	<80
Pre-HTN	120–139	Or	80–89
HTN, stage 1	140–159	Or	90–99
HTN, stage 2	≥ 160	Or	≥ 100

Table 8.2 American heart association criteria for dyslipidemia.

	Desirable	Near optimal	Borderline high	High risk	Very high risk
Total cholesterol/ mg dL^{-1}	<200		200–239		≥240
LDL-C/mg dL^{-1}	<100	100–129	130–159	160–189	≥190
Triglycerides/ mg dL^{-1}	<100	100–149	150–199	200–499	≥500
HDL-C/mg dL^{-1}		>60 or higher			<40 for men <50 for women

Each component of dyslipidemia contributes to coronary heart disease (CHD), but increased low-density lipoprotein (LDL) and triglycerides (TG) are the greatest risk factors. Although LDL-C levels are strongly associated with CVD risk at high concentrations, the predictive value is weaker at low concentrations. Apolipoprotein B (apoB) concentrations are believed to more accurately measure the number of atherogenic lipoprotein particles and therefore to more accurately predict CVD risk. Higher apoB and lower apoA-1 levels are associated with CVD. Results of several large-scale cohort studies (AMORIS, Quebec Cardiovascular Study, Northwick Park Heart Study, INTERHEART) suggest that apoB levels are a stronger predictor of CVD risk than any cholesterol ratio. Other studies (ARIC, Nurses Health Study, Ingelsson *et al.*) however, found that apoB/apoA-1 measurements added no significant information to other lipid measurements.[9,10]

According to a recent meta-analysis, statin use lowered LDL cholesterol by 28% and was associated with a 30% decrease in CHD events. CVD death rates in the US dropped 58% between 1972 and 2004, due in large part to the use of statins.[11]

Hypertriglyceridemia is an independent risk factor for CHD. The Multiple Risk Factor Intervention Trial and the 7 Countries Study demonstrated a 2% increase in risk for every 1% increase in total cholesterol (TC). The relationship between TG and CHD, however, is not as strong as for LDL-C. This may be due, in part, to the greater variability in measuring TG. Furthermore, high levels of TG are usually found in conjunction with other risk factors, *e.g.*, obesity and diabetes, and are more predictive of CVD for women than men. Six large cohort studies showed a 37% increase in risk for women, and a 14% increase in risk for men after adjustment for other lipid levels.[11] Two recent, large scale Mendelian studies have shown that loss of function mutations in the APOC3 gene, which encodes apolipoprotein C3, reduces total triglyceride levels and risk of ischemic vascular disease and ischemic heart disease by 41% and 36% respectively.[12,13]

High-density lipoprotein cholesterol (HDL-C) levels, on the other hand, are inversely correlated with CVD risk, independent of LDL-C. The Framingham study was the first to reveal that low levels of HDL-C (<40 mg dL^{-1} for men and <50 mg dL^{-1} for women) conferred an increased CVD risk. These results were supported by findings from multiple epidemiological studies including PROCAM, which showed a 70% reduction in CVD risk over 6 years with HDL-C >35 mg dL^{-1}, and the ARIC study which supported the inverse association over a 10 year follow-up. Every 1.0 mg dL^{-1} increase in HDL decreases CVD risk by 2–3%.[9] There is a similar relationship between the levels of apolipoprotein A-1 (apoA-1), the major protein component of HDLs, and CVD risk.

8.2.3 Obesity and Overweight

According to the Center for Disease Control and Prevention, an individual with a BMI of 25–29.9 kg m^{-2} is considered overweight, and a BMI of 30 kg m^{-2} or greater is obese. The American Heart Association estimates that 68.2% of adults over the age of 20 years are overweight or obese, and 34.6% are obese.[1] Current caloric intake in the US is 22% higher than between 1971–2004, whereas physical activity is decreased. 31.8% of children aged 2–19 years are currently considered overweight and obese which, according to the American Heart Association, confers a 16 times greater risk of adult obesity.[1] Obesity is strongly related to hypertension, glucose intolerance, type II diabetes, and dyslipidemia. The prevalence of insulin resistance and diabetes is also increasing in relation to obesity. It is estimated that 382 million individuals worldwide, 8.3% of the population, have diabetes and more than 90% of these have type II diabetes, which is strongly related to obesity.[14] Individuals with centripetal or abdominal obesity are at greater cardiovascular risk than individuals with peripheral obesity, and the waist-to-hip ratio is a stronger predictor of CVD risk than BMI. Each additional cm of waist circumference increases risk after controlling for BMI.[15] The Global Burden of Disease 2010 study estimated that 23% of ischemic heart disease DALYs are attributable to high BMI. Childhood obesity is a strong predictor of adult hypertension.

8.2.4 Diabetes (Table 8.3)[16]

Diabetics have a 2–4-fold increase in risk for CHD, and a 57% greater risk of death from myocardial infarction (MI) with unstable angina. Not only is diabetes an independent risk factor for CVD, but also it amplifies the effects of other risk factors. According to data from the Framingham study, the CVD

Table 8.3 American Diabetes Association criteria for the diagnosis of pre-diabetes and diabetes.

Parameter	Pre-diabetes	Diabetes
Hemoglobin A1c	5.7–6.4%	≥6.5%
8 h fasting plasma glucose/mg dL^{-1}	100–125	≥126
2 h plasma glucose in oral glucose tolerance test/mg dL^{-1}	140–199	≥200
Random plasma glucose in patients with symptoms of hyperglycemia/mg dL^{-1}		≥200

Table 8.4 Biological thresholds for metabolic syndrome defined by 2009 joint scientific statement of the International Diabetes Federation Task Force on Epidemiology and Prevention; National Heart, Lung, and Blood Institute; American Heart Association; World Heart Federation; International Atherosclerosis Society; and International Association for the Study of Obesity. For adults ≥20 years old; three or more criteria establishes a diagnosis of Metabolic Syndrome.

Clinical component	Criteria for metabolic syndrome
Central obesity	Waist circumference ≥102 cm for men ≥88 cm for women
Hyperglycemia/ insulin resistance	Fasting plasma glucose ≥100 mg dL^{-1} Or previously prescribed anti-hyperglycemic
Hypertension	Blood pressure ≥130/85 mmHg Or previously prescribed anti-hypertensive
Dyslipidemia	TG≥150 mg dL^{-1} HDL-C <40 mg dL^{-1} for men <50 mg dL^{-1} for women Or previously prescribed drug for lipid modification

burden in diabetic individuals exists up to 30 years before the diagnosis, even after adjustment for BMI.[17] Even before the development of overt diabetes, insulin resistance is associated with increased CVD risk.

Additionally, insulin resistance is often associated with a constellation of other risk factors, termed the "metabolic syndrome" (Table 8.4). Although there are slight variations among specific criteria by various professional groups, regardless of how it is defined, approximately 25% of the world's population meets criteria for the metabolic syndrome. It is unclear if the "syndrome" has greater predictive value than its individual components, however, the recognition that risk factors cluster is a clinically useful concept.

8.2.5 Diet

Compared to a typical Western diet (high in red and processed meats, carbohydrates, fried food, refined grains), the Mediterranean diet (high consumption of olive oil, nuts and seeds, fruits, vegetables, fish, and poultry) is associated with a lower risk of CVD. Ten percent of the global DALYs in 2010 were attributed to dietary risk factors and physical inactivity.

8.2.6 Tobacco and Air Pollution (Table 8.5)

Active smoking increases the risk of coronary artery disease (CAD) by 80%, and passive smoking (second hand smoke) by 30%.[18] Reportedly, there are 967 million smokers worldwide, with countless more individuals affected by

Table 8.5 Major environmental pollutants as risk factors for cardiovascular disease.[39–41]

Pollutant	Major source	Major CV effect
Particulate matter Fine <2.5 μm	Combustion of fossil fuels forest and grass fires, tobacco smoke, fireplaces and space heaters	Decreased HRV, HTN, ischemic CVD, HF, arrhythmias, vascular disease
Coarse 2.5 μm–10 μm	Industrial processes, dust, agriculture	Autonomic imbalance, decreased HRV
Ground level O_3	Urban smog Reaction of hydrocarbons, nitro compounds and organic compounds with sunlight	Oxidation of biomolecules leading to oxidative stress and inflammation. Acute arterial vasoconstriction, changes in HRV, *in vitro* lipid peroxidation
Nitrogen oxides NO, NO_2, N_2O_4, N_2O_5, nitrites, nitroso compounds, nitrous acid, nitrous oxide	Fossil fuel combustion, electrical utilities, coal burning, tobacco smoke, unvented furnaces and stoves	Increase in angina, MI arrhythmias, atherogenic risk and platelet activation
SO_2	Fossil fuel combustion, smelting	Increase in blood viscosity seen with an increase in PM also. Increase in P-selectin in circulating platelets, HTN
CO	Vehicle exhaust, tobacco smoke, industrial processes, fossil fuel combustion	Hypoxia leading to oxidative stress and atherosclerosis
CS_2	Primarily industrial processes of rayon, cellophane, CCl_4, rubber, chemicals, pesticides	Increase in atherosclerosis with increases in serum cholesterol and disturbances in lipid metabolism
Toxic metals Pb, Cd, As, Hg	Old pipes, dyes, paint, fish, tobacco smoking, pesticides, treated wood, contaminated food	HTN, lipid peroxidation, ROS, inflammation and atherosclerosis

second-hand smoke.[19] Tobacco smoke accounts for greater than 10% of the global disease burden across all age groups, second only to hypertension. Smoking cessation reduces the risks to near non-smoking levels within five years.[18] Additionally, environmental pollutants (*e.g.*, pollutants from solid fuels) have recently been recognized as an emerging risk factor. Studies done in Europe (Air Pollution and Health: a European Approach Project 2) and the United States by the American Cancer Society have shown a significant association between particulate matter pollution and overall cardiovascular mortality. Pollutants released from household solid fuels are considered to be the third leading risk factor for global DALYs.[3] The WHO estimates 8 00 000 deaths annually are due to environmental pollutant exposure.[20]

8.2.7 Obstructive Sleep Apnea (OSA) and Chronic Obstructive Pulmonary Disease (COPD)

These have been identified as risk factors for CVD. OSA is also associated with obesity, the metabolic syndrome and insulin resistance, especially in individuals under 60 years of age.[18] Airflow limitation, as seen in COPD, is an independent risk factor for CVD. CAD is present in 10–38% of patients with COPD, and a decreased forced expiratory volume is a strong predictor of CVD morbidity and mortality.[21]

8.2.8 Psycho-social Stress (Depression, Job-related Stress, Social Isolation, Post-traumatic Stress Disorder, and Lower Socioeconomic Status)

This is another recognized risk factor. Notably, these risk factors frequently co-exist with other confounding risk factors. The INTERHEART case-controlled study was the largest to examine the relationship between long-term stress and CHD. A cohort of 15 152 patients who had suffered a MI and 14 820 non-CHD individuals were compared. The odds ratio of MI was more than double among those who were stressed, even after controlling for other risk factors. The population attributable risk for all psychosocial risk factors was 28.8%, with general stress accounting for 12%.[22] Meta-analysis has shown that job-stress increases the risk for CHD by 50%, and the response is dose-related. Stress from social isolation results in a pooled relative risk of 1.51 in CHD free populations, according to a meta-analysis of 9 studies done through 2011.[23]

Depression is also strongly linked to CVD and is more common in patients with CAD than the general population. Multiple studies and meta-analysis have linked the development of CVD with depression, resulting in a reported pooled effect size between 1.5 and 2.7.[22] A sample of Vietnam veterans suffering from Post Traumatic Stress Disorder (PTSD) followed over 30 years showed a higher rate of mortality from CVD. Women with PTSD in a 14 year prospective study developed CHD at a rate 3 times higher than controls. Males with PTSD are at higher risk for non-fatal MI and fatal CHD.[24]

8.3 Mechanisms of Vascular Injury

Increasing evidence suggests that chronic inflammation, oxidative stress, and autonomic imbalance contribute to the underlying vascular pathology of apparently diverse risk factors. The process of atherosclerosis may be apparent in childhood, with the accumulation of lipids and fibrous elements in the endothelium, which appear as fatty streaks. As lipids and cholesterol build up, the inflammatory process begins, which contributes to further growth and instability of the plaque. LDLs are found in atherosclerotic lesions, and the deposition of cholesterol in the plaque is proportional to the concentration of plasma LDL-C. ApoB rich lipoproteins especially have an increased affinity for the extracellular matrix of the intima. Once retained in the artery wall, the LDL-C is oxidized, both as a result of, and further inducing, oxidative stress and inflammation.

The inflammatory process can be direct, occurring at the site of an atherosclerotic lesion, or systemic and contribute to the development of the plaque.[25] Vascular injury is not due to lipid accumulation alone, however. Oxidative stress and autonomic imbalance both contribute to the modification of the lipids and subsequent inflammatory response. Identification of the many signal molecules and pathways involved may help to identify inflammatory triggers, and therefore new risk factors for cardiovascular disease, as well possible therapeutic targets.

8.3.1 Inflammation

The inflammatory process can lead to the development of plaque formation, and reciprocally, the plaque itself may trigger an inflammatory cascade. The immune response is both innate, non-specific and inflammatory, and adaptive, with the activation of T cells contributing to destabilization of the plaque. Figure 8.1 outlines the relationship between inflammation and atherosclerosis. The entry of plasma LDLs into the vascular wall is passive, but as lipids accumulate in the intima, they become trapped by interaction between the $ApoB_{100}$ moiety and extracellular matrix glycoproteins. Intimal dendritic cells may be responsible for active uptake of the LDLs, which

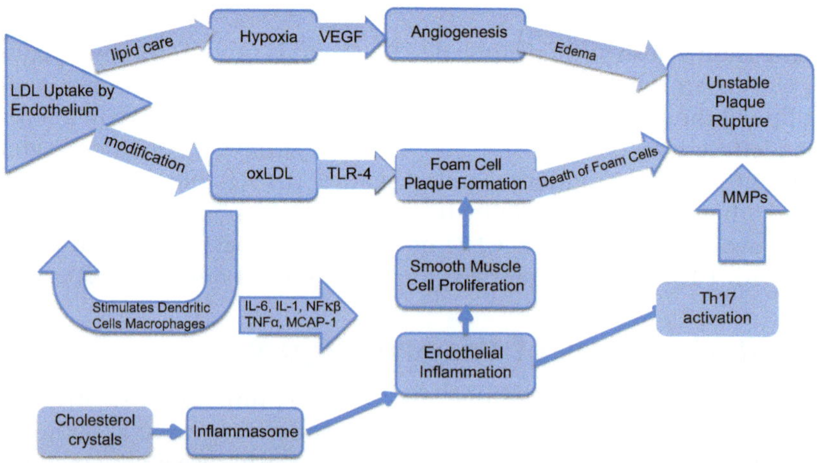

Figure 8.1 Inflammation and atherosclerosis. LDL-C uptake and oxidation leads to foam cell and plaque formation. Subsequent release of inflammatory mediators further stimulates plaque growth and activation of the adaptive immune response. As the lipid core thickens, hypoxia develops and angiogenesis occurs. The release of matrix metalloproteases from Th_{17} cells and edema due to leaky new vessels causes the plaque to become unstable and rupture.

under conditions of oxidative stress are modified, oxidized or glycolated by exposure to reactive oxygen (ROS) or reactive nitrogen species (RNS).[26] While the initial uptake is probably protective, lowering serum cholesterol, modification to oxidized LDL contributes to the cascade of inflammatory events causing monocyte recruitment and differentiation (Figure 8.1). Macrophage uptake of oxidatively modified LDLs (oxLDL) occurs through the Toll-like receptor 4 (TLR-4) resulting in the formation of foam cells, which form the core of the atherosclerotic plaque. TLR-4 may represent a potential therapeutic target, as TLR-4 knockout mice have reduced levels of plaque formation and the expression of these receptors in monocytes can be down-regulated by statins.[27]

In response to the irritation caused by plaque formation, the endothelial lining increases its expression of adhesion molecules. P-selectin, intercellular adhesion molecules (ICAM), and vascular cell adhesion molecules (VCAM) expression allow platelets to adhere which augments the growth of the atherosclerotic lesion. The oxLDL also stimulates the expression and release of a number of inflammatory mediators, primary among them interleukins 1 and 6 (IL-1, IL-6), nuclear factor kappa beta (NFκβ) and tumor necrosis factor (TNF). IL-6 is a proinflammatory cytokine released by a variety of cells, including endothelial cells and leukocytes, driving the production of downstream markers of inflammation, including C-reactive protein (CRP) and fibrinogen, and CVD. Although the exact role of IL-6 in plaque formation is unclear, a meta-analysis has shown a causal association between IL-6 receptor related pathways and CVD.[28] CRP is an acute phase protein which is involved in endothelial dysfunction, influencing the expression of adhesion molecules,[29] decreasing the activity of nitric oxide synthase (NOS),[30] activating macrophages and up-regulating cytokine secretion *via* protein kinase C and NFκβ.[31] While CRP levels correlate with, and are predictive of, the risk of future cardiac events, whether the relationship is causal or simply correlative is unclear.[32,33]

NFκβ is a transcription factor that promotes the up-regulation of other pro-inflammatory cytokines by endothelial cells, including monocyte chemoattractant protein-1 (MCP-1) and IL-1. MCP-1 is one of the principal chemokines responsible for the infiltration of leukocytes into the growing lesion and IL-1 release causes the release of TNF-α. Free cholesterol may also activate inflammasomes, which help maintain the inflammatory response.[27] Plaque elaboration is enhanced by the migration of smooth muscle cells into the intima in response to these growth factors and IL-1. Under the influence of TGF-β, as well as angiotensin, smooth muscle cells proliferate and produce collagen and elastin, which form a fibrous cap over the plaque.

TNF-α contributes to an adaptive response by the immune system. Recruitment of T-lymphocytes contributes to the inflammatory response and destabilization of the plaque. The Th17 subset release IFN-γ, which inhibits collagen and extracellular matrix production by smooth muscle cells and stimulates the production of matrix metalloproteases by macrophages. MMPs are collagenases, which, along with other gelatinases and proteases, contribute to plaque rupture and thrombosis.[34] It has been suggested that MMP inhibitors may be a possible therapy for CVD.

Foam cell formation of the lipid core, and the migration of smooth muscle cells into the intima contribute to hypoxic conditions by blocking the diffusion of oxygen. This not only stimulates the production of hypoxia inducible factors (HIF-1α), but also the expression of pro-angiogenic genes. Vascular endothelial growth factor (VEGF) promotes the growth of new blood vessels from the vasa vasorum into the area of the lesion. The vessels, however, tend to be permeable and the resultant edema contributes to the accumulation of immune mediators and cells within the vessel wall, adding to plaque instability.[27]

8.3.2 Oxidative Stress

Oxidative stress is defined as an imbalance between the production of reactive oxygen species (ROS), largely superoxide anions (O_2^-), and antioxidative processes. The half-life of superoxide anion is very short, one microsecond, being rapidly converted to hydrogen peroxide (H_2O_2) or peroxynitrile by dismutases. H_2O_2 and O_2^- contribute to vascular injury by influencing monocyte adhesion, platelet aggregation, inflammatory gene induction, and loss of vasorelaxation.[35]

ROS can be generated *via* a number of mechanisms, including the leakage of electrons from the mitochondrial respiratory chain, NADPH oxidase (NOX), xanthine oxidase, myeloperoxidase (MPO), lipoxygenase (LO), and uncoupled NO synthase (NOS). The NOX complexes, activated in monocytes and neutrophils, are also found in vascular endothelial cells, smooth muscle cells, fibroblasts and cardiomyocytes, and are a major source of O_2^-. Increased NOX activity has been associated with clinical risk factors for CVD and hypertension, and plays a role in the formation of plaque in animal models.[36]

Myeloperoxidases generate ROS that modify lipids and proteins, and play a role in destabilizing plaques. High levels of MPO activity have been found in macrophages of ruptured lesions[37] possibly also activating MMPs in the artery wall.[38] The levels of MPO activity correlate well with the extent of atherosclerosis in CAD, and are predictive of future events.[39]

Lipoxygenase catalyzes the oxidation of polyunsaturated fatty acids. The isoforms 12/15-LO and 5-LO are expressed in inflammatory and endothelial cells with 5-LO activity found in macrophages, foam cells, mast cells and dendritic cells of human plaques. The activity increases as the plaque grows and becomes unstable and may contribute to the production of leukotrienes and increased vascular permeability.[40] Drugs that inhibit 5-LO activity decrease leukotriene production and plaque formation after acute coronary syndrome.[41]

Concomitant with the production of ROS, especially the superoxide anion, the production and the bioavailability of nitric oxide (NO) decreases. The activity of the endothelial isoform of NO synthase (eNOS) is constitutive in most tissues, with L-arginine being the substrate for eNOS mediated reduction. However, the production of oxLDLs by the enzymes described above can upregulate arginase enzymes, decreasing the availability of eNOS substrate and uncoupling electron transfer. Similar uncoupling is seen with reduction in the tetrahydroxy biopterin (BH_4) cofactor.

ROS has both extracellular and intracellular effects. Upon release, ROS activates macrophages and mast cells, thereby increasing the inflammatory process. The intracellular response to ROS is *via* the activation of transcription factors, primarily NFκβ and activation protein 1 (AP-1) which leads to the production of chemokines, cytokines and adhesion molecules. This chain of events is amplified as the production of oxLDLs can further stimulate the production of ROS through the uncoupling of endothelial NO synthase and stimulation of NADPH oxidases. The loss of NO availability also contributes directly to endothelial dysfunction. NO is cardioprotective in that it inhibits platelet aggregation, the expression of adhesion molecules and smooth muscle cell proliferation, as well as modulating vascular tone. Therefore, it is reasonable to conclude that conditions that result in oxidative stress increase the production of ROS and RNS, which both cause, and are markers of, vascular injury.

8.3.3 Autonomic Imbalance

Autonomic imbalance is marked by a decrease in vagal tone, as indexed by changes in heart rate variability (HRV). HRV is a measure of beat-to-beat heart rate fluctuation over time, which reflects the balance between sympathetic and parasympathetic output. Although the heart receives innervation from both, the parasympathetic influence dominates at rest. Depressed vagal tone is associated with both traditional and emerging risk factors for CVD, in part because of the resultant predominant sympathetic tone in these states. Numerous studies have linked decreased vagal activity and HRV with morbidity and mortality, in both at risk and healthy populations.[42]

Autonomic imbalance is associated with inflammation and oxidative stress, although the directionality of the relationships is unclear (Figure 8.2). Decreased vagal tone correlates with immune dysfunction and inflammation, as indicated by an increase in pro-inflammatory cytokines. Serum IL-6 and CRP levels have primarily established the relationship between HRV and inflammation. Results from multiple studies of both healthy and CVD populations show an increase in these biomarkers with a decrease in HRV.

Changes to the inflammatory response may be mediated through what has been termed the cholinergic anti-inflammatory pathway. The vagal afferents can be activated by cytokines causing release of corticotropin-releasing hormone into the hypothalamic-pituitary-axis (HPA) followed by ACTH stimulation of the adrenals and glucocorticoid secretion. In culture, acetylcholine, the parasympathetic nervous system neurotransmitter, also decreases TNF-α production by macrophages in culture, and vagal nerve

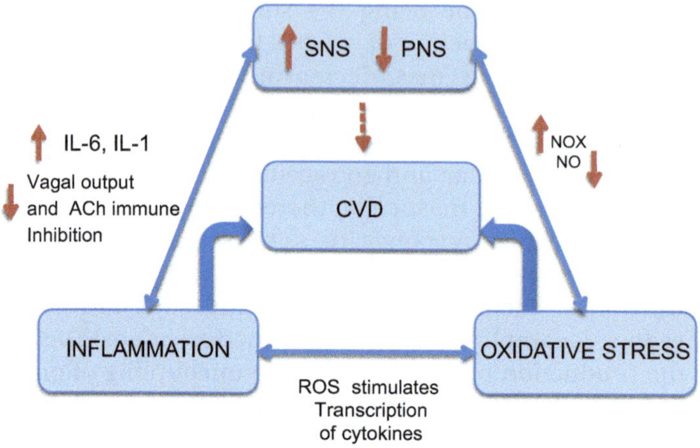

Figure 8.2 Autonomic imbalance. The relationship between inflammation, oxidative stress and autonomic imbalance is reciprocal. As oxidative stress develops, the ROS stimulate the transcription of proinflammatory cytokines IL-6 and IL-1. Output from the SNS is also proinflammatory, while PNS output, primarily vagal, is anti-inflammatory. An imbalance favoring the SNS leads to increased inflammation and oxidative stress, both of which contribute directly to CVD.

stimulation in dogs decreases the release of TNF-α and IL-6. Long-term stimulation normalizes the levels of these proteins in both the myocardium and the plasma.[43–45] The overall effect is immunosuppressive.[42] In addition to contributing to the inflammatory mechanism, depressed vagal and parasympathetic tone are associated with increased oxidative stress and the production of ROS, particularly in cardiomyocytes. Conversely, sympathetic nervous system activation is pro-inflammatory. The question of whether the connection between autonomic dysfunction and inflammation is due to increased sympathetic stimulation of the immune response, or a decreased cholinergic anti-inflammatory influence is unclear.

8.4 Clinical Translation

8.4.1 Hypertension

Markers of oxidative stress and inflammation have been described not only in hypertensive patients, but also prior to the actual development of overt hypertension. A cohort study, which examined 1514 men and 1528 women ages 18–89 years old, who were CVD free, showed an association between pre-hypertensive status and oxidative stress. Specifically, both systolic and diastolic blood pressures were inversely correlated with total antioxidant capacity and positively correlated with serum levels of oxLDL-C, even after adjustment for other risk factors. Normotensive individuals with a positive family history of hypertension have higher plasma levels of hydrogen superoxide than normotensive individuals with no family history, suggesting that oxidative stress both precedes the development of hypertension, and is, in part, genetically controlled.[46] In addition, CRP level, a marker of inflammation predicts the development of hypertension, independent of blood pressure.[47]

Both hypertension-related production of ROS and inflammation are mediated in part through activation of the renin–angiotensin–aldosterone system (RAAS). This system is active in a variety of tissues, *e.g.* the adrenals, brain, heart, and blood vessels, as well as the kidneys. Angiotensin II (angII) is a vasoactive octapeptide that binds to G-coupled receptors AT_1R and AT_2R. The majority of the actions of angII, cell growth, oxidative stress and vasoconstriction, are mediated *via* the AT_1R, found throughout the cardiovascular system. AngII contributes directly to hypertension by vasoconstriction and by stimulating production and release of the adrenal mineralocorticoid, aldosterone.

The indirect contribution of angII to hypertension and CVD is through stimulation of oxidative stress and inflammation. Hypertensive patients show increased LDL peroxidation and elevated titers of autoantibodies against oxLDLs, which can be reduced by angiotensin converting enzyme (ACE) inhibitors.[47] AngII contributes to this process of lipid modification by up-regulating several NADPH oxidase isoforms, NOX1 and 2, through binding of the AT_1R. Ang II also up-regulates the lectin-like oxidized low-density lipoprotein receptor-1 (LOX-1), a scavenger receptor found on vascular endothelial cells. This up-regulation has been seen not only in hypertension, but also diabetes mellitus, hyperlipidemia, and atherosclerosis, and can be blocked by the angiotensin receptor blocker (ARB) losartan.[48]

Both angII and heightened oxidative stress may influence the autonomic control of blood pressure. Hypertension is associated with decreased baroreceptor sensitivity and augmented chemoreceptor sensitivity (reciprocal dysautonomia).[49] Increased RAAS activity in the hypothalamus, or the brainstem may result in production of ROS leading to impaired baroreceptor function and a compensatory hyperstimulation of the sympathetic nervous system *via* increased chemoreceptor activity.

AngII induced autonomic imbalance may stimulate inflammation and T-cell activation in hypertension. Induced hypertension *via* angII infusion, or salt administration leads to increased levels of chemokines, cytokines, adhesion molecules and inflammatory cells *via* activation of the transcription factor NFκβ. ACE inhibitors have been shown to reduce NFκβ activity and inflammation, and irbesartan and candelsartan administration in hypertensive patients lowers serum levels of adhesion molecules TNF-α and MCP-1.[50] Moreover, angII-induced hypertension is blocked in IL-6 knockout mice.[49]

Animal models also suggest a role for the adaptive arm of the immune system in hypertension. Immune deficient $RAG^{-/-}$ mice have no pressor effect with angII infusion, but the pressor response is restored with T lymphocyte, not B lymphocyte, transfer. However, T-cells lacking AT_1R, or functional NADPH oxidases, only showed blunted hypertensive responses.[51] One hypothesis suggests that vascular damage caused by the prehypertensive state causes the formation of "neo-antigens" that result in the activation of the adaptive immune system.[49] Regardless of the mechanism of activation, there is abundant experimental evidence for the role of the T-cell in the development of hypertension. T-lymphocytes have AT_1R and proliferate in response to angII, and in addition, they contain NADPH oxidases. Thymectomy prevents hypertension in experimental models; infusion of allo-activated T-cells in cancer treatment causes a rise in blood pressure; and suppression of the adaptive immune system inhibits hypertension in experimental animals and humans.[51]

8.4.2 Dyslipidemia

LDL uptake and oxidation contributes to plaque formation, while oxidative stress directly contributes to plaque rupture and thrombosis. The cardioprotective actions of HDL/apoA-1 extend beyond reverse cholesterol transport. HDLs have anti-inflammatory, anti-thrombotic, and anti-oxidative properties. In addition, they support endothelial integrity by preventing endothelial cell death and stimulating proliferation and migration. HDLs exert anti-inflammatory and anti-thrombotic effects *via* inhibition of adhesion molecule expression, therefore preventing monocyte, macrophage and platelet accumulation and activation. The anti-platelet effects are further enhanced by the increased NO bioavailability that results from the anti-oxidative effects. HDLs prevent the oxidation of LDL-Cs, which stimulates foam cell formation and uncouples eNOS, as discussed above. HDLs prevent this uncoupling and upregulate eNOS expression and activity. Finally, HDLs inhibit the apoptotic actions of TNF-α, oxLDL and angII.[9]

8.4.3 Obesity, Insulin Resistance, and Diabetes

Overweight children with hypertension and adults with metabolic syndrome have higher measures of oxidative stress. Further, individuals with abdominal obesity show higher levels of oxidative stress and inflammation than lean individuals, or those with subcutaneous adiposity. Subcutaneous adipocytes have higher insulin sensitivity and adiponectin production, which prevents the inflammatory cascade and protects against obesity. Conversely, visceral adipocytes have low insulin sensitivity, higher rates of lipolysis and triglyceride and glucose storage. The ROS and proinflammatory adipokines produced in visceral adipose tissue may contribute to the association of obesity with hypertension and diabetes. In children, CRP levels correlate with obesity, based on waist circumference, even before the onset of puberty.

Obesity leads to adipocyte dysfunction, which is manifest by the infiltration of inflammatory cells into adipose tissue, and the release of proinflammatory cytokines and adipokines. Excess glucose leads to the production of advanced glycation end products which bind to receptors on macrophages and induces NFκβ. Macrophages found in adipose tissue of obese children, adults and mice express TNF-α, and other cytokines in proportion to the obesity.[52] The stimulus for adipocyte inflammation may be increased oxidative stress, or the increased levels of angII produced in this tissue. Hyperglycemia and hyperlipidemia also stimulate adipocyte hypertrophy, hyperplasia and dysfunction leading to a persistent elevation in circulating triglycerides. *In vitro*, ROS generation is increased by exposure of adipocytes, endothelial cells and vascular smooth muscle cells to high concentrations of glucose, and in mice adipocytes, this effect can be blocked with antioxidants.

The complications of diabetes, impaired circulation leading to limb loss, cerebrovascular disease and CAD, are associated with changes to the endothelium as a result of dyslipidemia, oxidative stress and inflammation. Insulin plays a role in metabolic signaling that is, in part, responsible for decreasing vascular tone, vascular smooth muscle cell (VSMC) proliferation and decreased adhesion of platelets and inflammatory cells to the endothelium. Consequently, insulin resistance contributes to the development of atherosclerosis.

Hyperinsulinemia and hyperglycemia, consequences of insulin resistance, result in depletion of HDL-C. Increased insulin levels in type II diabetes also stimulate mitogenic and growth factors within smooth muscle. Additionally, in response to hyperglycemia, adipose tissue releases proinflammatory adipokines, as well as angiotensinogen, which also causes inflammation (Figure 8.3). The endothelium becomes dysfunctional as NO bioavailability is lost with increasing oxidative stress, and VSMCs proliferate under the influence of high insulin levels.

Increased levels of F_2-isoprostanes, markers of lipid oxidation, have been found in both type I and type II diabetic patients, and correlate with the level of glycemic control. Additionally, F_2-isoprostanes production can be induced in VSMC in culture using high glucose concentrations.[53] There is ample experimental evidence to support the increase in ROS and RNS production by endothelial cells, VSMC and adipocytes with high concentration of fatty acids or glucose. Coupled with the loss of insulin induced eNOS activation, the oxidative stress results in loss of bioavailability of NO and endothelial dysfunction. The formation and binding of advanced glycation end products occurs in diabetes through the same mechanisms, and with the same results, as seen with hyperglycemia and obesity. The heightened oxidative stress seen in these patients has the end result of inflammation and atherosclerosis.

There is also evidence of autonomic imbalance in diabetics indicated by a decreased vagal tone, even after adjustment for other factors. High blood glucose levels result in the formation of O-linked *N*-acetylglucosamine, which activates calcium-calmodulin dependent protein kinase II (CaMKII). This enzyme regulates calcium levels in cardiomyocytes, and therefore the electrical activity in the heart. Consequently, overstimulation of CaMKII may lead to arrhythmias.[54]

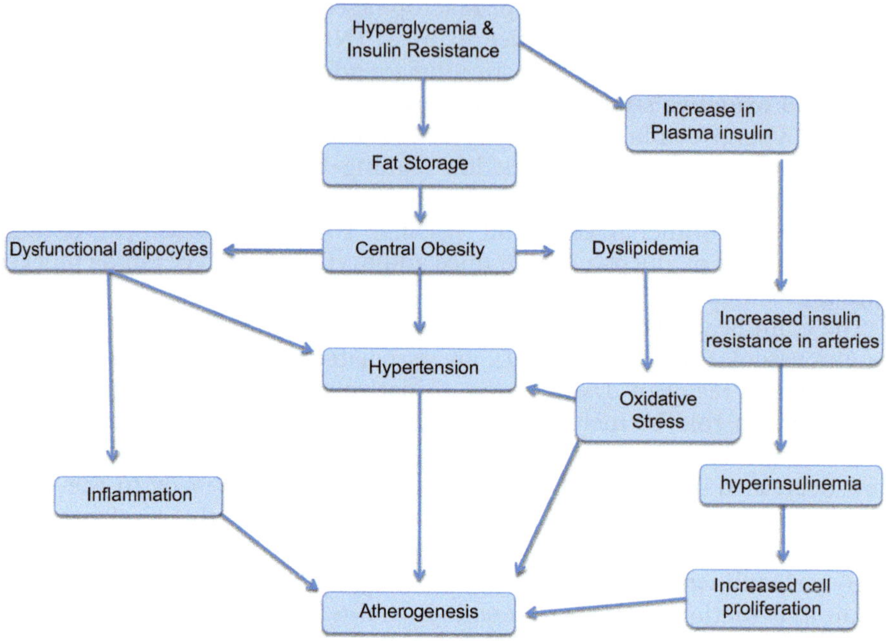

Figure 8.3 Hyperglycemia, insulin resistance and atherosclerosis. Hyperglycemia and insulin resistance lead to an increase in fat storage, contributing to central obesity, hypertension, and atherogenesis. Hypertension is further heightened by the resulting dyslipidemia, oxidative stress and inflammation. Plaque development is enhanced by the high plasma insulin levels stimulating cell proliferation.

8.4.4 Diet

A typical "Western diet" is associated with increases in pro-inflammatory cytokines and markers of endothelial dysfunction. Conversely, the Mediterranean diet characterized by low consumption of red and processed meats, dairy and sweets, and high consumption of olive oil, nuts and seeds, fruits, vegetables, fish and poultry, as well as small amounts of red wine, is associated with lowered CVD risk. Specifically, diets low in fruits, nuts and seeds, whole grains, vegetables and omega-3 fatty acids, while high in sodium were considered to account for a tenth of the global disease burden.[3,5]

Olive oil, nuts and seeds, the components of the Mediterranean diet that primarily mediate the cardioprotective effects, are rich in omega-3 polyunsaturated fatty acids (n-3 PUFAs). In contrast, the Western diet is rich in saturated fatty acids, trans fatty acids and omega-6 polyunsaturated fatty acids (n-6 PUFAs). These latter compounds activate pathways leading to inflammation, arrhythmias, vasoconstriction and platelet activation, and have been associated with increased plasma levels of TNF-α, IL-1, IL-6, CRP, E-selectin, ICAM-1 and VCAM-1. Conversely, n-3 PUFAs have been shown to decrease levels of these inflammatory biomarkers.[55] Consumption of eicosapentaenoic acid (EPA 20: 5n-3) and docosahexaenoic acid (DHA 22: 6n-3) also lowers the level of arachidonic acid, which is the precursor to atherosclerotic eicosanoids. Omega 3 PUFAs, when incorporated into lipids, increase membrane fluidity, and alter ion channel function, especially Ca^{2+} channels. This may be the mechanism through which EPA and DHA decrease arrhythmias and lower blood pressure and heart rate.[56] A Mediterranean diet supplemented with nuts has also been shown to decrease waist circumference by 5% and serum triglycerides by 10%, as well as shift the lipoprotein sub fractions to a less atherogenic profile.[57]

8.4.5 Tobacco and Air Pollution

The pollutants found in both tobacco smoke as well as the environment induce oxidative stress and inflammation (Figure 8.4). Active smoking places a greater oxygen demand on the heart through both stimulation of the

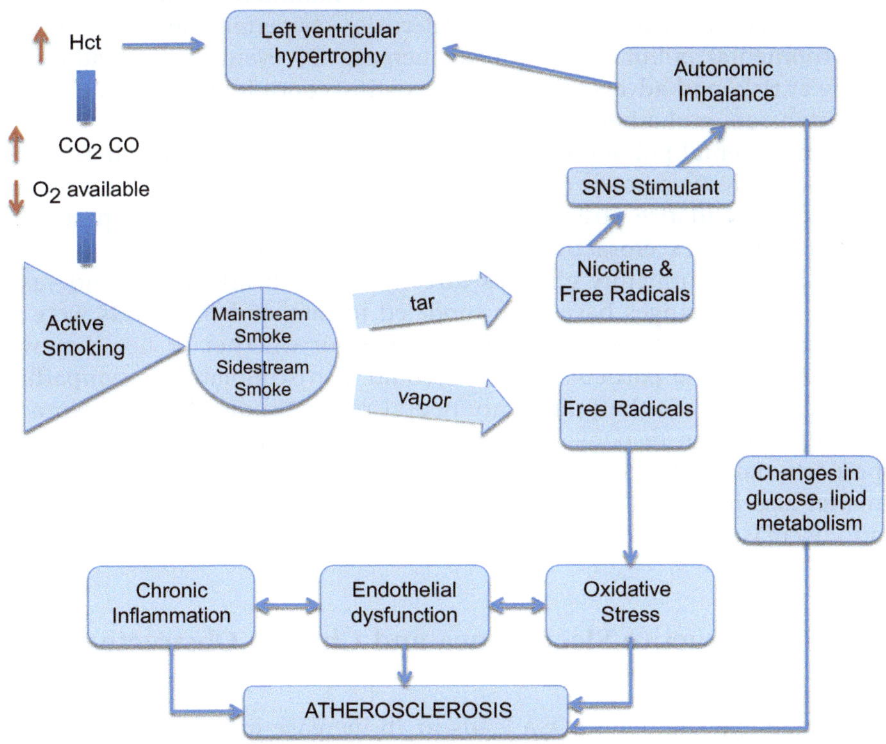

Figure 8.4 The effect of tobacco smoking. The tar and vapors produced by tobacco smoking contain free radicals that contribute to oxidative stress which injures the endothelium and leads to inflammation and atherosclerosis. Stimulation of the SNS by nicotine heightens autonomic imbalance resulting in changes to metabolism, increased heart rate and oxygen demand, which when coupled with an increased hematocrit, results in left ventricular hypertrophy.

sympathetic nervous system and a decrease in oxygen availability. The presence of free radicals from smoking both increases the ROS produced, as well as inhibits catalase, superoxide dismutase, and glutathione peroxidase, which are protective enzymes normally active in the cell. The end result is the decreased expression and uncoupling of eNOS, and the depletion of BH_4, the eNOS cofactor.[58] The loss of NO production leads to changes in vasomotor function, and increased platelet activation, leading to a prothrombotic state.

Smokers also show evidence of chronic inflammation, with 20–25% increased peripheral leukocyte counts, and increased fibrinogen levels correlating with the number of cigarettes smoked.[59] Other inflammatory markers CRP, IL-6 and TNF-α are also elevated. Smokers also tend to be insulin resistant and have increased LDL-C, decreased HDL-C and apoA-1 levels. Cohort studies have shown that smoking increases the risk of type II diabetes in a dose-dependent fashion.[58]

Environmental pollutants contribute to inflammation, vasoconstriction, thrombosis, arrhythmias, atherosclerosis, hypertension and CAD.[60] Environmental pollutants may also increase CV events in individuals already at risk, and adversely affect pre-existing conditions, such as diabetes, hypertension, and atherosclerosis.

Environmental pollutants are thought to affect the cardiovascular system either directly, being toxic to the exposed tissues, or indirectly *via* autonomic imbalance, oxidative stress and inflammation. Both ultrafine particulate matter and gases cross the pulmonary epithelium and induce oxidative stress and inflammatory changes directly on the heart muscle and vasculature. These changes have been modeled in rats, and the effects were the same regardless of the age of the animals, or whether or not they were hypertensive. The particulate matter could also stimulate the sympathetic nervous system *via* the pulmonary neural reflex. Studies have shown a decrease in HRV in response to particulate matter exposure that is inversely proportional to the concentration.[60] Exposure to concentrated particulate matter also increased plasma levels of fibrinogen, IL-6, IL-1β, CRP and ROS and decreased NO.[60]

8.4.6 Obstructive Sleep Apnea and Chronic Obstructive Pulmonary Disease

Nocturnal hypoxia due to OSA leads to an increase in sympathetic nervous system activity. Markers of inflammation and oxidative stress are elevated in OSA, as are P-selectin and coagulation factors, further increasing the risk of CV events. Lipid peroxidation is also heightened with increased expression of LOX-1 leading to atherosclerosis.

In patients with COPD, severe airflow obstruction results in higher cardiac infarction injury scores, and increased levels of CRP, leukocytes, platelets, and fibrinogen. Carotid intima thickness is also increased in patients with

COPD.[61] The inflammatory mediators are produced initially in the lungs by alveolar macrophages, bronchial epithelial cells and lymphocytes. Dyspnea and increased respiratory drive in COPD also lead to sympathetic nervous system stimulation, and decreased vagal tone, both contributing to systemic inflammation and oxidative stress.

8.4.7 Psychosocial Stress

The effects of stress, depression, and low socioeconomic status on CVD are difficult to study due to confounding factors. Chronic stress and depression activate the hypothalamic-pituitary-adrenocorticotropic axis resulting in increased levels of cortisol and immune dysfunction. This is reflected in the increased serum levels of inflammatory cytokines and angII. The role of inflammation in depression is strengthened by clinical data showing some efficacy of anti-inflammatory therapy in depressed patients. The CARDIA study investigated the levels of CRP and IL-6 as related to socioeconomic status, and found in general, that they are inversely related. While inflammation may be both a cause and a consequence of depression, a study done in 2006 which followed patients for 6 years showed increased levels of IL-6 coincided with severity of depressive symptoms. However, IL-6 was not a predictor of change in depressive symptoms, suggesting that while the relationship is bidirectional, depression is more a cause, than a result of inflammation.[62] Similarly, individuals suffering from PTSD show increased levels of inflammation and risk for CVD, even after accounting for depression, which often is a coexisting condition.

HRV is also decreased in patients exposed to chronic stress, or suffering from depression, reflecting a loss of parasympathetic nervous system output and vagal tone.

8.4.8 Infection

Although the evidence is not conclusive, results of epidemiological and serological studies have suggested a link between infection and atherosclerosis. Antibody titers against cytomegalovirus (CMV), herpes simplex virus types 1 and 2 (HSV), hepatitis A virus (HAV), human immunodeficiency virus (HIV), human parvovirus B19, human papilloma virus (HPV), *Chlamydia pneumonia* (Cp) and *Helicobacter pylori* (Hp) and oral pathogens have all been associated with increased rates of cardiovascular events.[63] PCR analysis has confirmed the presence of Cp, Hp, HSV, CMV and periodontal pathogens in plaques, although only HSV and Cp were found intact. CMV, Cp and gingival bacteria were found in 46% of atheromatous arteries, but less than 1% of healthy vessels.[59] Acute respiratory infection has been shown to be a risk factor for acute MI, even when other CVD risk factors are not present.

Pathogens may trigger an inflammatory response which contributes to vascular injury. Vascular injury could also result as part of an autoimmune response to human heat shock proteins (HSP). Heat shock proteins are highly conserved and are expressed on the surface of cells in response to stress, infection, oxidative stress, shear stress or inflammation. Because these proteins are highly conserved, a previous infection may lead to cross reactivity between the infectious agent and self HSPs. The resultant inflammation would be chronic and self-maintaining independent of current infection. Chronic Cp infection and human HSP-60 antibody titers have been predictive of coronary events, especially when CRP levels are high.[64] The release of cytokines as a result of infection may also activate a subset of self-reactive T-cells which could also lead to autoimmunity.

Thus far, there have been no long- or short-term benefits for CVD prevention gained from antimicrobial therapy. Questions remain as to the role of infectious agents in plaques. Do they precede or follow formation? Are they actively involved in the process of atherogenesis? Many of the epidemiological studies have not accounted for socioeconomic confounders.

8.5 Conclusion

Cardiovascular disease is the number one cause of mortality worldwide, conferring a huge burden on the global population both in terms of life lost and economics. Epidemiological studies have shown that this burden is increasing as the disease becomes more prevalent in less developed countries. Many of the risk factors for CVD have been well established and investigated. While there has been some success in treating and managing certain of the traditional risk factors, others, such as obesity and diabetes, are growing at an alarming rate. Moreover, novel risk factors such as pollution, respiratory disorders and psychosocial stress, continue to emerge. In each instance, inflammation and oxidative stress contribute to vascular injury. An understanding of these underlying mechanisms may help lead to the development of novel therapies.

Abbreviations

ACE	angiotensin converting enzyme
angII	angiotensin II
ARB	angiotensin receptor blocker
CVD	Cardiovascular disease
CAD/CHD	coronary artery/heart disease
COPD	chronic obstructive pulmonary disease
CRP	C-reactive protein
HDL	high-density lipoprotein
HRV	heart rate variability
ICAM	intercellular adhesion molecule
LDL	low density lipoprotein
NFκβ	nuclear factor kappa beta
NO	nitric oxide

NOS	nitric oxide synthase (eNOS, endothelial form)
NOX	NADPH oxidase
OSA	obstructive sleep apnea
PTSD	post-traumatic stress disorder
PUFA	polyunsaturated fatty acid
RAAS	renin–angiotensin–aldosterone system
ROS	reactive oxygen species
RNS	reactive nitrogen species
TG	triglycerides
TNF	tumor necrosis factor
VCAM	vascular cell adhesion molecule
VEGF	vascular endothelial growth factor
VSMC	vascular smooth muscle cell

References

1. A. S. Go, D. Mozaffarian, V. L. Roger, E. J. Benjamin, J. D. Berry, W. B. Borden, D. M. Bravata, S. Dai, E. S. Ford, C. S. Fox, *et al.*, *Circulation*, 2013, **127**, 143.
2. P. A. Heidenreich, J. G. Trogdon, O. A. Khavjou, J. Butler, K. Dracup, M. D. Ezekowitz, E. A. Finkelstein, Y. Hong, S. C. Johnston, A. Khera, *et al.*, *Circulation*, 2011, **123**, 933.
3. R. Lozano, M. Naghavi, K. Foreman, S. Lim, K. Shibuya, *et al.*, *Lancet*, 2012, **380**, 2095.
4. T. R. Dawber, F. E. Moore, Jr. and G. V. Mann, *Am. J. Public Health Nation's Health*, 1957, **47**, 4.
5. S. S. Lim, T. Vos, A. D. Flaxman, G. Danaei and K. Shibuya, *Lancet*, 2012, **380**, 2224.
6. A. V. Chobanian, G. L. Bakris, H. R. Black, W. C. Cushman, L. A. Green, J. L. Izzo, Jr., D. W. Jones, B. J. Materson, S. Oparil, J. T. Wright, Jr., E. J. Roccella and the National High Blood Pressure Education Program Coordinating Committee, *Hypertension*, 2003, **42**, 1206.
7. G. Santulli, *J. Cardiovasc. Dis.*, 2013, **1**, 1.
8. M. Miller, N. J. Stone, C. Ballantyne, V. Bittner, M. H. Criqui, H. N. Ginsberg, A. C. Goldberg, W. J. Howard, M. S. Jacobson and P. M. Kris-Etherton, *Circulation*, 2011, **123**, 2292.
9. D. J. Hausenloy and D. M. Yellon, *Postgrad. Med. J.*, 2008, **84**, 590.
10. E. Ingelsson, E. J. Schaefer, J. H. Contois, J. R. McNamara, L. Sullivan, M. J. Keyes, M. J. Pencina, C. Schoonmaker, P. W. F. Wilson, R. B. D'Agostino and R. S. Vasan, *J. Am. Med. Assoc.*, 2007, **298**, 776.
11. A. S. Leon and U. G. Bronas, *Am. J. Lifestyle Med.*, 2009, **3**, 257.
12. The TG and HDL Working Group of the Exome Sequencing Project, *N. Engl. J. Med.*, 2014, **371**, 22.
13. A. B. Jørgensen, R. Frikke-Schmidt, B. G. Nordestgaard and A. Tybjærg-Hansen, *N. Engl. J. Med.*, 2014, **371**, 32.

14. C. Manrique, G. Lastra and J. R. Sowers, *Ann. N. Y. Acad. Sci.*, 2014, **1311**, 138.
15. B. Arsenault, P. Pibarot and J. Despres, *Cardiol.*, 2009, **113**, 35.
16. American Diabetes Association, *Diabetes Care*, 2014, **37**, S14.
17. S. R. Preis, M. J. Pencina, R. B. D'Agostino, Sr., P. J. Savage and C. S. Fox, *Diabetes Care*, 2013, **36**, 1590.
18. J. A. Ambrose and R. S. Barua, *J. Am. Coll. Cardiol.*, 2004, **43**, 1731.
19. M. Ng, M. K. Freeman, T. D. Fleming, M. Robinson, L. Dwyer-Lindgren, B. Thomson, A. Wollum, E. Sanman, S. Wulf, A. D. Lopez, C. J. L. Murray and E. Gakidou, *J. Am. Med. Assoc.*, 2014, **311**, 183.
20. R. C. Anakwue and A. C. Anakwue, *Cardiovasc. Toxicol.*, 2014, DOI: 10.1007/s12012-013-9242-y.
21. K. Ghoorah, A. DeSoyza and V. Kunadian, *Cardiol. Rev.*, 2013, **21**, 196.
22. A. Neylon, C. Canniffe, S. Anand, C. Kreatsoulas, G. J. Blake, D. Sugrue and C. McGorrian, *Prog. Cardiovasc. Dis.*, 2013, **55**, 574.
23. A. Steptoe and M. Kivimäki, *Annu. Rev. Public Health*, 2013, **34**, 337.
24. B. A. Wentworth, M. B. Stein, L. S. Redwine, Y. Xue, P. R. Taub, P. Clopton, K. R. Nayak and A. S. Maisel, *Cardiol. Rev.*, 2013, **21**, 16.
25. M. E. Rosenfeld, *Curr. Opin. Pharmacol.*, 2013, **13**, 154.
26. K. Paulson, S. Zhu, M. Chen, S. Nurmohamed, J. Jongstra-Bilen and M. I. Cybulsky, *Circ. Res.*, 2010, **106**, 383.
27. B. W. Wong, A. Meredith, D. Lin and B. M. McManus, *Can. J. Cardiol.*, 2012, **28**, 631.
28. The Interleukin-6 Receptor Mendelian Randomisation Analysis Consortium, *Lancet*, 2012, **379**, 1214.
29. V. Pasceri, J. Willerson and E. Yeh, *Circulation*, 2000, **102**, 2165.
30. J. Chen, J. Jin, M. Song, H. Dong, G. Zhao and L. Huang, *Gene*, 2012, **496**, 128.
31. S. Devaraj, U. Singh and I. Jialal, *Am. Assoc. Clin. Chem.*, 2009, **55**, 229.
32. J. Danesh, J. G. Wheeler, G. M. Hirschfield, S. Eda, G. Eiriksdottir, A. Rumley, G. D. O. Lowe, M. B. Pepys and V. Gudnason, *N. Engl. J. Med.*, 2004, **350**, 1387.
33. E. Grad and H. D. Danenberg, *Blood Rev.*, 2013, **27**, 23.
34. R. R. Packard, A. H. Lichtman and P. Libby, *Semin. Immunopathol.*, 2009, **31**, 5.
35. A. V. Sima, *Cell Tissue Res.*, 2009, **335**, 191.
36. K. Bedard and K. Krause, *Physiol. Rev.*, 2007, **87**, 245.
37. A. Daugherty, J. Dunn, D. L. Rateri and J. W. Heinecke, *J. Clin. Invest.*, 1994, **94**, 437.
38. X. Fu, S. Y. Kassim, W. C. Parks and J. W. Heinicke, *J. Biol. Chem.*, 2001, **276**, 41279.
39. L. Stoner, A. A. Lucero, B. R. Palmer, L. M. Jones, J. M. Young and J. Faulkner, *Clin. Biochem.*, 2013, **46**, 1353.
40. O. Radmark and B. Samuelsson, *Prostaglandins Other Lipid Mediators*, 2007, **83**, 162.

41. J. Tardif, P. L. L'Allier, R. Ibrahim, J. C. Grégoire, A. Nozza, M. Cossette, S. Kouz, M. Lavoie, J. Paquin, T. M. Brotz, R. Taub and J. Pressacco, *Circ. Cardiovasc. Imaging*, 2010, **3**, 298.
42. J. Thayer and R. D. Lane, *Biol. Psychol.*, 2007, **74**, 224.
43. V. Papaioannou, I. Pneumatikos and N. Maglaveras, *Front. Physiol.*, 2013, **4**, 1.
44. M. De Jong and D. C. Randall, *J. Cardiovasc. Nursing*, 2005, **20**, 186.
45. H. Sabbah, I. Ilsar, A. Zaretsky, S Rastogi, M. Wang and R. Gupta, *Heart Failure Rev.*, 2011, **16**, 171.
46. C. Chrysohoou, D. B. Panagiotakos, C. Pitsavos, J. Skoumas, M. Economou, L. Papadimitriou and C. Stefanadis, *Atherosclerosis*, 2007, **192**, 169.
47. H. D. Sesso, J. E. Buring, N. Rifai, G. J. Blake, J. M. Gaziano and JP. M. Ridker, *J. Am. Med. Assoc.*, 2003, **290**, 2945.
48. L. Wang, X. Hu, W. Zhang and F. Tian, *Inflammation Res.*, 2013, **62**, 219.
49. F. M. Abboud, S. C. Harwani and M. W. Chapleau, *Hypertension*, 2012, **59**, 755.
50. R. M. Touyz, *Curr. Opin. Nephrol. Hypertens.*, 2005, **14**, 125.
51. T. J. Guzik, N. E. Hoch, K. A. Brown, L. A. McCann, A. Rahman, S. Dikalov, J. Goronzy, C. Weyland and D. G. Harrison, *J. Exp. Med.*, 2007, **204**, 2449.
52. C. Nathan, *Mol. Med.*, 2008, **14**, 485.
53. G. L. Milne, E. S. Musiek and J. D. Morrow, *Biomarkers*, 2005, **10**, 10.
54. J. R. Erickson, L. Pereira, L. Wang, G. Han, A. Ferguson, K. Dao, R. J. Copeland, F. Despa, G. W. Hart, C. M. Ripplinger and D. M. Bers, *Nature*, 2013, **502**, 372.
55. Y. Adkins and D. S. Kelley, *J. Nutr. Biochem.*, 2010, **21**, 781.
56. W. S. Harris, M. Miller, A. P. Tighe, M. H. Davidson and E. J. Schaefer, *Atherosclerosis*, 2008, **197**, 12.
57. N. R. Damasceno, A. Sala-Vila, M. Cofán, A. M. Pérez-Heras, M. Fitó, V. Ruiz-Gutiérrez, M. Martínez-González, D. Corella, F. Aros, R. Estruch and E. Ros, *Atherosclerosis*, 2013, **230**, 347.
58. M. Kitami and M. K. Ali, *Global Heart*, 2012, **7**, 121.
59. J. A. Ambrose and R. S. Barua, *J. Am. Coll. Cardiol.*, 2004, **43**, 1731.
60. R. C. Anakwue and A. C Anakwue, *Cardiovasc. Toxicol.*, 2014, **14**, 193.
61. K. Ghoorah, A. De Soyza and V. Kunadian, *Cardiol. Rev.*, 2013, **21**, 196.
62. J. C. Stewart, K. L. Rand, M. F. Muldoon and T. W. Kamarck, *Brain, Behav., Immun.*, 2009, **23**, 936.
63. D. Pedicino, A. F. Giglio, V. A. Galiffa, P. Cialdella, F. Trotta, F. Graziani and G. Liuzzo, *Int. J. Cardiol.*, 2013, **166**, 572.
64. M. Leinonen and P. Saikku, *Lancet Infect. Dis.*, 2002, **2**, 11.

CHAPTER 9

Gene–Environment Interaction and Oxidative Stress in Cardiovascular Disease

JEFFREY W. STEPHENS* AND SARAH L. PRIOR

Diabetes Research Group, Institute of Life Science, College of Medicine, Swansea University, Swansea, Wales, UK, SA2 8PP
*Email: j.w.stephens@swansea.ac.uk

9.1 Introduction

During the past 2–3 decades numerous biochemical risk factors have been implicated to play a role in cardiovascular disease (CVD). However, our basic knowledge of the underlying pathophysiological mechanisms within vascular biology remains unclear. Although patients with CVD commonly have one or more identifiable risk factor,[1] many cardiovascular events occur in the absence of classical risk factors.[2] For example, in the United States, cholesterol screening fails to identify almost 50% of the 1.3 million individuals who develop a myocardial infarction each year and have no other classical cardiovascular risk factor.[3] Furthermore, no more than 25% of the excess coronary heart disease (CHD) risk associated with diabetes can be accounted for by classical risk factors.[4] This illustrates the complexity of CVD, particularly in high risk states such as diabetes, and may explain the inaccuracy of algorithms such as the Framingham equation and the PROCAM calculation to predict CHD risk in diabetes.[5] Many of the CVD risk factors are associated with increased inflammation and oxidative stress and these are not routinely incorporated into risk algorithms. Furthermore, the

RSC Drug Discovery Series No. 45
Cardiovascular and Metabolic Disease: Scientific Discoveries and New Therapies
Edited by Philip Peplow, James Adams and Tim Young

Published by the Royal Society of Chemistry, www.rsc.org

process of atherosclerosis and CVD involves many different pathophysiological processes and these factors are determined by both genetic and environmental factors and the interaction between them. The available risk prediction tools do not take into account genetic variation (nor gene–environment interaction) which may alter an individual's susceptibility to CVD.[6] Furthermore, there is now considerable evidence that oxidative stress and inflammation play key roles in the initiation and progression of CVD. This chapter will examine the role of oxidative stress in CVD and how gene–environment interaction might contribute to this.

9.2 The Biological Relevance of Oxidative Stress

Free radical-mediated damage (or oxidative stress) contributes to many chronic disorders, in particular CHD, diabetes and cancer.[7] Free radicals are atoms or molecules with one or more unpaired electrons in their atomic structure and as a consequence the molecules are highly reactive. Oxygen is the most ubiquitous of all biological molecules and provides a major source of reactive oxygen species (ROS). It has been estimated that up to 5% of inhaled oxygen becomes an active oxygen species, which is approximately 40×10^{21} molecules per person per day or approximately 25 000 000 000 molecules per cell per day.[8] Oxidative stress results when there is an imbalance between oxidant production and antioxidant defences.[9] ROS include the superoxide anion (O_2^-), hydrogen peroxide (H_2O_2), the hydroxyl radical ($OH^\bullet$) and the peroxynitrite radical ($OONO^\bullet$). As illustrated in Figure 9.1, increased oxidative stress is associated with many of the classical risk factors involved in atherosclerosis including diabetes, hypercholesterolaemia, renal

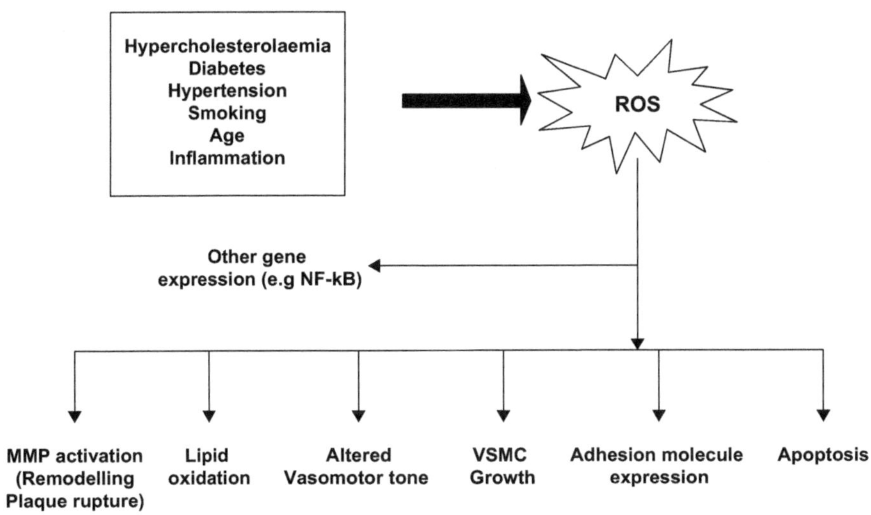

Figure 9.1 Causes & consequences of oxidative stress. ROS: reactive oxygen species; NFκ-B: nuclear factor kappa beta.

failure, ageing, hypertension and smoking.[10] Redox imbalance may result in cell death (apoptosis) following widespread macromolecule oxidation or more subtle changes in a range of signal transduction pathways involved in atherosclerosis.[11] All molecules are potential targets for ROS related damage (proteins, lipids and DNA), but because of their ubiquitous distribution within cell membranes and their propensity to contain double bonds, unsaturated lipids are often targeted.[12] Endogenous antioxidant systems exist to reduce oxidative stress (for review see ref. 9). They include (a) vitamins (A, B1, B2, B6, B12, C, E); (b) enzymes such as mitochondrial superoxide dismutase (SOD), which catalyses the conversion of O_2^- to H_2O_2, which may then be detoxified further in the cytoplasm to form H_2O by the action of glutathione peroxidase (mitochondria), glutathione *s*-transferase (cytoplasm) or catalase (peroxisomes); (c) minerals (copper, zinc, selenium); (d) mitochondrial proteins (uncoupling proteins); and (e) other dietary sources (bioflavanoids, carotenoids). There are a number of factors which determine the bio-availability of these antioxidants, including genetic variation, dietary consumption, basal *in vivo* levels of oxidation and inflammation, and other co-existing chronic disease and associated risk factors.

9.3 Sources of Reactive Oxygen Species

9.3.1 Glucose Dependent Sources of Oxidative Stress

Diabetes mellitus is intimately associated with increased oxidative stress and several studies have demonstrated that hyperglycaemia, *per se*, is associated with increased oxidative stress.[13–16] This is supported by the observation that plasma from diabetes subjects contains increased levels of end-products of oxidative damage, including thiobarbituric acid reactive substances (TBARS), lipid peroxides and urinary F_2-isoprostanes compared to healthy controls.[17] F_2-isoprostanes are the end-product of free radical attack on cell membrane phospholipids, and are often considered to be the 'gold standard' measure of plasma oxidative stress. Further evidence for a direct role of hyperglycaemia in increasing oxidative stress, comes from the observation that with improved glycaemic control, urinary F_2-isoprostane levels decrease.[14] There are several mechanisms by which hyperglycaemia may induce ROS formation.[9,12] Firstly, hyperglycaemia may result in the glucose-mediated non-enzymatic glycosylation of proteins (the *Maillard* reaction) which results in the formation of advanced glycosylation end products (AGEs). These not only increase ROS production but may also initiate a cascade of events, which have harmful effects on the vascular system. Secondly, glucose may undergo auto-oxidation to form a highly reactive enediol radical. This not only increases AGE formation but may also catalyse the conversion of molecular oxygen to O_2^- (hence increasing ROS). Thirdly, excess cellular glucose enters the citric acid cycle, resulting in the generation of excess mitochondrial nicotinamide adenine dinucleotide (NADH). When excessive NADH cannot be dissipated by oxidative phosphorylation the mitochondrial proton gradient increases, and single electrons are transferred to molecular oxygen,[18] resulting in the formation of superoxide and hence ROS.[12] Pancreatic β-cells, neural tissue and endothelial cells may be particularly affected by 'overfeeding' with excess glucose as these cells are not dependent on insulin for glucose uptake, and therefore if 'overfed' will be more prone to oxidative damage. Many studies have suggested that β-cell dysfunction results from prolonged exposure to high glucose, elevated free fatty acids (FFA) or both.[19] β-Cells are particularly sensitive to ROS because they are low in antioxidants such as glutathione peroxidase, catalase and SOD.[20,21] Therefore increased oxidative stress may not only result from hyperglycaemia associated with diabetes, but may also have an important causal role in β-cell failure and the development of insulin resistance, type 2 diabetes, endothelial dysfunction and CVD.

9.3.2 Glucose Independent Sources of Oxidative Stress

As illustrated in Figure 9.2, there are also glucose-independent sources of oxidative stress.[10] These include enzymes such as nicotinamide adenine dinucleotide phosphate (NADPH) oxidase and xanthine oxidase. Both of

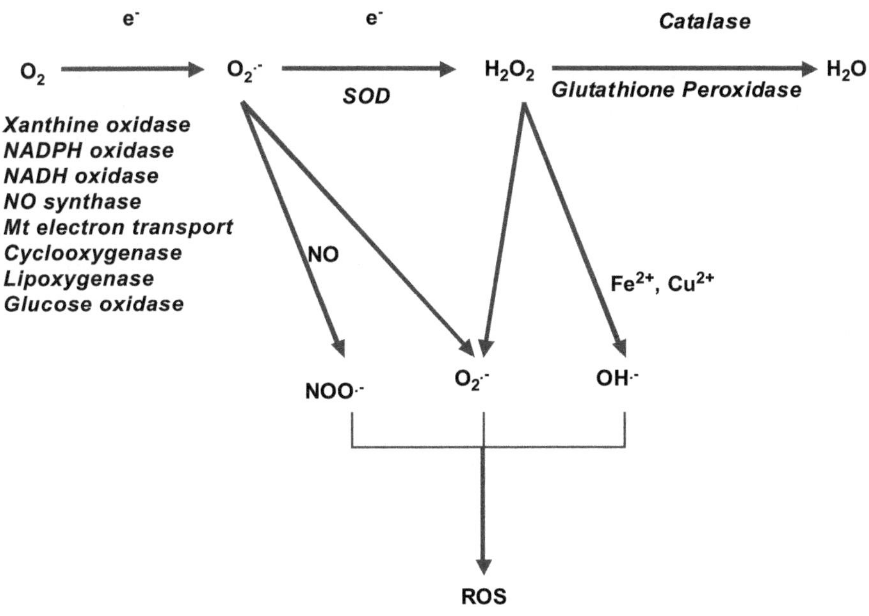

Figure 9.2 Glucose independent sources of ROS. SOD: Mitochondrial superoxide dismutase; O_2^-: superoxide anion; H_2O_2: hydrogen peroxide; OH: hydroxyl radical; OONO: peroxynitrite radical.
Reprinted from J. W. Stephens, S. C. Bain and S. E. Humphries, The biological relevance and measurement of plasma markers of oxidative stress in diabetes and cardiovascular disease, *Atherosclerosis*, 2009, **202**, 321–329, with permission from Elsevier.

these catalyse the conversion of molecular oxygen to O_2^-. Under the influence of mitochondrial SOD, this is converted to hydrogen peroxide. Cytoplasmic catalase or glutathione peroxidase then detoxifies this further to form water. Therefore, if there is increased NADPH activity (or any of the enzymes generating O_2^- as a by-product) or reduced SOD or glutathione activity, ROS generation will be increased. For example, the peroxynitrite ($OONO^\bullet$) radical results from the reaction between O_2^- and nitric oxide (NO). The result would be impaired NO signalling within the vasculature in response to endothelium-dependent vasodilators, which might lead to increased platelet aggregation, expression of cell adhesion molecules,[22] neutrophil migration to endothelial cells and proliferation of smooth muscle cells.[23] Clearly, this will have a major effect in patients with diabetes and CVD.

9.4 Oxidative Stress and Coronary Heart Disease

Several studies have demonstrated that plasma markers of oxidative stress are elevated in CHD or in the presence of its classical risk factors.[10,24,25] As summarised in Figure 9.1, numerous adverse effects on the vascular system are associated with increased oxidative stress. The oxidation of vulnerable cell membrane unsaturated lipids[12] may modulate diverse signal transduction pathways, leading to numerous adverse effects implicated in the pathogenesis of atherosclerosis.[10,11] These include increased expression of cell adhesion molecules, induction of pro-inflammatory pathways, matrix metalloproteinase activation, vascular smooth muscle cell proliferation, endothelial dysfunction and lipid peroxidation (LDL-oxidation). *In vivo* animal studies have shown increased oxidative stress during experimental hypoxia and during cardiac ischaemia.[26] In humans, increased oxidative stress has also been demonstrated during coronary by-pass grafting, post-myocardial infarction and in congestive cardiac failure.[26] A cross-sectional (total antioxidant status, F_2-isoprostane) and prospective (total antioxidant status) association between baseline plasma measures and CHD risk has also been observed, independent of baseline classical risk factors.[27]

9.4.1 Oxidation of LDL and Coronary Heart Disease

Apart from the global effects associated with increased oxidative stress described above, more specific effects also occur. Low density lipoprotein-cholesterol (LDL) is an important target of oxidation, and oxidative modification of LDL is a key step in atherosclerosis.[28] The original interest in oxidised-LDL (Ox-LDL) came from two basic observations. The first was that Ox-LDL was cytotoxic to endothelial and smooth muscle cells.[29] The second was that uptake of native LDL by macrophages occurs at a sufficiently low rate to prevent foam cell formation, but uptake of Ox-LDL was unregulated and led to macrophage foam cell formation and subsequent atherosclerosis.[30,31] Elevated Ox-LDL is independently associated with increased

Table 9.1 Consequences of elevated Ox-LDL. MCP-1: Monocyte chemotactic protein-1; IL-1: Interleunkin-1; ICAMs: Intercellular adhesion molecules; SAA: Serum amyloid A; PPARγ: Peroxisome proliferator-activated receptor-gamma.

Consequence	Ref.
Increased foam cell formation	113, 114
Increased monocytes and T-cells chemotaxis	115, 116
Increased vascular smooth muscle and macrophage production	117, 118
Altered gene expression (MCP-1, IL-1, ICAMs)	119
The induction of proinflammatory genes (PPARγ, Haemoxygenase, SAA, Ceruloplasmin)	120
Increased Immunogenicity (elicits autoantibody formation and activated T-cells)	116
Enhanced procoagulant pathways	121,122
Altered arterial vasomotor properties	123
Increased extracellular matrix synthesis	124
Increased LDL binding affinity	125

atherosclerosis and CHD risk.[32–34] It is now clear that Ox-LDL, with its many oxidatively-modified lipids and degradation products, contributes to the pathophysiology of both the initiation and progression of atherosclerosis (as summarised in Table 9.1). Clearly the contributions described in Table 9.1 are complex and dependent on a number of genetic and environmental contributions. Ox-LDL concentrations are strongly correlated with plasma LDL concentrations, and the latter is thus a key factor in determining absolute plasma Ox-LDL concentration. To overcome this, several researchers in the field have considered the ratio of the Ox-LDL (units per L) to LDL-cholesterol (LDL-C; mmol L^{-1}).[35] This gives an Ox-LDL : LDL-C ratio (units per mmol of LDL-C). A previous study has shown a cross-sectional association between Ox-LDL : LDL-C ratio and CHD risk.[27]

9.5 Genes and the Environment

Health homeostasis depends on maintaining the optimum levels of functional proteins within an organism, which are themselves synthesised and metabolised by enzymes, all of which are encoded by genes. However, cellular functions clearly adapt and develop in relation to the surrounding environment, and failure to adapt results in altered physiological homeostasis which usually has a detrimental effect. Complex diseases such as CVD may arise due to failure at the genetic level (*e.g.* gene transcription) or at the protein level (*i.e.* translation) due to environmental exposure to a toxin (*e.g.* tobacco smoke) or usually due to an interaction between the two. Rarely do these conditions arise from a single mutation in a single important homeostatic gene. Although an example would be single mutations in the gene coding for the LDL-receptor that causes familial hypercholesterolemia which results in greater LDL oxidation and premature CVD.[36] In general, an individual at risk of CVD will have a modest 'genetic' input from minor mutations or variation (functional gene polymorphisms) in several different genes modulating risk together with an environmental insult (*e.g.* cigarette smoking or a high fat diet) which cannot be compensated at the gene or protein level. The importance of the gene–environment interaction is that only when an individual with a high-risk genetic profile enters into a high-risk environment will the effect be so great that disease will develop. In this way, a gene variant may be compared to the analogy of a 'loaded gun' and the environment to a 'firing trigger'. The result is 'Genes load the gun, but the environment pulls the trigger' (Elliot Joslin 1921). There are several examples where gene–environment interaction has been studied both in relation to CVD risk and biochemical intermediate risk phenotypes.[37] Within this chapter, we focus on functional gene polymorphisms associated with changes in plasma oxidative stress in relation to CVD. There has been a paucity of research within the literature in this area, which is likely to be related to the lack of a specific biomarker for oxidative damage that can be measured with relative ease in a large number of samples.[27] Part of this is related to the fact that, by definition, free radicals are highly reactive and are difficult to measure in any biological sample, especially in easily accessible specimens such as serum or plasma.

9.6 Common Gene Variants which may Influence Oxidative Stress

Several candidate gene variants (Table 9.2) relating to oxidative stress have been examined in cross-sectional studies with respect to CVD risk. These include those implicated in the generation of cellular ROS and those with anti-oxidant properties. However, association with intermediate biochemical risk factors and interaction with pro-oxidant environments (*e.g.* cigarette smoking, obesity) are limited within the available literature. This chapter will focus on the association between plasma markers of oxidative stress and LDL-oxidation with variants from three specific candidate genes implicated in oxidative stress. These candidate genes encode proteins that have anti-oxidant effects in three different surroundings: apolipoprotein E (a plasma lipoprotein), mitochondrial uncoupling protein 2 (responsible for uncoupling the electrochemical proton gradient generated across the mitochondrial membrane by the electron transport chain) and glutathione-*s*-transferase (a cellular antioxidant protein).

9.6.1 Variation in the Apolipoprotein E (*APOE*) Gene

Apolipoprotein E (apoE) is synthesised in the liver and intestine, and is found in association with triglyceride-rich lipoproteins,[38,39] facilitating the removal of these particles from the circulation. This gene and protein has been comprehensively studied in relation to plasma lipid levels and CHD risk.[37] The human *APOE* gene is located on chromosome 19, and genetic

Table 9.2 Other gene variants studied in relation to plasma markers of oxidative stress. RAGE: Receptor for advanced glycation end products; PON1: Paraoxonase 1; AGTR1: Angiotensin II AT1 receptor gene; eNOS: Endothelial nitric oxide synthase; TBARS: Thiobarbituric acid-reactive species; ApoA-IV: Apolipoprotein A-IV; TAOS: Total antioxidant status; IL-6: Interleukin-6.

Gene	Variant(s)	Genotype associated with higher levels	Sample size	Plasma marker studied	Interaction examined	Ref.
RAGE	1704G>T 2184A>G	Haplotype of T+ & G+	371	α-tocopherol, caratenoids	No	126
PON1	Q192R	QR/RR	22	*Ex vivo* LDL oxidation lag time	Tomato juice*	127
PON1	Q192R	QQ	244	PON1 activity	Smoking	128
AGTR1	A153G	AA	100	Urinary F_2-isoprostane	No	129
AGTR1	A1166C	CC	239	Protein carbonyls/ Myeloperoxides	Heart failure*	130
apoA-IV	T347S	SS	705	TAOS	CVD	131
IL-6	−174G>C	GC/CC	552	TAOS	Urinary protein excretion	132
Catalase	rs7947841	A allele	1746	Catalase activity	None	133
UCP2	−866G>A	A allele	1190	Telomere length	None	134

*No interaction checked statistically.

variation in exon 4 results in three common gene variants ε2, ε3 and ε4 which have strong and consistent influences on plasma lipids,[40] and CHD risk.[41] Plasma apoE levels differ by genotype such that ε4ε4 subjects have the lowest levels and ε2ε2 the highest.[42] ε3 is the most common isoform with a frequency of 0.77 in Caucasian populations, while ε4 and ε2 are seen at frequencies of 0.15 and 0.08 respectively.[37]

9.6.1.1 *Association with Lipids and CHD Risk*

Carriers of the ε2 allele have plasma total cholesterol levels approximately 10% lower than ε3 homozygotes, whilst ε4 carriers have cholesterol levels approximately 5% higher than ε3 homozygotes. This is almost entirely due to the differential metabolism of apoB-containing lipoproteins, with parallel effects on fasting plasma LDL levels. Compared to men homozygous for the ε3 allele, those carrying ε2 (ε2+) appear to be protected from both CHD and stroke,[43] whereas those with the ε4 allele (ε4+) have a relatively higher risk.[44] With respect to CHD, two meta-analyses have demonstrated increased risk in ε4+,[41,45] with ε4+ subjects having a 42% higher CHD risk compared to ε3ε3 and ε2+ subjects. The role of the gene–environment interaction is important as the risk associated with ε4+ is greater in smokers, and with a modest or non-significant increase in risk in ε4+ non-smokers. In the 10 year prospective UK-based NPHSII study, compared to the risk in the combined group of never smokers (irrespective of genotype), smoking was observed to increase risk of CHD significantly in ε4+ smokers (odds ratio [OR]: 3.17), with much more modest effect in ε3ε3 subjects (OR: 1.38), and in the ε2+ smokers (OR: 1.18). Interestingly, while ε3ε3 ex-smokers still had a high risk (OR: 1.75), the risk in ε4+ ex-smokers was lower (OR: 0.84), suggesting a particular benefit in this group.

This effect was confirmed by Talmud *et al.*,[46] who re-examined data from 1668 men (316 with CVD events) from the Framingham Offspring Study focusing on *APOE*:smoking interactions. Overall the hazard ratio (HR) for CVD events in smokers was 1.95 (95% confidence interval: 1.52–2.50) compared to non-smokers. When the ε3ε3 subjects were used as the reference group, in non-smokers there was no evidence for significant CVD risk associated with ε2+ (HR:1.04 [95% CI: 0.61–1.76]) or ε4+ (HR:1.04 [95% CI: 0.70–1.54]), but in smokers, HRs were 1.96 (95% CI: 1.26–2.78) in ε3ε3 men, 3.46 (95% CI: 2.14–5.60; $P=0.09$ for interaction) in ε2+ and 3.81 in ε4+ (95%CI: 2.49–5.84; $P=0.01$ for interaction), with a significant interaction between daily cigarette consumption and *APOE* genotype on risk ($P=0.03$). While the greater ε2+ smoking risk seen in Framingham was not seen in NPHSII, the greater than multiplicative effect on risk in ε4+ smokers seen in Framingham and NPHSII middle aged men, strongly suggests this is not a chance finding. Importantly, the risk associated with the *APOE* genotype in NPHSII and Framingham was independent of classical risk factors including lipid levels,[47] suggesting that these genotype effects on risk must be acting through a novel mechanism. Another more recent study which looked at non-smokers and smokers has also observed a protective effect in ε2+ smokers (1.45 [95% CI: 1.00–2.10]) compared to ε3ε3 (2.25 [95% CI: 1.90–2.68]) and ε4+ subjects (2.37 [95% CI: 1.85–3.04]).[48]

9.6.1.2 Association with Biochemical Markers of Oxidative Stress

The different isoforms of apoE have different antioxidant activity and studies have shown differential effects on susceptibility of apoE containing lipoproteins to oxidation in plasma. *In vitro*, the antioxidant efficacy of apoE lipoproteins in relation to hydrogen peroxide cytotoxicity is apoE2 > apoE3 > apoE4.[49,50] Furthermore, serum lipid peroxides are higher in ε4+ subjects,[51] and in subjects with Alzheimer's disease products of lipid peroxidation in plasma and urine are lower in ε2+ compared to ε4+ subjects.[52] Plasma lipoproteins from apoE-deficient mice are more prone to *in vitro* oxidation than lipoproteins from wild-type mice.[53] This effect may also be reversed with supplementation with vitamin E.[54] The antioxidant effect associated with the ε2 allele is likely to be due to the fact that apoE2 has two free –SH groups, apoE3 one –SH group and apoE4 none.

The potential anti-oxidant mechanism for this *APOE* ε4:smoking interaction was examined in a group of 728 Caucasian patients with diabetes (UDACS),[46] where markers of plasma oxidative stress were measured including Ox-LDL and total antioxidant status (TAOS),[55] where a lower plasma TAOS is a result of increased oxidative burden. There was no association between *APOE* genotype and plasma Ox-LDL or TAOS in non-smokers. However as illustrated in Figure 9.3, in smokers, ε4+ had 26% higher plasma Ox-LDL than other genotypes (*APOE*:smoking interaction $P = 0.04$), while ε2+ had 18% higher plasma TAOS than ε3ε3 and ε4+ combined (*APOE*:smoking interaction $P = 0.026$). Although direct extrapolation must be made with caution, these results suggest that the increased CVD risk of ε4+ on oxidative stress will be confined to smokers, and a feasible mechanism is presented by the reduced antioxidant capacity and increased Ox-LDL of apoE4.

9.6.2 Variation in the Mitochondrial Uncoupling Protein 2 (*UCP2*) Gene

The mitochondrial electron transport chain (ETC) is associated with substantial ROS generation during the process of aerobic metabolism.[56,57] Uncoupling proteins (UCPs 1–3)[58] dissipate the inner mitochondrial membrane proton electrochemical gradient that drives ATP synthesis.[59] UCP1 expression is restricted to brown adipose tissue,[60] whilst UCP3 is predominantly expressed in skeletal muscle.[61] UCP2 is ubiquitous and is likely to be the ancestral UCP sharing 59% and 73% sequence homology with UCP1 and UCP3 respectively.[62–64] The ubiquity and electrochemical actions

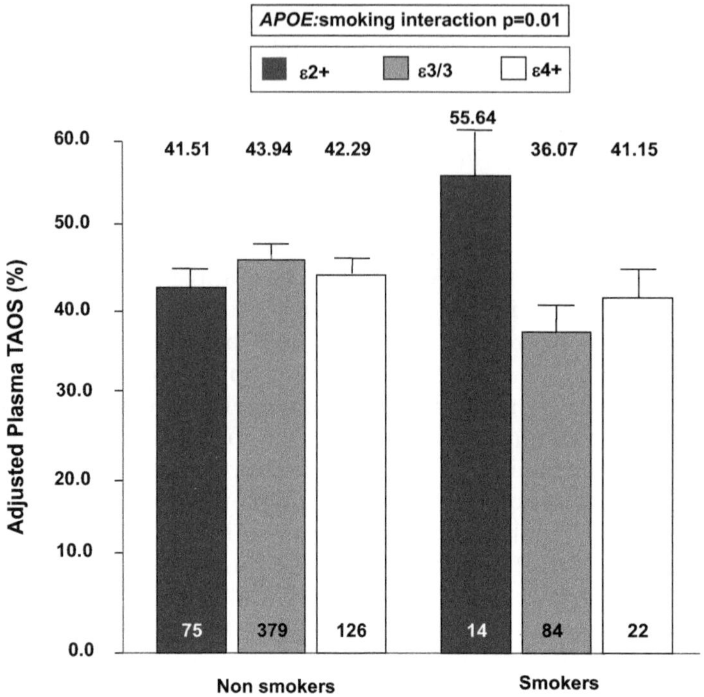
APOE:smoking interaction p=0.01
ε2+
ε3/3
ε4+
Adjusted Plasma TAOS (%)
60.0
50.0
40.0
30.0
20.0
10.0
0.0
41.51
43.94
42.29
55.64
36.07
41.15
75
379
126
14
84
22
Non smokers
Smokers

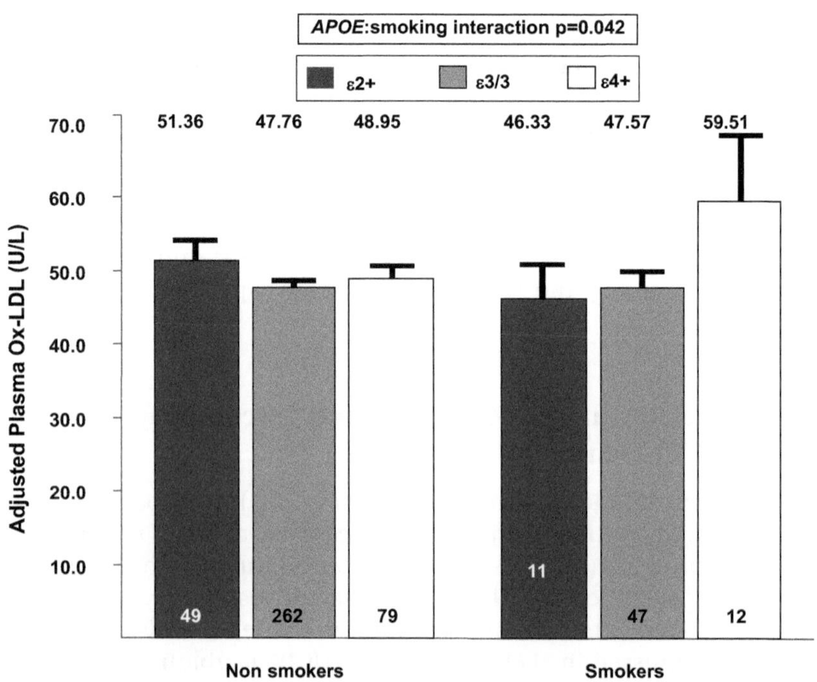
APOE:smoking interaction p=0.042
ε2+
ε3/3
ε4+
Adjusted Plasma Ox-LDL (U/L)
70.0
60.0
50.0
40.0
30.0
20.0
10.0
51.36
47.76
48.95
46.33
47.57
59.51
49
262
79
11
47
12
Non smokers
Smokers

of UCP2 make it a plausible negative regulator of ROS production.[65,66] Uncoupling leads to a higher flow rate though the electron chain and a reduced half-life of the radical intermediates. This leads to a reduction in the formation of ROS, at the cost of wasting energy as heat. Therefore, it is plausible that UCP2 has a role in protecting the cell from mitochondrial generated oxidative stress. Interestingly, UCP2 activity itself is induced by by-products of lipid peroxidation[67] and intra-mitochondrial superoxide,[68] which may serve to protect against further ROS generation. Excess nutrient intake (glucose, FFA) lead to increased ROS production *via* the electron transport chain, and of note UCP2 is over-expressed in animal models and humans under these conditions.[69,70] The role and mode of action of UCP2 is currently under debate.[71] Furthermore, UCP2 has recently been proposed to play a major role in mitochondrial calcium transport and downstream signaling events.[72]

The human *UCP2* gene is located on chromosome 11 and consists of 8 exons.[73] The transcription site is preceded by a region containing a strong *cis*-acting positive regulatory element (−141 to − 65) which may underlie the ubiquitous expression of *UCP2*. Several consensus sequences exist for transcription control elements, such as C/EBP-β (important for interleukin-6 (IL-6) gene expression), CREB-1 (cAMP response binding protein 1), 2 peroxisome proliferator-activated receptors-gamma (PPARγ) responsive elements, 2 thyroid hormone response elements (TREs) and nuclear factor kappa beta (NFκ-B). Four common gene variants have been described in the *UCP2* gene. The − 866G > A promoter variant is in 100% linkage disequilibrium with another − 2723T > A promoter variant. The − 866G > A variant is also at the junction between a negative and positive *cis*-acting DNA regions, and within a region that contains binding sites for hypoxia, aromatic hydrocarbons and inflammatory cytokines.[74] Thus, based on its location, the − 866G > A change is likely to influence *UCP2* transcription, and there is good *in vitro* data to support this.[74,75] In exon 8, a 3′UTR 45 base pair deletion/insertion (D/I) variant exist, and is in moderate linkage disequilibrium (75%) with the − 866G > A variant.[74,76] In exon 4, a +164C > T variant results

Figure 9.3 Association of *APOE* genotype in non-smokers and smokers with biochemical markers of oxidative stress in UDACS. (A) Plasma total antioxidant status (TAOS). TAOS was adjusted for triglyceride, HDL and glucose. (B) Oxidised-LDL (Ox-LDL). Ox-LDL was adjusted for age, LDL and HDL. Mean and standard error are shown. Numbers of subjects are shown at the base of each column.
Reprinted from (i) J. W. Stephens, S. C. Bain and S. E. Humphries, Gene–environment interaction and oxidative stress in cardiovascular disease, *Atherosclerosis*, 2008, **200**, 229–238, with permission from Elsevier. (ii) P. J. Talmud, J. W. Stephens, E. Hawe, S. Demissie, L. A. Cupples, S. J. Hurel, S. E. Humphries and J. M. Ordovas, The significant increase in cardiovascular disease risk in APOEepsilon4 carriers is evident only in men who smoke: potential relationship between reduced antioxidant status and ApoE4, *Ann. Hum. Genet.*, 2005, **69**, 13–22, with permission of John Wiley and Sons.

in an amino acid change of alanine to valine (Ala55Val). This is in strong linkage disequilibrium with the – 866G>A variant (97%) and with the D/I variant (97%). The – 866G>A variant has been reported to explain 71% of the variation in mRNA transcript ratio of the exon 8 D/I variant.[74] Although all 4 sites may themselves be directly functional, the – 866G>A is the most likely candidate, with reported effects associated with the other single nucleotide polymorphisms being due to linkage disequilibrium with this site.

9.6.2.1 *Previous Associations with UCP2 Gene Variants*

Considerable interest has developed in the role that the *UCP2* – 866G>A promoter variant may play in determining obesity and insulin sensitivity.[74–79] These studies have shown an association between the – 866G allele with higher BMI and also relative protection from type 2 diabetes.[74–77,80] Furthermore, the – 866A allele is associated with β-cell dysfunction.[75] O'Rahilly has previously described this association as 'UCP2 being an adiposity angel and diabetes devil'.[78] The mechanism for this can be understood in terms of mitochondrial physiology. In adipose tissue, the A allele relative to the G allele, is associated with higher UCP2 expression.[74] Therefore, in relation to obesity, the G allele would be associated with less uncoupling of the proton gradient across the mitochondrial membrane. This would hypothetically results in increased Adenosine Triphosphate (ATP) synthesis and storage. Conversely with the A allele, increased UCP2 expression would increase uncoupling of the proton gradient, leading to reduced ATP synthesis and storage and increased energy loss as heat. With respect to pancreatic β-cell dysfunction in rat cells, insulin secretion is an active process dependent on ATP.[75] Therefore, increased UCP2 expression in these cells would result in increased uncoupling of the proton gradient and reduced ATP synthesis and subsequent insulin secretion. For these reasons it is understandable why much of the previous work relating to variation in the *UCP2* gene has been related to energy expenditure, obesity and insulin secretion. However, if this gene has such a 'major' effect on the mitochondrial proton gradient and hence in allowing the neutralisation of superoxide, a valid hypothesis would be that this variant may alter cellular ROS production and hence play an important antioxidant role and protect against CHD. Reports have been published which have shown an association between the – 866A allele with poor outcome post myocardial infarction in subjects with diabetes,[81] and with prospective CHD risk in healthy males.[82]

9.6.2.2 *Evidence for the Role of UCP2 in Modulating Oxidative Stress*

The *UCP2* knockout mouse is able to generate more ROS during infection with *T. gonadii,* an infection normally fatal in wild type mice.[83] The knockout mouse has higher levels of ROS and a greater burden of arterial disease. The *UCP2* knockout mouse is also more susceptible to pathogen activation-induced apoptosis and high levels of ROS.[84] Decreased *UCP2* expression (through endothelial anti-sense strategies,[64] or macrophage gene-deletion)[83] increases ROS generation, and bone marrow transplant from *UCP2* knockout donor mice increases markers of oxidative stress and lesion size in athero-sclerotic-prone mice.[85]

9.6.2.3 Gene–Environment Interaction with UCP2 Gene Variants

Previously,[82] amongst 465 men with diabetes, an association of the *UCP2* – 866A allele with significantly lower plasma TAOS in those without CHD has been observed. Interestingly, in those subjects with CHD (associated with a higher oxidative burden), the association was more pronounced (TAOS 30.1 ± 16.1% *vs.* 41.6 ± 12.4% for AA *vs.* GG; $P = 0.016$). Therefore the G-allele appeared to be associated with better antioxidant defense in a pro-oxidant environment. In a sample of 20 men with diabetes selected for homozygosity for the *UCP2* – 866G>A variant, matched for baseline characteristics, plasma markers of oxidative stress (TAOS and esterified F_2-isoprostane) in those with CHD were significantly different in AA men (TAOS 31.7 ± 7.3% *vs.* 52.6 ± 6.3%; $P = 0.001$ and F_2-isoprostanes 220.6 ± 37.2 pg mL^{-1} *vs.* 109.9 ± 51.1 pg mL^{-1}; $P = 0.005$ for AA *vs.* GG). A further study examined the association between this gene variant with prospective CHD risk amongst 2695 healthy men (from the NPHSII study) over a median period of 10.2 years. AA homozygosity was associated with a doubling in CHD risk after adjustment for established risk factors (HR 1.99 [1.37–2.90]; $P = 0.002$) and the risk associated with this genotype was substantially increased by the presence of other risk factors (obesity, hypertension and diabetes).

Another study among 453 Caucasian men with diabetes observed a significant interaction ($P = 0.001$) between genotype and smoking in determining plasma TAOS.[86] As shown in Figure 9.4, current smokers with the – 866AA genotype had the lowest TAOS (indicating higher oxidative stress) of all subjects. In a sub-sample of 20 subjects (10 GG, 10 AA) matched for baseline characteristics, plasma markers of oxidative stress in current smokers were significantly higher in AA compared to GG subjects (TAOS 36.8 ± 9.5% *vs.* 51.4 ± 9.5%, $P = 0.04$; F_2-isoprostanes 1133.6 ± 701.2 pg mL^{-1} *vs.* 500.8 ± 64.7 pg mL^{-1}, $P = 0.04$).

9.6.3 Variation in the Glutathione *s*-Transferase (*GSTT1*) Gene

Glutathione *s*-transferases (GSTs) are important cellular antioxidant defenses as they catalyse the reduction of potentially harmful peroxides.[87] GSTs are a family of detoxification enzymes responsible for the metabolism

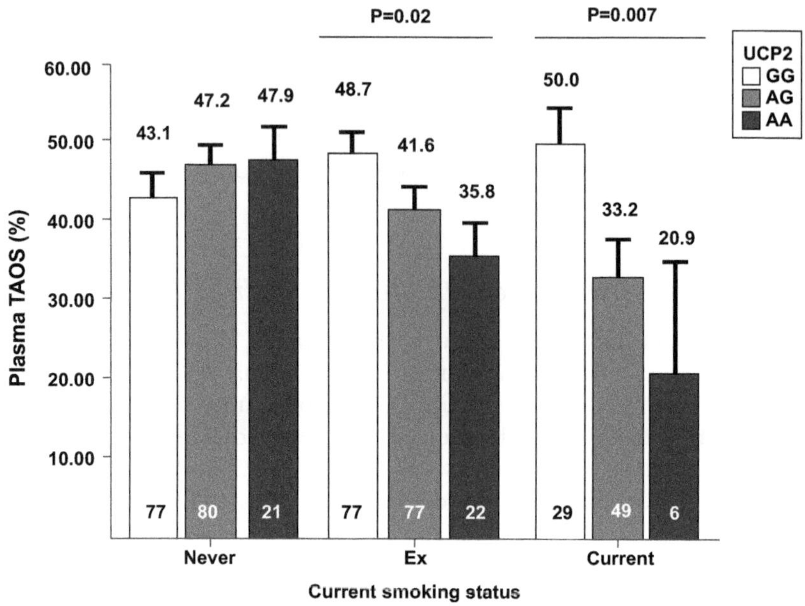

Figure 9.4 Plasma TAOS in relation to the UCP2 −866G>A gene variant and smoking status in UDACS. Mean and standard error are shown. Plasma TAOS adjusted for correlates for plasma TAOS (HDL, and negatively with triglyceride, glucose and HbA1c). Numbers of subjects are shown at the base of each column. In ex-smokers and current smokers there was a linear association between genotype and plasma TAOS, $P=0.02$ and $P=0.007$ respectively. There was a significant interaction between genotype and smoking status in determining plasma TAOS ($P<0.001$).
Reprinted from (i) J. W. Stephens, S. C. Bain and S. E. Humphries, Gene–environment interaction and oxidative stress in cardiovascular disease, *Atherosclerosis*, 2008, **200**, 229–238, with permission from Elsevier. (ii) J. W. Stephens, S. S. Dhamrait, A. R. Mani, J. Acharya, K. Moore, S. J. Hurel and S. E. Humphries, Interaction between the uncoupling protein 2 −866G>A gene variant and cigarette smoking to increase oxidative stress in subjects with diabetes, *Nutr. Metab. Cardiovasc. Dis.*, 2008, **18**, 7–14, with permission from Elsevier.

of many xenobiotics and carcinogens.[88,89] These enzymes catalyse the interaction of glutathione with a wide variety of endogenous and exogenous electrophilic compounds including potential atherogenic substances, such as lipid peroxides, to form thioesters (Figure 9.5).[90] During this process, reduced glutathione is oxidised and subsequently recycled back to reduced glutathione by the action of NADPH-dependent glutathione reductase. As seen in Figure 9.5, the anti-oxidant role of glutathione peroxidase is also dependent on the oxidation of glutathione. Therefore, these two enzymes compete for reduced glutathione availability to exert their anti-oxidant role. Thus, if the anti-oxidant activity of GST is increased, it might be that glutathione peroxidase would be less efficient in detoxification (alternatively the reverse may apply).

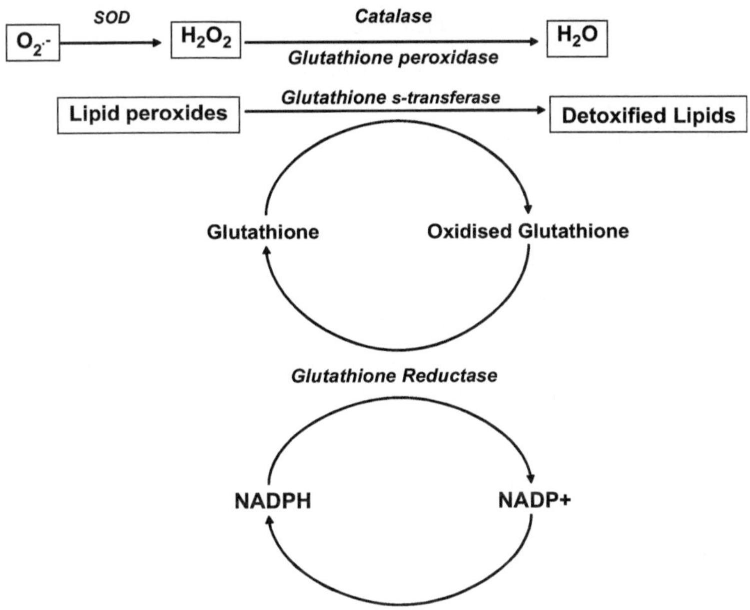

Figure 9.5 The role of glutathione *s*-transferase. NADP: Nicotinamide adenine dinucleotide phosphate.
Reprinted from J. W. Stephens, S. C. Bain and S. E. Humphries, Gene–environment interaction and oxidative stress in cardiovascular disease, *Atherosclerosis*, 2008, **200**, 229–238, with permission from Elsevier.

GST expression is induced by molecules known to result in chemical stress and carcinogenesis such as aromatic compounds. *GST* genes are highly polymorphic. Previous work has focused on the *GSTM1* and *GSTT1* deletion variants. *GSTM1* (chromosome 1p13.3) and *GSTT1* (chromosome 22q11.2) encode the cytosolic enzymes GST-μ and GST-θ respectively.[91] Approximately 50% of the Caucasian population are homozygous for the deletion of *GSTM1* and 20% homozygous for the null variant gene for *GSTT1*.[90] In both variants homozygosity for the deletion allele results in deletion of the entire gene.[91] Polymorphic deletion variants in the *GSTM1* and *GSTT1* genes therefore produce either a functional protein (non-deletion alleles or heterozygous deletion, *GSTM1-1* and *GSTT1-1*) or result in the complete absence of the protein (homozygous deletion alleles, *GSTM1-0* and *GSTT1-0*).[91] Considerable interest has focused on the association of these gene variants in relation to tobacco related cancers,[90] since *GSTM1* and *GSTT1* enzymes detoxify not only products of oxidative stress but also carcinogenic compounds such as polycyclic aromatic hydrocarbons, a major constituent of tobacco smoke.[90] Several epidemiological studies have suggested that *GSTM1-0* and *GSTT1-0* status are associated with increased risk of smoking related cancers including lung, bladder, ovarian and colorectal cancers.[92–96]

9.6.3.1 Association with CHD Risk

In contrast to the extensive investigation of the *GSTM1* and *GSTT1* gene variants in relation to malignancy, their role in CVD risk and the pathophysiology of atherosclerosis, particularly in relation to exposure to cigarette smoke, remains unclear.[97–102] Recent studies have suggested that variation in the *GSTM1* or *GSTT1* genes may alter the susceptibility to atherosclerosis.[97–102] Previous association studies looking at these gene variants have focused on the interaction with cigarette smoking, therefore in the setting of a high ROS environment. Previously, studies in smokers have shown that subjects with the *GSTM1-0* compared to *GSTM1-1* variant had a higher risk for CHD, but not of lower extremity arterial disease.[97,98] In another study focusing on smokers and using carotid intima media thickness as an outcome, *GSTM1-0* compared to *GSTM1-1* had increased carotid atherosclerosis over a 2 year period.[100] Another study has also observed increased CHD risk in smokers with the GSTM1-0 compared to non-smokers with this genotype (OR: 3.54, $P = 0.079$).[103] Contradictory to these findings, two further studies focusing on a sample of UK Caucasian smokers and secondly a UK sample of subjects with a South Asian origin, showed a reduced risk of acute myocardial infarction in *GSTM1-0* subjects.[99,104] The literature is also unclear with respect to role of the *GSTT1* variant in relation to CVD risk. The Atherosclerosis Risk In Communities (ARIC) study showed *GSTT1-1* smokers to be at an increased risk of CHD, lower extremity arterial disease and carotid artery atherosclerosis.[97,98]

9.6.3.2 Association with Biochemical Markers of Oxidative Stress

Surprisingly, despite the interest between the *GSTT1* and *GSTM1* variants and CVD risk, within the literature there are limited reports looking at the association between these genotypes and a plasma marker of inflammation and oxidative stress.[105] One study examined 90 subjects with severe acute pancreatitis (AP) and 253 healthy controls. The *GSTT-1* was more prevalent in those with AP compared to controls (OR: 6.6 [95%CI: 2.3–18.7]) and was associated with increased plasma C-reactive protein (CRP). Although the levels of reduced glutathione were higher in those with severe AP, no *GST* genotype effect on the levels was observed.

9.6.3.3 Gene–Environment Interaction with GSTT1 Gene Variants

Previously, interactions between the *GSTT1* variant and cigarette smoking in relation to intermediate plasma markers of oxidative stress have been observed.[106] In a sample of subjects with diabetes (UDACS), *GSTT1-1* compared to *GSTT1-0* subjects had significantly higher CRP ($P = 0.001$), Ox-LDL ($P = 0.004$) and smaller LDL particles ($P = 0.01$). The association with Ox-LDL is illustrated in Figure 9.6. In subjects without CVD, there was a significant

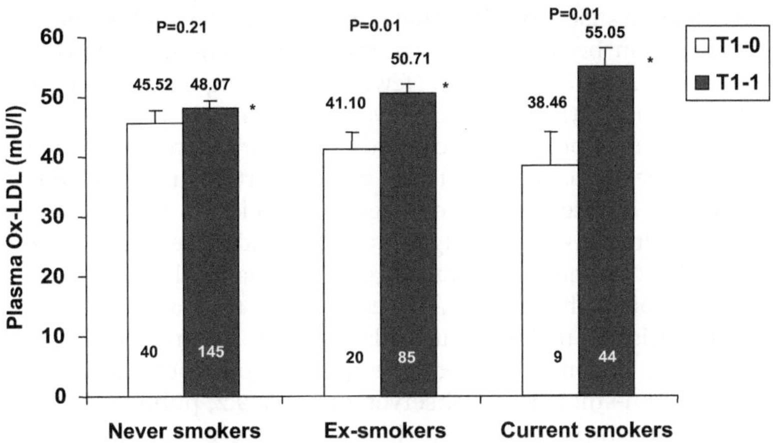

Figure 9.6 Plasma Ox-LDL by GSTT1 genotype & smoking status in subjects without CVD from UDACS. Mean (± standard error) shown. Numbers of subjects are shown at the base of each column. The interaction between genotype & smoking status was significant, $p = 0.042$. After adjustment the interaction remained significant ($p = 0.02$). *$P = 0.04$, linear association for Ox-LDL between current, ex and never smokers of the *GSTT1-1* genotype. Reprinted from (i) J. W. Stephens, S. C. Bain and S. E. Humphries, Gene–environment interaction and oxidative stress in cardiovascular disease, *Atherosclerosis*, 2008, **200**, 229–238, with permission from Elsevier. (ii) T. Hayek, J. W. Stephens, C. S. Hubbart, J. Acharya, M. J. Caslake, E. Hawe, G. J. Miller, S. J. Hurel and S. E. Humphries, A common variant in the glutathione *S* transferase gene is associated with elevated markers of inflammation and lipid peroxidation in subjects with diabetes mellitus, *Atherosclerosis*, 2006, **184**, 404–12, with permission from Elsevier.

interaction between the *GSTT1-1* variant and smoking in determining Ox-LDL ($P = 0.04$). Furthermore, CVD risk was higher in smokers compared to non-smokers with *GSTT1-1*. No significant associations were observed by *GSTM1*. Few other studies have examined this within the literature to date.

9.7 Conclusions and Other Gene Variants Studied in Relation to Plasma Markers of Oxidative Stress

Within this chapter, we have focused on three genetic examples which may influence oxidative stress in three different cellular components. Clearly, intracellular and extracellular mechanisms involved in ROS generation and antioxidant defenses are complex, with numerous genes encoding proteins involved in atherosclerosis and CHD. Furthermore, there are numerous ingested environmental toxins and antioxidants which may influence gene transcription, translation and post-translational modification of proteins. This is further complicated by possible interactions between different antioxidant mechanisms (*e.g.* GST, UCP2) and how the environment may influence disease susceptibility. There are relatively few publications looking at the association between gene variants and biochemical markers of oxidative stress *in vivo* and even fewer examining the effects of gene–environment interaction on oxidative stress. These are summarised in Table 9.2. This may be related to the difficulty and expense of measuring a surrogate marker of oxidative burden *in vivo*, particularly in a large number of samples, as is often required in gene association studies. However, a previous quantitative study[107] suggests that the additive effects of genes may explain 50% of the phenotypic variance in plasma total antioxidant capacity by metmyoglobin. This study also comments that as much as 83% of phenotypic variance in total antioxidant status in smokers is due to the additive effects of genes, whereas the genetic contribution is significantly less (49%) in non-smokers. As observed in Table 9.2, published studies have relatively small sample numbers, which raises the possibility that these associations may be chance observations. Also the effects of these gene variants appear to be less than those discussed within this review. Furthermore, to our knowledge the first four listed variants have not been associated with prospective CVD risk.

All of the gene variants listed may be mechanistically associated with increased oxidative stress. With respect to the receptor for AGEs (RAGE), the formation of AGEs and oxidative stress is intimately linked (glycoxidation). Since RAGE is present on endothelial cells, mononuclear phagocytes, smooth muscle cells, mesangial cells and neurons, it is understandable why it may be considered in the aetiology of CVD and diabetes complications.[108] Paraoxonase (PON) is an enzyme bound to high density lipoprotein-cholesterol (HDL), which has the ability to reduce LDL oxidation, and therefore may reduce atherosclerosis and CVD risk.[109] The angiotensin type-1 receptor (AT1R) provides a site for the action of angiotensin II. This may result in the generation of ROS by activating NAD(P)H oxidases, a family of membrane-associated enzyme complexes expressed in many cell types implicated in atherosclerosis including cardiomyocytes, vascular endothelial cells, and smooth muscle cells.[110] Apolipoprotein A-IV (apoA-IV) is a glycoprotein that circulates freely or in association with chylomicrons and HDL. The presence of the *APOA-IV* transgene in apoE-deficient mice (*APOA4/apoe*$^{-/-}$) supports the anti-atherogenic role of apoA-IV and, furthermore, these mice have reduced oxidative markers, including aldehyde-modified LDL, suggesting an antioxidant effect of apoA-IV.[111] Increased IL-6 is associated with numerous adverse effects on the cardiovascular system, including increased synthesis of ROS *via* NAD(P)H oxidases,[10] and hence increased oxidative stress. Previous studies have shown the −174C allele to be associated with prospective CHD risk and hypertension.[112]

Therefore, it appears that the literature is lacking in studies looking at the associations between variants within candidate genes and intermediate phenotypic markers of oxidative stress. Furthermore, the effect of gene–environment interaction and translation to prospective CHD risk needs further investigation.

References

1. P. W. Wilson, *Am. J. Hypertens.*, 1994, 7, 7S.
2. L. G. Futterman and L. Lemberg, *Am. J. Crit. Care*, 1998, 7, 240.
3. N. Rifai and P. M. Ridker, *Clin. Chem.*, 2001, **47**, 403.
4. K. Pyorala, M. Laakso and M. Uusitupa, *Diabetes/Metab. Rev.*, 1987, **3**, 463.
5. J. W. Stephens, G. Ambler, P. Vallance, D. J. Betteridge, S. E. Humphries and S. J. Hurel, *Eur. J. Cardiovasc. Prev. Rehabil.*, 2004, **11**, 521.
6. S. E. Humphries, J. A. Cooper, P. J. Talmud and G. J. Miller, *Clin. Chem.*, 2007, **53**, 8.
7. J. W. Baynes, *Diabetes*, 1991, **40**, 405.
8. S. L. Prior, J. Javid, G. V. Gill, S. C. Bain and J. W. Stephens, *Kidney Int.*, 2008, **74**, 1361.
9. A. C. Maritim, R. A. Sanders and J. B. Watkins, 3rd, *J. Biochem. Mol. Toxicol.*, 2003, **17**, 24.
10. D. Harrison, K. K. Griendling, U. Landmesser, B. Hornig and H. Drexler, *Am. J. Cardiol.*, 2003, **91**, 7A.
11. Y. J. Suzuki, H. J. Forman and A. Sevanian, *Free Radical Biol. Med.*, 1997, **22**, 269.
12. J. L. Evans, I. D. Goldfine, B. A. Maddux and G. M. Grodsky, *Endocr. Rev.*, 2002, **23**, 599.
13. P. Odetti, S. Garibaldi, G. Noberasco, I. Aragno, S. Valentini, N. Traverso and U. M. Marinari, *Acta Diabetol.*, 1999, **36**, 179.
14. G. Davi, G. Ciabattoni, A. Consoli, A. Mezzetti, A. Falco, S. Santarone, E. Pennese, E. Vitacolonna, T. Bucciarelli, F. Costantini, F. Capani and C. Patrono, *Circulation*, 1999, **99**, 224.
15. A. Ceriello, L. Quagliaro, B. Catone, R. Pascon, M. Piazzola, B. Bais, G. Marra, L. Tonutti, C. Taboga and E. Motz, *Diabetes Care*, 2002, **25**, 1439.
16. A. Telci, U. Cakatay, R. Kayali, C. Erdogan, Y. Orhan, A. Sivas and T. Akcay, *Horm. Metab. Res.*, 2000, **32**, 40.
17. J. Nourooz-Zadeh, J. Tajaddini-Sarmadi, S. McCarthy, D. J. Betteridge and S. P. Wolff, *Diabetes*, 1995, **44**, 1054.
18. P. Maechler, L. Jornot and C. B. Wollheim, *J. Biol. Chem.*, 1999, **274**, 27905.
19. J. L. Evans, I. D. Goldfine, B. A. Maddux and G. M. Grodsky, *Diabetes*, 2003, **52**, 1.
20. M. Tiedge, S. Lortz, J. Drinkgern and S. Lenzen, *Diabetes*, 1997, **46**, 1733.
21. R. P. Robertson, J. Harmon, P. O. Tran, Y. Tanaka and H. Takahashi, *Diabetes*, 2003, **52**, 581.
22. M. W. Radomski, R. M. Palmer and S. Moncada, *Br. J. Pharmacol.*, 1987, **92**, 639.
23. U. C. Garg and A. Hassid, *J. Clin. Invest.*, 1989, **83**, 1774.
24. H. Cai and D. G. Harrison, *Circ. Res.*, 2000, **87**, 840.

25. G. M. Chisolm and D. Steinberg, *Free Radical Biol. Med.*, 2000, **28**, 1815.
26. C. Ceconi, A. Boraso, A. Cargnoni and R. Ferrari, *Arch. Biochem. Biophys.*, 2003, **420**, 217.
27. J. W. Stephens, D. R. Gable, S. J. Hurel, G. J. Miller, J. A. Cooper and S. E. Humphries, *Clin. Chem.*, 2006, **52**, 446.
28. J. L. Witztum and D. Steinberg, *Trends Cardiovasc. Med.*, 2001, **11**, 93.
29. J. R. Hessler, D. W. Morel, L. J. Lewis and G. M. Chisolm, *Arteriosclerosis*, 1983, **3**, 215.
30. J. W. Heinecke, H. Rosen and A. Chait, *J. Clin. Invest.*, 1984, **74**, 1890.
31. U. P. Steinbrecher, S. Parthasarathy, D. S. Leake, J. L. Witztum and D. Steinberg, *Proc. Natl. Acad. Sci. U. S. A.*, 1984, **81**, 3883.
32. S. Toshima, A. Hasegawa, M. Kurabayashi, H. Itabe, T. Takano, J. Sugano, K. Shimamura, J. Kimura, I. Michishita, T. Suzuki and R. Nagai, *Arterioscler., Thromb., Vasc. Biol.*, 2000, **20**, 2243.
33. S. Ehara, M. Ueda, T. Naruko, K. Haze, A. Itoh, M. Otsuka, R. Komatsu, T. Matsuo, H. Itabe, T. Takano, Y. Tsukamoto, M. Yoshiyama, K. Takeuchi, J. Yoshikawa and A. E. Becker, *Circulation*, 2001, **103**, 1955.
34. T. Weinbrenner, M. Cladellas, M. Isabel Covas, M. Fito, M. Tomas, M. Senti, J. Bruguera and J. Marrugat, *Atherosclerosis*, 2003, **168**, 99.
35. P. G. Scheffer, G. Bos, H. G. Volwater, J. M. Dekker, R. J. Heine and T. Teerlink, *Diabet. Med.*, 2003, **20**, 563.
36. S. E. Humphries, R. A. Whittall, C. S. Hubbart, S. Maplebeck, J. A. Cooper, A. K. Soutar, R. Naoumova, G. R. Thompson, M. Seed, P. N. Durrington, J. P. Miller, D. J. Betteridge and H. A. Neil, *J. Med. Genet.*, 2006, **43**, 943.
37. J. W. Stephens and S. E. Humphries, *J. Intern. Med.*, 2003, **253**, 120.
38. S. C. Rall, Jr., K. H. Weisgraber and R. W. Mahley, *J. Biol. Chem.*, 1982, **257**, 4171.
39. P. J. Talmud and S. E. Humphries, *Curr. Opin. Lipidol.*, 2002, **13**, 149.
40. J. Davignon, R. E. Gregg and C. F. Sing, *Arteriosclerosis*, 1988, **8**, 1.
41. P. W. Wilson, E. J. Schaefer, M. G. Larson and J. M. Ordovas, *Arterioscler., Thromb., Vasc. Biol.*, 1996, **16**, 1250.
42. M. Smit, P. de Knijff, M. Rosseneu, J. Bury, E. Klasen, R. Frants and L. Havekes, *Hum. Genet.*, 1988, **80**, 287.
43. C. Kessler, C. Spitzer, D. Stauske, S. Mende, J. Stadlmuller, R. Walther and R. Rettig, *Arterioscler., Thromb., Vasc. Biol.*, 1997, **17**, 2880.
44. L. U. Gerdes, C. Gerdes, K. Kervinen, M. Savolainen, I. C. Klausen, P. S. Hansen, Y. A. Kesaniemi and O. Faergeman, *Circulation*, 2000, **101**, 1366.
45. Y. Song, M. J. Stampfer and S. Liu, *Ann. Intern. Med.*, 2004, **141**, 137.
46. P. J. Talmud, J. W. Stephens, E. Hawe, S. Demissie, L. A. Cupples, S. J. Hurel, S. E. Humphries and J. M. Ordovas, *Ann. Hum. Genet.*, 2005, **69**, 613.
47. S. E. Humphries, P. J. Talmud, E. Hawe, M. Bolla, I. N. Day and G. J. Miller, *Lancet*, 2001, **358**, 115.

48. J. Gustavsson, K. Mehlig, K. Leander, E. Strandhagen, L. Bjorck, D. S. Thelle, L. Lissner, K. Blennow, H. Zetterberg and F. Nyberg, *Atherosclerosis*, 2012, **220**, 486.
49. C. Jolivalt, B. Leininger-Muller, P. Bertrand, R. Herber, Y. Christen and G. Siest, *Free Radical Biol. Med.*, 2000, **28**, 129.
50. M. Miyata and J. D. Smith, *Nat. Genet.*, 1996, **14**, 55.
51. J. D. Smith, M. Miyata, S. E. Poulin, L. M. Neveux and W. Y. Craig, *Int. J. Clin. Lab. Res.*, 1998, **28**, 116.
52. M. A. Fernandes, M. T. Proenca, A. J. Nogueira, M. M. Grazina, L. M. Oliveira, A. I. Fernandes, B. Santiago, I. Santana and C. R. Oliveira, *Int. J. Mol. Med.*, 1999, **4**, 179.
53. T. Hayek, J. Oiknine, J. G. Brook and M. Aviram, *Biochem. Biophys. Res. Commun.*, 1994, **201**, 1567.
54. W. Palinski, V. A. Ord, A. S. Plump, J. L. Breslow, D. Steinberg and J. L. Witztum, *Arterioscler. Thromb.*, 1994, **14**, 605.
55. M. J. Sampson, N. Gopaul, I. R. Davies, D. A. Hughes and M. J. Carrier, *Diabetes Care*, 2002, **25**, 537.
56. M. K. Shigenaga, T. M. Hagen and B. N. Ames, *Proc. Natl. Acad. Sci. U. S. A.*, 1994, **91**, 10771.
57. J. F. Turrens, *Biosci. Rep.*, 1997, **17**, 3.
58. F. Bouillaud, E. Couplan, C. Pecqueur and D. Ricquier, *Biochim. Biophys. Acta*, 2001, **1504**, 107.
59. K. D. Garlid, M. Jaburek and P. Jezek, *Biochem. Soc. Trans.*, 2001, **29**, 803.
60. C. S. Lin and M. Klingenberg, *FEBS Lett.*, 1980, **113**, 299.
61. O. Boss, S. Samec, A. Paoloni-Giacobino, C. Rossier, A. Dulloo, J. Seydoux, P. Muzzin and J. P. Giacobino, *FEBS Lett.*, 1997, **408**, 39.
62. C. Fleury, M. Neverova, S. Collins, S. Raimbault, O. Champigny, C. Levi-Meyrueis, F. Bouillaud, M. F. Seldin, R. S. Surwit, D. Ricquier and C. H. Warden, *Nat. Genet.*, 1997, **15**, 269.
63. R. E. Gimeno, M. Dembski, X. Weng, N. Deng, A. W. Shyjan, C. J. Gimeno, F. Iris, S. J. Ellis, E. A. Woolf and L. A. Tartaglia, *Diabetes*, 1997, **46**, 900.
64. C. Duval, A. Negre-Salvayre, A. Dogilo, R. Salvayre, L. Penicaud and L. Casteilla, *Biochem. Cell Biol.*, 2002, **80**, 757.
65. A. Negre-Salvayre, C. Hirtz, G. Carrera, R. Cazenave, M. Troly, R. Salvayre, L. Penicaud and L. Casteilla, *FASEB J.*, 1997, **11**, 809.
66. L. Casteilla, M. Rigoulet and L. Penicaud, *IUBMB Life*, 2001, **52**, 181.
67. K. S. Echtay, D. Roussel, J. St-Pierre, M. B. Jekabsons, S. Cadenas, J. A. Stuart, J. A. Harper, S. J. Roebuck, A. Morrison, S. Pickering, J. C. Clapham and M. D. Brand, *Nature*, 2002, **415**, 96.
68. K. S. Echtay, M. P. Murphy, R. A. Smith, D. A. Talbot and M. D. Brand, *J. Biol. Chem.*, 2002, **277**, 47129.
69. N. Lameloise, P. Muzzin, M. Prentki and F. Assimacopoulos-Jeannet, *Diabetes*, 2001, **50**, 803.
70. G. Patane, M. Anello, S. Piro, R. Vigneri, F. Purrello and A. M. Rabuazzo, *Diabetes*, 2002, **51**, 2749.

71. F. Criscuolo, J. Mozo, C. Hurtaud, T. Nubel and F. Bouillaud, *Biochim. Biophys. Acta*, 2006, **1757**, 1284.
72. M. Trenker, R. Malli, I. Fertschai, S. Levak-Frank and W. F. Graier, *Nat. Cell Biol.*, 2007, **9**, 445.
73. C. Pecqueur, M. C. Alves-Guerra, C. Gelly, C. Levi-Meyrueis, E. Couplan, S. Collins, D. Ricquier, F. Bouillaud and B. Miroux, *J. Biol. Chem.*, 2001, **276**, 8705.
74. H. Esterbauer, C. Schneitler, H. Oberkofler, C. Ebenbichler, B. Paulweber, F. Sandhofer, G. Ladurner, E. Hell, A. D. Strosberg, J. R. Patsch, F. Krempler and W. Patsch, *Nat. Genet.*, 2001, **28**, 178.
75. F. Krempler, H. Esterbauer, R. Weitgasser, C. Ebenbichler, J. R. Patsch, K. Miller, M. Xie, V. Linnemayr, H. Oberkofler and W. Patsch, *Diabetes*, 2002, **51**, 3331.
76. H. Wang, W. S. Chu, T. Lu, S. J. Hasstedt, P. A. Kern and S. C. Elbein, *Am. J. Physiol.: Endocrinol. Metab.*, 2004, **286**, E1.
77. G. Sesti, M. Cardellini, M. A. Marini, S. Frontoni, M. D'Adamo, S. Del Guerra, D. Lauro, P. De Nicolais, P. Sbraccia, S. Del Prato, S. Gambardella, M. Federici, P. Marchetti and R. Lauro, *Diabetes*, 2003, **52**, 1280.
78. S. O'Rahilly, *Nat. Med.*, 2001, **7**, 770.
79. G. Andersen, L. T. Dalgaard, J. M. Justesen, S. Anthonsen, T. Nielsen, L. W. Thorner, D. Witte, T. Jorgensen, J. O. Clausen, T. Lauritzen, J. Holmkvist, T. Hansen and O. Pedersen, *Int. J. Obes.*, 2013, **37**, 175.
80. D. R. Gable, J. W. Stephens, S. S. Dhamrait, E. Hawe and S. E. Humphries, *Diabetes, Obes. Metab.*, 2007, **9**, 130.
81. B. R. Palmer, C. L. Devereaux, S. S. Dhamrait, T. J. Mocatta, A. P. Pilbrow, C. M. Frampton, L. Skelton, T. G. Yandle, C. C. Winterbourn, A. M. Richards, H. E. Montgomery and V. A. Cameron, *Cardiovasc. Diabetol.*, 2009, **8**, 31.
82. S. S. Dhamrait, J. W. Stephens, J. A. Cooper, J. Acharya, A. R. Mani, K. Moore, G. J. Miller, S. E. Humphries, S. J. Hurel and H. E. Montgomery, *Eur. Heart J.*, 2004, **25**, 468.
83. D. Arsenijevic, H. Onuma, C. Pecqueur, S. Raimbault, B. S. Manning, B. Miroux, E. Couplan, M. C. Alves-Guerra, M. Goubern, R. Surwit, F. Bouillaud, D. Richard, S. Collins and D. Ricquier, *Nat. Genet.*, 2000, **26**, 435.
84. T. Cao, Y. Dong, R. Tang, J. Chen, C. Y. Zhang and K. Zen, *Cell. Immunol.*, 2013, **286**, 39.
85. J. Blanc, M. C. Alves-Guerra, B. Esposito, S. Rousset, P. Gourdy, D. Ricquier, A. Tedgui, B. Miroux and Z. Mallat, *Circulation*, 2003, **107**, 388.
86. J. W. Stephens, S. S. Dhamrait, A. R. Mani, J. Acharya, K. Moore, S. J. Hurel and S. E. Humphries, *Nutr., Metab. Cardiovasc. Dis.*, 2007, **18**, 7.
87. J. A. Berliner, M. Navab, A. M. Fogelman, J. S. Frank, L. L. Demer, P. A. Edwards, A. D. Watson and A. J. Lusis, *Circulation*, 1995, **91**, 2488.

88. B. Mannervik and U. H. Danielson, *CRC Crit. Rev. Biochem.*, 1988, **23**, 283.
89. B. Ketterer, J. M. Harris, G. Talaska, D. J. Meyer, S. E. Pemble, J. B. Taylor, N. P. Lang and F. F. Kadlubar, *Environ. Health Perspect.*, 1992, **98**, 87.
90. T. R. Rebbeck, *Cancer Epidemiol., Biomarkers Prev.*, 1997, **6**, 733.
91. S. Pemble, K. R. Schroeder, S. R. Spencer, D. J. Meyer, E. Hallier, H. M. Bolt, B. Ketterer and J. B. Taylor, *Biochem. J.*, 1994, **300**(Pt 1), 271.
92. D. A. Bell, J. A. Taylor, D. F. Paulson, C. N. Robertson, J. L. Mohler and G. W. Lucier, *J. Natl. Cancer Inst.*, 1993, **85**, 1159.
93. G. Chenevix-Trench, J. Young, M. Coggan and P. Board, *Carcinogenesis*, 1995, **16**, 1655.
94. J. E. McWilliams, B. J. Sanderson, E. L. Harris, K. E. Richert-Boe and W. D. Henner, *Cancer Epidemiol., Biomarkers Prev.*, 1995, **4**, 589.
95. J. Brockmoller, I. Cascorbi, R. Kerb and I. Roots, *Cancer Res.*, 1996, **56**, 3915.
96. S. S. Coughlin and I. J. Hall, *Genet. Med.*, 2002, **4**, 250.
97. R. Li, E. Boerwinkle, A. F. Olshan, L. E. Chambless, J. S. Pankow, H. A. Tyroler, M. Bray, G. S. Pittman, D. A. Bell and G. Heiss, *Atherosclerosis*, 2000, **149**, 451.
98. R. Li, A. R. Folsom, A. R. Sharrett, D. Couper, M. Bray and H. A. Tyroler, *Atherosclerosis*, 2001, **154**, 729.
99. M. H. Wilson, P. J. Grant, L. J. Hardie and C. P. Wild, *FASEB J.*, 2000, **14**, 791.
100. F. G. de Waart, F. J. Kok, T. J. Smilde, A. Hijmans, H. Wollersheim and A. F. Stalenhoef, *Atherosclerosis*, 2001, **158**, 227.
101. X. L. Wang, M. Greco, A. S. Sim, N. Duarte, J. Wang and D. E. Wilcken, *J. Cardiovasc. Risk*, 2002, **9**, 25.
102. S. Masetti, N. Botto, S. Manfredi, M. G. Colombo, A. Rizza, C. Vassalle, A. Clerico, A. Biagini and M. G. Andreassi, *J. Mol. Med.*, 2003, **81**, 488.
103. N. J. Martin, A. C. Collier, L. D. Bowen, K. L. Pritsos, G. G. Goodrich, K. Arger, G. Cutter and C. A. Pritsos, *Mutat. Res.*, 2009, **674**, 93.
104. M. H. Wilson, P. J. Grant, K. Kain, D. P. Warner and C. P. Wild, *Biomarkers*, 2003, **8**, 43.
105. S. H. Rahman, K. Ibrahim, M. Larvin, A. Kingsnorth and M. J. McMahon, *Gastroenterology*, 2004, **126**, 1312.
106. T. Hayek, J. W. Stephens, C. S. Hubbart, J. Acharya, M. J. Caslake, E. Hawe, G. J. Miller, S. J. Hurel and S. E. Humphries, *Atherosclerosis*, 2006, **184**, 404.
107. X. L. Wang, D. L. Rainwater, J. F. VandeBerg, B. D. Mitchell and M. C. Mahaney, *Arterioscler., Thromb., Vasc. Biol.*, 2001, **21**, 1190.
108. A. M. Schmidt, O. Hori, R. Cao, S. D. Yan, J. Brett, J. L. Wautier, S. Ogawa, K. Kuwabara, M. Matsumoto and D. Stern, *Diabetes*, 1996, **45**(Suppl 3), S77.

109. B. Mackness, M. I. Mackness, S. Arrol, W. Turkie, K. Julier, B. Abuasha, J. E. Miller, A. J. Boulton and P. N. Durrington, *Atherosclerosis*, 1998, **139**, 341.
110. K. K. Griendling, D. Sorescu and M. Ushio-Fukai, *Circ. Res.*, 2000, **86**, 494.
111. M. A. Ostos, M. Conconi, L. Vergnes, N. Baroukh, J. Ribalta, J. Girona, J. M. Caillaud, A. Ochoa and M. M. Zakin, *Arterioscler., Thromb., Vasc. Biol.*, 2001, **21**, 1023.
112. S. E. Humphries, L. A. Luong, M. S. Ogg, E. Hawe and G. J. Miller, *Eur. Heart J.*, 2001, **22**, 2243.
113. S. Tomono, S. Kawazu, N. Kato, T. Ono, C. Ishii, Y. Ito, M. Shimizu, M. Shimoyama, T. Nakano and K. Nakajima, *J. Atheroscler. Thromb.*, 1994, **1**, 98.
114. A. Kawakami, M. Tani, T. Chiba, K. Yui, S. Shinozaki, K. Nakajima, A. Tanaka, K. Shimokado and M. Yoshida, *Arterioscler., Thromb., Vasc. Biol.*, 2005, **25**, 424.
115. S. D. Cushing, J. A. Berliner, A. J. Valente, M. C. Territo, M. Navab, F. Parhami, R. Gerrity, C. J. Schwartz and A. M. Fogelman, *Proc. Natl. Acad. Sci. U. S. A.*, 1990, **87**, 5134.
116. T. B. Rajavashisth, A. Andalibi, M. C. Territo, J. A. Berliner, M. Navab, A. M. Fogelman and A. J. Lusis, *Nature*, 1990, **344**, 254.
117. A. Kawakami, A. Tanaka, T. Nakano, A. Saniabadi and F. Numano, *Horm. Metab. Res.*, 2001, **33**, 67.
118. A. Kawakami, A. Tanaka, T. Chiba, K. Nakajima, K. Shimokado and M. Yoshida, *Circulation*, 2003, **108**, 2679.
119. H. K. Shin, Y. K. Kim, K. Y. Kim, J. H. Lee and K. W. Hong, *Circulation*, 2004, **109**, 1022.
120. P. Delerive, C. Furman, E. Teissier, J. Fruchart, P. Duriez and B. Staels, *FEBS Lett.*, 2000, **471**, 34.
121. A. Stiko-Rahm, A. Hultgardh-Nilsson, J. Regnstrom, A. Hamsten and J. Nilsson, *Arterioscler. Thromb.*, 1992, **12**, 1099.
122. A. R. Saniabadi, K. Umemura, M. Shimoyama, M. Adachi, M. Nakano and M. Nakashima, *Thromb. Haemost.*, 1997, **77**, 996.
123. K. Kugiyama, S. A. Kerns, J. D. Morrisett, R. Roberts and P. D. Henry, *Nature*, 1990, **344**, 160.
124. D. D. Roh, V. S. Kamanna and M. A. Kirschenbaum, *Am. J. Nephrol.*, 1998, **18**, 344.
125. J. E. Murphy, P. R. Tedbury, S. Homer-Vanniasinkam, J. H. Walker and S. Ponnambalam, *Atherosclerosis*, 2005, **182**, 1.
126. K. Kankova, I. Marova, J. Zahejsky, J. Muzik, A. Stejskalova, V. Znojil and J. Vacha, *Metabolism*, 2001, **50**, 1152.
127. A. Bub, S. W. Barth, B. Watzl, K. Briviba and G. Rechkemmer, *Br. J. Nutr.*, 2005, **93**, 291.
128. C. C. Bortolasci, H. O. Vargas, A. Souza-Nogueira, D. S. Barbosa, E. G. Moreira, S. O. Nunes, M. Berk, S. Dodd and M. Maes, *J. Affect. Disord.*, 2014, **159**, 23.

129. O. Ormezzano, J. L. Cracowski, J. M. Mallion, O. Poirier, J. Bessard, S. Briancon, P. Francois and J. P. Baguet, *Free Radical Biol. Med.*, 2005, **38**, 583.
130. V. A. Cameron, T. J. Mocatta, A. P. Pilbrow, C. M. Frampton, R. W. Troughton, A. M. Richards and C. C. Winterbourn, *Hypertension*, 2006, **47**, 1155.
131. W. M. Wong, J. W. Stephens, J. Acharya, S. J. Hurel, S. E. Humphries and P. J. Talmud, *J. Lipid Res.*, 2004, **45**, 1565.
132. J. W. Stephens, S. J. Hurel, G. D. Lowe, A. Rumley and S. E. Humphries, *Mol. Genet. Metab.*, 2007, **90**, 422.
133. K. Mohammedi, T. A. Patente, N. Bellili-Munoz, F. Driss, M. B. Monteiro, R. Roussel, E. J. Pavin, N. Seta, F. Fumeron, M. J. Azevedo, L. H. Canani, S. Hadjadj, M. Marre, M. L. Correa-Giannella and G. Velho, *Diabetologia*, 2013, **56**, 2733.
134. K. D. Salpea, P. J. Talmud, J. A. Cooper, C. G. Maubaret, J. W. Stephens, K. Abelak and S. E. Humphries, *Atherosclerosis*, 2010, **209**, 42.

CHAPTER 10

New Drug Design and Therapeutic Development for Heart Failure

ELNAZ MENHAJI-KLOTZ* AND THOMAS V. MAGEE

Pfizer, Inc., 610 Main St., Cambridge, MA 02139, USA
*Email: Elnaz.menhaji-klotz@pfizer.com

10.1 Introduction

Cardiovascular diseases (CVDs) are the number one cause of death globally according to the World Health Organization.[1] They account for approximately 30% of all global deaths and incur great medical costs. Chronic or congestive heart failure (CHF) remains the most prevalent diagnosis requiring hospitalization.[2] Broadly speaking, CHF refers to the situation in which the heart cannot pump enough blood to the body's organs. Major causes of heart failure (HF) include ischemia, hypertension and valve disease. Symptoms include shortness of breath (dyspnea), fluid buildup (edema), fatigue and increased heart rate among others. HF progresses through a chronic phase where compensatory mechanisms allow the patient to function. Acute heart failure (AHF) is defined by the International Acute Heart Failure Syndromes (AHFS) Working Group as a "gradual or rapid onset of new or worsening signs and/or symptoms of HF requiring urgent therapy, usually resulting in hospitalization",[3] also known as acute decompensation.

Diet, exercise and lifestyle choices remain the best way to prevent CVD especially since current treatments aim to improve quality of life by treating signs and symptoms. Despite pharmacological interventions, patient

RSC Drug Discovery Series No. 45
Cardiovascular and Metabolic Disease: Scientific Discoveries and New Therapies
Edited by Philip Peplow, James Adams and Tim Young

Published by the Royal Society of Chemistry, www.rsc.org

outcomes remain poor. In fact, after hospitalization the six month prognosis of death or recurrent hospitalization approaches 50%.[3] An aging population means that the associated costs and burden on society will only grow. Discovery and development of new HF therapies is an active research area seeking to address this unmet medical need. For the purposes of this chapter, we review (1) new drugs in the clinic for existing targets and (2) new targets that are currently being evaluated in the clinic. In general, targets with compelling preclinical data that have not reached the clinic have been omitted due to space constraints.

10.2 Modulators of the Renin–Angiotensin–Aldosterone System (RAAS) Pathway

The renin–angiotensin–aldosterone system (RAAS) plays a central physiological role in controlling vasoconstriction and cardiorenal function.[4] The current standards of care for the treatment of hypertension and HF are drugs that modulate the RAAS pathway, namely the angiotensin I converting enzyme inhibitors (ACEi), the angiotensin II receptor blockers (ARB), and the mineralocorticoid receptor antagonists (MRA).[5,6] Direct renin inhibitors (DRi) are an old concept but relatively new as a drug class, exemplified by aliskerin. Aliskerin clinical trials have raised some concerns about its safety, in particular hyperkalemia and hypotension when combined with an ACEi or an ARB.[7]

10.2.1 Neprilysin Inhibitors (NEPi)

The endogenous natriuretic peptides oppose the effects of the RAAS and stimulate vasodilation, natriueresis, diuresis, and inhibition of fibrosis. Neprilysin is a vasopeptidase (neutral endopeptidase) that has broad activity, including the degradation of natriuretic peptides (NPs),[8] and therefore inhibitors of neprilysin (NEPi) would be expected to effectively counteract the effects of RAAS activation.[9] Novartis is currently developing LCZ696, a 1 : 1 fixed dose combination of an ARB (valsartan) and a neprilysin inhibitor (ethyl ester prodrug AHU-377), also called "ARNi," which has shown great promise (Figure 10.1).

In a phase III outcomes trial (PARADIGM-HF) LCZ696 demonstrated a statistically significant reduction in cardiovascular (CV) mortality when compared to the ACEi enalapril, and thus represents a promising therapy for the treatment of HF with reduced ejection fraction (HFrEF).[10] Previous combination of NEPi and ACEi had led to safety concerns, specifically angioedema due to bradykinin accumulation, however this was not observed with LCZ696. Combinations of RAAS modulators with other classes, including triple combination therapies of a RAAS modulator, a diuretic, and a calcium channel blocker have been argued for recently in the management of hypertension.[6]

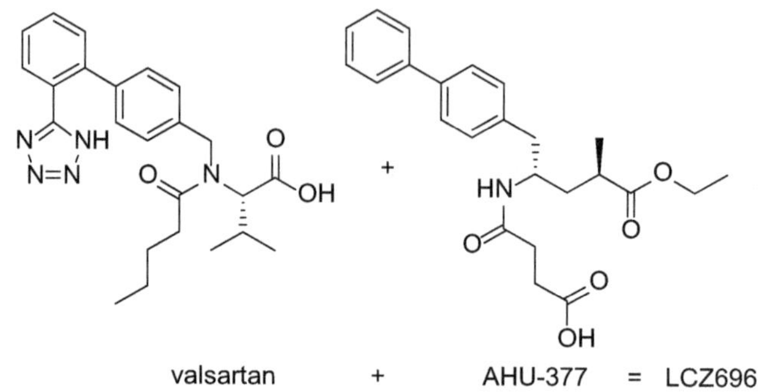

Figure 10.1 LCZ696: fixed-dose valsartan and AHU-377.

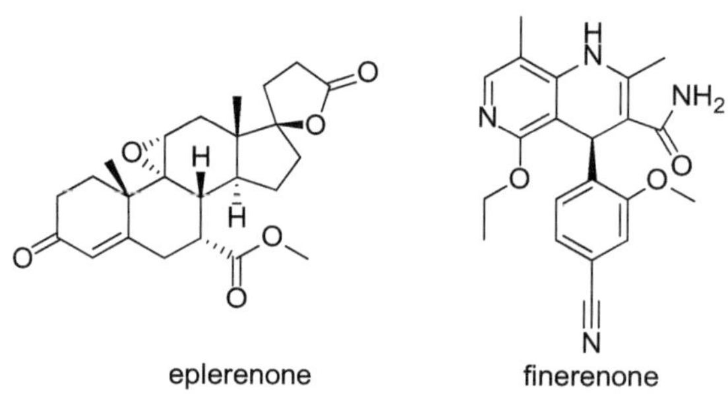

Figure 10.2 Mineralcorticoid receptor antagonists.

10.2.2 Mineralcorticoid Receptor Antagonists (MRA)

Angiotensin II, in addition to promoting vasoconstriction, triggers the release of the mineralocorticoid (aldosterone) in the adrenal cortex which in turn stimulates the renal tubular uptake of sodium and water and a subsequent increase in blood pressure. Thus, the mineralocorticoid receptor antagonist (MRA) class of drugs is also under the umbrella of the RAAS pathway. The MRA eplerenone (Figure 10.2), already marketed by Pfizer for HF, recently demonstrated a statistically significant improvement in CV death and hospitalization rate for HF in the EMPHASIS-HF phase III outcomes trial when compared to placebo.[11,12] Eplerenone also showed a significant reduction in the incidence of new-onset atrial fibrillation in this study, pointing to the possibility of mineralocorticoid receptor (MR) involvement in cardiac remodeling and fibrosis. This represents a significant advance for the use of eplerenone and potentially other drugs in this class and have prompted a modification in treatment guidelines for MRAs,[13] in spite of concerns about safety, in particular hyperkalemia.[14] Bayer has discovered a novel, next-generation non-steroidal mineralocorticoid antagonist, BAY 94-8862 or finerenone (Figure 10.2),[15] which is substantially more selective than steroidal MRAs such as spironolactone or eplerenone *versus* other members of the oxosteroid family of receptors. A phase II comparison to spironolactone in HFrEF patients (ARTS trial) demonstrated efficacy comparable to spironolactone but with a lower incidence of hyperkalemia.[16] Bayer plans to advance finerenone to phase III trials in 2015.

10.2.3 Aldosterone Synthase Inhibitor

One downside to antagonizing the mineralocorticoid receptor is that it leads to an increase in circulating aldosterone, which can have detrimental effects independent of the MR. An alternative approach to MR antagonism is inhibition of aldosterone synthase, or CYP11B2.[17–19] Phase II trials in patients with hypertension were run on the aldosterone synthase inhibitor LCI699 (Figure 10.3),[20] and although it demonstrated antihypertensive efficacy it also repressed cortisol synthesis due to its poor selectivity *versus* 11-beta-hydroxylase, or CYP11B1, and its therapeutic value was therefore limited.[21] More selective chemical lead matter has been reported recently that might help overcome this issue.[22]

10.2.4 Biased Signaling

A novel approach to modulating the RAAS axis is utilizing beta-arrestin biased signaling at the angiotensin II receptor type 1 (AT1R).[23] A parentally dosed peptide being developed by Trevena, TRV027 (also known as TRV120027, Figure 10.4) competitively blocks binding of angiotensin II to AT1R, but is differentiated in its pharmacology in that it is driven by beta-arrestin recruitment and signaling and does not activate G-protein coupling. The selective stimulation of the beta-arrestin pathway downstream of AT1R leads to up-regulation of CV tone by way of nitric oxide (NO) release and also

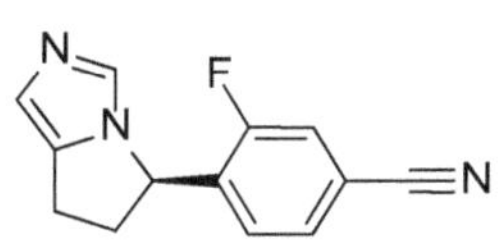

Figure 10.3 LCI699.

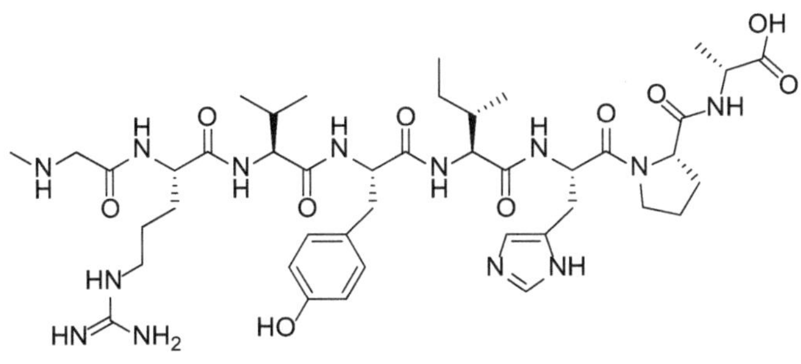

Figure 10.4 TRV027.

improves myocyte contractility.[24] In a canine model of HF in combination with the diuretic furosemide, TRV027 was also shown to both preserve the furosemide diuresis and reduce cardiac preload and afterload.[25] TRV027 was further shown to promote cardiac protection in preclinical models in contrast to an ARB, suggesting that this mode of action has differential effects from conventional therapy in the context of acute cardiac injury.[26] In a phase IIa clinical trial for AHF, TRV027 dosed by IV infusion demonstrated a rapid decrease in blood pressure while preserving cardiac output and renal function.[27]

10.2.5 Angiotensin II Receptor Type 2 (AT2R) Activation

An alternative strategy is activation of the angiotensin II receptor type 2 (AT2R), which has been shown to have anti-fibrotic, anti-inflammatory, and anti-hypertensive effects preclinically.[28,29] Vicore has discovered the small molecule AT2R agonist C21 (Figure 10.5),[30] which has shown efficacy in a rodent model of HF and could potentially be paired with an ACEi or ARB.[31]

10.2.6 Mas Receptor

Other recent developments in this area have shown that angiotensins I and II are substrates for the carboxypeptidase ACE2, and that a soluble form of this enzyme (sACE2) is elevated following myocardial infarction (MI).[32] ACE2 cleaves angiotensin I to angiotensin-(1-9) and angiotensin II to angiotensin-(1-7), which have been shown to counter cardiac myocyte hypertrophy in preclinical models.[33] Angiotensin-(1-9) is cleaved by ACE or neprilysin (NEP) to angiotensin-(1-7), which in turn signals through the Mas receptor.[34] Mas receptor signaling has shown evidence of counteracting RAAS signaling, and the ACE2-Angiotensin-(1-7)-Mas receptor axis has been proposed as a novel therapeutic approach for CVD.[35] Interestingly, angiotensin II can also be generated independently of the renin pathway. Angiotensin-(1-12), which is derived from angiotensinogen in a renin-independent process, is cleaved in cardiac myocytes to angiotensin II primarily by chymase, not by ACE, suggesting alternative mechanisms outside the traditional RAAS modulators for addressing cardiomyopathy.[36,37]

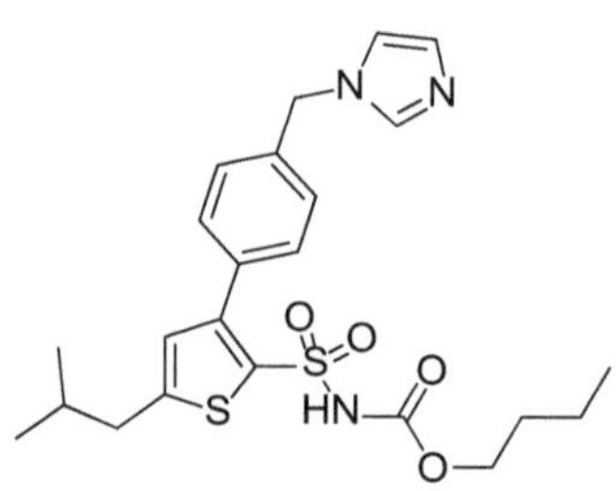

Figure 10.5 C21.

10.2.7 (Pro)Renin–Renin Receptor ((P)RR)

Renin can mediate pro-fibrotic effects independently of angiotensins I and II *via* the (pro)renin–renin receptor, ((P)RR)[38] and was further observed to be elevated in both rodent models of HF and in human patients with dilated cardiomyopathy.[39] Thus, (P)RR may offer a new point of intervention in CVD given its relationship to renin at the top of the RAAS cascade, although the fundamental biology of this receptor is still being uncovered.[40]

10.3 Natriuretic Peptides

Natriuretic peptides (NPs) are circulating hormones known to antagonize the RAAS pathway and are elevated in patients with HF and are part of a compensatory response, and thus can serve as both biomarkers[41] and as potential therapeutics.[42,43] Inhibitors of NP degradation combined with RAAS antagonism is a promising treatment for HF as described above. The NPs activate membrane-bound guanylate cyclase (GC) receptors (subtypes-A and -B, also known as NPR-1 and -2, respectively) to produce the vasodilator cyclic guanosine monophosphate (cGMP), affecting vascular tone and cardiac remodeling.[44,45] Recombinant versions of human endogenous NPs, such as atrial (ANP), B-type (BNP), and C-type (CNP), offer potential as therapeutics for the treatment of HF. ANP and BNP are selective for the GC-A subtype, while CNP activates GC-B.[46] One prominent example is the recombinant BNP nesiritide, which recently completed a large phase III trial in HF patients (ASCEND-HF). Nesiritide however failed to improve mortality or re-hospitalization rates in this study.[47] Furthermore, the hypotensive effects and worsening renal function by reducing the glomerular filtration rate pose a considerable safety concern[48] and create the clear need for improved NP-related therapies. Novel synthetically engineered peptides such as CD-NP (cenderitide, Figure 10.6),[49,50] a chimeric hybrid of CNP and *Dendroaspis* NP (DNP), derived from the green mamba, represent a first generation of NPs engineered to overcome some of the obstacles faced by the native NPs, including their poor pharmacokinetic properties.[51–53] Cenderitide consists

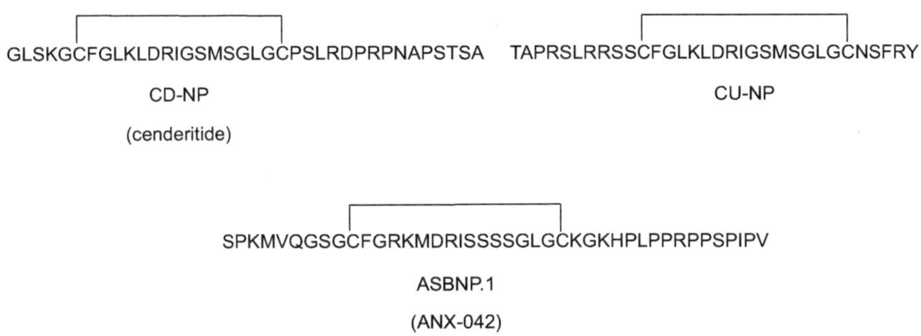

Figure 10.6 Natriuretic peptides.

of the 5 amino acid (AA) N-terminus and 17 AA cyclic peptide core of CNP linked to the 15 AA C-terminus of DNP. Cenderitide is more resistant to proteolytic degradation than endogenous NPs.[54] Furthermore it has expanded pharmacology, with pan-activity on both GC-A and -B and robust efficacy in rodent anti-fibrotic models of HF.[55] Cenderitide elevated circulating cGMP, lowered serum aldosterone, and increased diuresis and natriuresis in phase I trials; however it still induced a small but significant reduction in mean arterial blood pressure.[56]

Cenderitide received fast-track status from the U.S. Food and Drug Administration (FDA) for treating post-acute HF in phase II trials utilizing chronic subcutaneous infusion technology for sustained exposure.[57,58] Another example of an engineered NP showing promise is CU-NP (Figure 10.6), although still in the preclinical stage. CU-NP is a hybrid of CNP and an ANP-related peptide hormone found in the urine of human patients with renal failure, urodilatin. The structure of CU-NP consists of the 17 amino acid cyclic peptide core of CNP, and the N- and C-termini derived from urodilatin, making it a fully human chimera.[59] CU-NP, like cenderitide, also activates both the GC-A and -B subtypes and appears to signal through the sodium-hydrogen exchanger-1/calcineurin pathways and exert anti-hypertrophic effects.[60] Another strategy for engineering novel NPs is utilizing alternative RNA splice variants found endogenously. This technique has been used to make AS-BNP, a C-terminally modified analogue of BNP found in patients with HF, and a corresponding C-terminal truncated form AS-BNP.1 (Figure 10.6).[61] AS-BNP.1, also called ANX-042, demonstrated no hypotensive effects and increased glomerular filtration rate and lowered renin plasma levels in a canine model of HF. ANX-042 is a weak activator of GC-A (4 μM) and does not appear to activate GC-B at all (>10 μM), so a detailed understanding of its molecular pharmacology has yet to be revealed. Nevertheless Anexon has initiated phase I trials as a potential renal-enhancing non-hypotensive agent for treating HF on the strength of the preclinical data observed for ANX-042.

10.4 Soluble Guanylate Cyclase Modulators

Natriuretic peptides stimulate the generation of cGMP *via* their membrane-bound GC receptors. However a soluble form of GC (sGC), stimulated by NO rather than by NPs, also plays an important role in CV function.[62,63] NO donor drugs suffer from a lack of specificity, but also from a loss of efficacy with chronic treatment.[64] Direct stimulators of sGC that act independently of NO may offer an alternative mode of increasing cGMP.[65] The leading drug in this class is riociguat (Figure 10.7) discovered by researchers at Bayer,[66] and recently approved for the treatment of pulmonary hypertension.

Both NO and riociguat bind to ferrous iron (Fe^{2+}) in the pendant heme domain present in sGC to stimulate the formation of cGMP. Resistance to NO develops when the iron is oxidized to the ferric (Fe^{3+}) state, as will happen in disease environments dominated by reactive oxygen species (ROS)

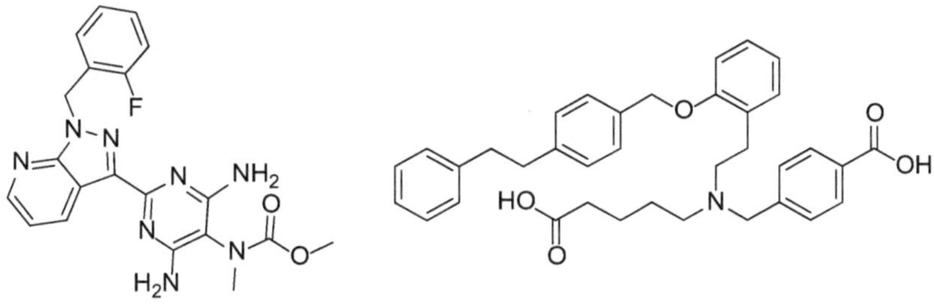

riociguat cinaciguat

Figure 10.7 Soluble guanylate cyclase modulators.

such as HF,[67,68] and ultimately the oxidized state results in heme cleavage from the sGC. Thus, activators that still function on the oxidized and heme-cleaved sGC may offer new modes in intervention.[69] Cinaciguat (Figure 10.7) is such an activator that showed good preclinical efficacy in HF models.[70,71] Cinaciguat was being developed by Bayer for acute decompensated HF; however it was recently discontinued in phase II for unspecified reasons.[72] Recent crystallographic results on sGC have been reported and may offer novel insights and modes of intervention in this pathway.[73–75]

10.5 Vasopressin Antagonists

HF can lead to chronic elevation in arginine vasopressin, a hormone that promotes renal reabsorption of free water, and consequently hypervolemia and hyponatremia. The vasopressin antagonists, or vaptans, have been reviewed recently.[76] There are three vasopressin receptors: V1a, V1b, and V2. The V1a receptor is associated with cardiac myocytes and vascular smooth muscle cells, whereas V1b is in the central nervous system and its function is poorly understood. The V2 receptor is found in the kidney and when activated stimulates translocation of the aquaporin ducts to the apical membrane of the collecting duct, which in turn leads to free water reabsorption and natriuresis.[77] Conivaptan,[78] was the first in class agent, and although it is peptidic it is dosed both orally and intravenously. However, its use is limited owing to an association with hepatic toxicity due to drug interactions with CYP3A4 as a victim, *i.e.* its exposure goes up to unsafe levels due to concomitant dosing with other drugs that inhibit CYP3A4. Conivaptan is a non-selective antagonist with activity at both V1a and V2. The leading agent in the vaptan class is also the first non-peptidic small molecule, tolvaptan (Figure 10.8),[79,80] which is selective for V2, however still suffers from the same CYP3A4-driven hepatic liabilities. Because of this safety risk, the FDA has issued restrictions which limit the dosing interval to 30 days.[81] A phase III trial (EVEREST) on HF patients with tolvaptan showed no reduction in all-cause mortality compared to placebo.[82,83] However, a retrospective analysis of the EVEREST trial showed a significant improvement in hyponatremia

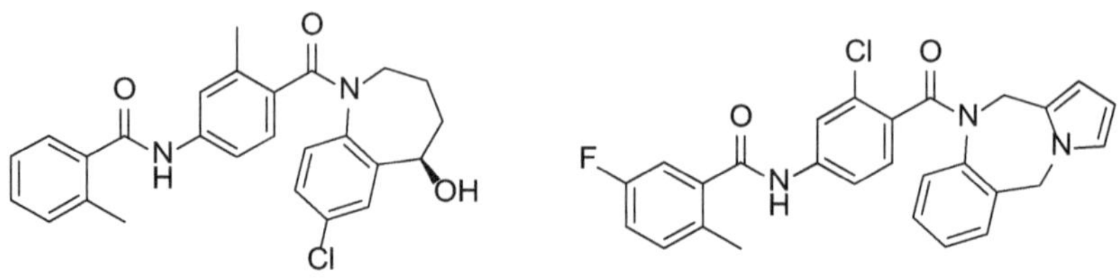

Figure 10.8 Vasopressin antagonists.

and also dyspnea symptoms albeit in a limited subgroup of (hyponatremic) patients, providing support for prospective trials.[84,85]

Another small molecule antagonist, lixivaptan, has shown improvements in hyponatremia in phase III trials (Figure 10.8),[86,87] however some of the claims to its efficacy has been called into question.[88] Furthermore, like the other vaptans, it is also metabolized by CYP3A4 and will likely suffer from a similar risk/benefit profile.

10.6 Electrolyte Polymers

Diuretic drugs for the reduction of hypervolemia remain a first line therapeutic option in the treatment of hypertension.[89,90] However, diuretics, in particular loop diuretics, are actually associated with worse outcomes.[91] Maintaining a neutral sodium balance is strongly implicated in the early treatment of HF.[92,93] Oral cation binders are an alternative strategy for achieving electrolyte balance without modulating hemodynamics or interfering with RAAS as conventional diuretics do. A recent phase II study[94] using cation-binding polymers administered orally[95] in patients with symptomatic HF with chronic kidney disease resulted in significant improvement in weight loss and lower NT-pro-BNP levels *versus* placebo. Cation-binding polymers thus represent a possible novel alternative to conventional diuretic therapy, however obviously more studies are needed in this area.

10.7 Transthyretin

Transthyretin (TTR) is a homo-tetrameric carrier protein that chaperones the thyroid hormone thyroxine (T4) and retinol binding protein in both plasma and cerebrospinal fluid. Tetramer dissociation and misfolding of TTR – in both wild type and in several reported protein mutant forms – are known to precipitate neurological and cardiac amyloidosis that can lead to serious pathologies, including cardiomyopathy and HF.[96,97] A recent review covered novel drug approaches to thwarting TTR-related disease, including stabilization of the TTR tetramer and its variants.[98]

Tafamidis (Figure 10.9)[99] is the leading drug in the stabilizer class for inhibiting the amyloid cascade and is currently on the market for slowing

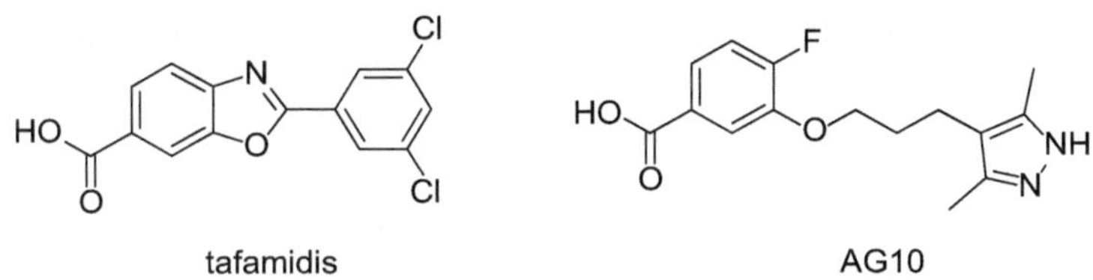

Figure 10.9 Transthyretin stabilizers.

the progression of familial amyloid polyneuropathy (FAP) caused by TTR dysfunction.[100] There are no drugs available to treat familial amyloid cardiomyopathy (FAC) currently, however recruiting for a phase III cardiomyopathy trial (ATTR-ACT) has begun recently. Different mutations in TTR promote FAC, however the V122I variant is one of the dominant ones, accelerating the rate of tetramer dissociation that leads to amyloid fibrile formation.[101] Although tafamidis does stabilize V122I, the pyrazole AG10 (Figure 10.9) was recently reported as more potent than tafamidis *versus* this important mutant.[102] An alternative to stabilization is blocking the synthesis of TTR using small interfering RNA (siRNA) or antisense oligonucleotide (ASO) strategies. TTR is primarily secreted from hepatocytes, and Alnylam Pharmaceuticals has developed a siRNA against both wild type and mutant TTR (patisiran or ALN-TTR02), formulated in nanoparticles that are tissue selective for the liver and dosed intravenously.[103] Patisiran is currently in a phase III trial (APOLLO) for treating FAP. An alternative subcutaneous formulation is also being developed (ALN-TTRsc), recently starting in phase II trials in patients with FAC and senile systemic amyloidosis. Isis is developing an ASO (ISIS-TTR$_{Rx}$) that targets synthesis of TTR, also both the mutant and wild type forms, to be dosed subcutaneously.[104] ISIS-TTR$_{Rx}$ is currently in a phase III trial for patients with FAP.

10.8 Inotropes

10.8.1 Sarcoplasmic Reticulum Calcium ATPase Cardiac Isoform 2a (SERCA2a)

The sarcoplasmic reticulum calcium ATPase cardiac isoform 2a (SERCA2a) is responsible for pumping cytosolic calcium into the sarcoplasmic reticulum in order to facilitate cardiac contraction and relaxation. Decreased SERCA2a activity and perturbation of cellular calcium handling is a hallmark of HF.[105] Increasing SERCA2a activity should increase cardiomyocyte calcium handling and improve cardiac contractility thereby improving systolic and diastolic function.[106]

Celladon are front runners in the SERCA2a field and have produced noteworthy clinical data on their gene therapy, Mydicar. Mydicar is a recombinant adeno-associated viral vector (AAVV) containing the human SERCA2a cDNA. Celladon's CUPID phase I/IIa trial in 39 CHF patients examined safety and efficacy of a single Mydicar coronary arterial infusion in comparison to placebo. At twelve months, the treatment group had an 88% reduction in CV events *versus* placebo.[107] No adverse events were associated with drug treatment and unlike traditional inotropes, there was no increase in patient heart rate. Furthermore, after three years, investigators found an 82% reduction in events in comparison to placebo.[108]

Mydicar has been granted fast track designation by the FDA and is enrolling for their CUPID-2b trial where the primary outcomes will be time to a recurrent event in the presence of terminal events up to 12 months.[109] One limitation of this approach is the prevalence of pre-existing antibodies in patients to the viral vector used to deliver Mydicar; approximately 50% of patients were excluded from the CUPID trial for this reason.[107] In addition, there is an uncertain regulatory path for gene therapies.

The SERCA2a gene therapy clinical data is compelling and a small molecule activator of the system would be of great value. While there are no known small molecule SERCA2a activators in the clinic, publications have put forth chemical matter in the public domain. Merck published an early report describing the effect of the flavonoid quercitin on SERCA2 activity: stimulatory at low concentrations and inhibitory at high concentrations of quercitin.[110] One approach to activate SERCA2a has been to modulate the interaction between SERCA2a and its co-expressed and inhibitory transmembrane protein, phospholamban (PLB or PLN). Celladon has published on such a strategy where a fluorescence resonance energy transfer (FRET) assay between labeled SERCA2a and PLB was used to identify activators.[111,112] Initial screening hits were tested in a functional calcium ATPase assay and a cardiomyocyte contractility assay where the most promising compounds were found to increase SERCA2a V_{max} by 50–60% in cardiac cells. Compound structures are not disclosed in the 2013 publication and graphical activity plots indicate EC_{50}'s in the micromolar range.

Researchers at Orion and the University of Helsinki have also published on small molecule inhibitors of PLB.[113,114] They used a phage display library to identify cyclic peptides with binding affinity for a soluble portion of PLB. Knowledge gained from the peptide library was used to define a pharmacophore and design small molecule inhibitors of PLB.

10.8.2 Ryanodine Receptor (RyR) Modulator

While SERCA2a is responsible for transporting calcium ions from the cytosol into the sarcolemma, the ryanodine receptor (RyR) allows calcium transport back into the cytosol and is required during systole. Published reports implicate 'leaky' RyR in aberrant membrane potential which can result in cardiac arrhythmias.[115] ARMGO have an approach where the closed form of the receptor is stabilized by a small molecule.[116,117] In 2006 ARMGO entered into a relationship with Servier licensing their RyR technology which is believed to currently be in phase II clinical trials.[116]

10.8.3 Myosin Activator

Mysosin acts as the molecular motor for cardiac muscle contractility and has been proposed as a promising therapy for HF.[118] Researchers at Cytokinetics have published the first, selective cardiac myosin small molecule activator.[119] A myosin ATPase activity assay was used for high throughput screening and the initial hit was elaborated to provide omecamtiv mecarbil (CK-1827452, Figure 10.10) a compound that increases fractional shortening in rats.

Cytokinetics and Amgen are currently developing omecamtiv mecarbil for HF in a phase II COSMIC-HF trial. The trial will look at pharmacokinetics as well as heart function over 20 weeks. This drug is being developed for both oral and IV formulations which will provide flexibility for inpatient IV formulation for AHF patients and the ability to continue outpatient care *via* oral dosing.

10.8.4 Calcium Sensitizer

Levosimendan (Figure 10.11), the active enantiomer of simendan is an IV drug marketed in the EU for HF since 2000 but is not currently available in the US. Levosimendan is a calcium sensitizer that binds troponin C in cardiac myocytes.[120] This action stabilizes the calcium-bound conformation of troponin C and prolongs the actin–myosin interaction thereby increasing cardiac contractility.

Levosimendan has been tested in several clinical trials and has shown mixed results. For example, REVIVE-II showed that worsening HF was slightly better with drug but was associated with hypotension, arrythmias and more deaths than placebo.[121] The CASINO trial, on the other hand, was terminated early due to the reduced mortality in the levosimendan treated

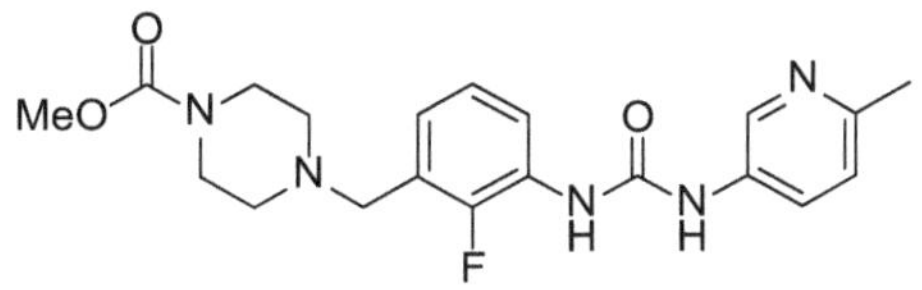

Figure 10.10 Omecamtiv mecarbil.

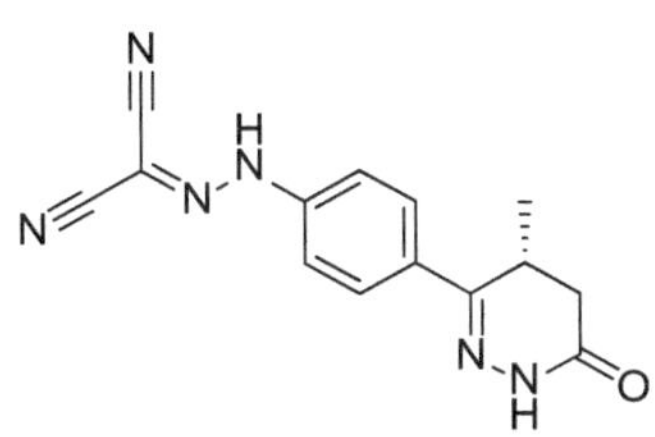

Figure 10.11 Levosimendan.

group. The investigators found that at one month, levosimendan significantly reduced mortality (6.0%) *versus* placebo (8.1%) and dobutamine (14.0%). After six month follow up, levosimendan reduced mortality (18.0%), *versus* placebo (28.3%) and dobutamine (42.0%).[122] The FDA has granted levosimendan fast track status and a Phase III trial (LEVO-CTS) is enrolling 760 patients with left ventricular systolic dysfunction undergoing cardiac surgery. The primary endpoints will look at all-cause death or use of medical assist device at 30 days.

10.8.5 Sodium Channel Antagonist

It has been shown that blocking the late sodium channel in cardiomyocytes can reduce sodium overload and minimize calcium overload.[123] Maintaining calcium homeostasis can preserve mitochondrial function and restore contractility and diastolic function. Ranolazine, an antagonist of the sodium, calcium and potassium channels, is a marketed Gilead drug for the treatment of stable angina (Figure 10.12).[123] It is now being tested as a treatment for cardiomyopathy in a phase III trial examining tachycardia, ventricular function or death over two years. It is also in phase II for heart failure with preserved ejection fraction (HFpEF), a patient population not served by current drugs, looking at six minute walk test (6MWT) after 28 weeks of treatment.

Gilead is also evaluating GS-6615 a selective, late sodium channel antagonist.[124] They are running a phase II trial in 210 subjects with arrhythmias, with implantable cardioverter defibrillator (ICD) or cardiac resynchronization therapy-defibrillator. The primary endpoints examined will be the overall occurrence of ICD interventions through 24 weeks.

10.8.6 Beta Adrenergic Receptor Antagonists

Beta adrenergic receptor antagonists, also commonly known as 'beta-blockers,' are a first line therapy for acute decompensated HF and are an accepted treatment for chest pain, arrhythmia, hypertension and hypertrophic cardiomyopathy. They act as negative inotropic agents that decrease heart rate and cardiac output which lowers blood pressure.[125] Importantly, they have also been shown to reduce CV events and deaths.[126] Drug development in this target class is challenging because one must differentiate new drugs from the surplus of available low cost generics and deal with typical beta blocker nuisance effects including fatigue and sexual dysfunction.

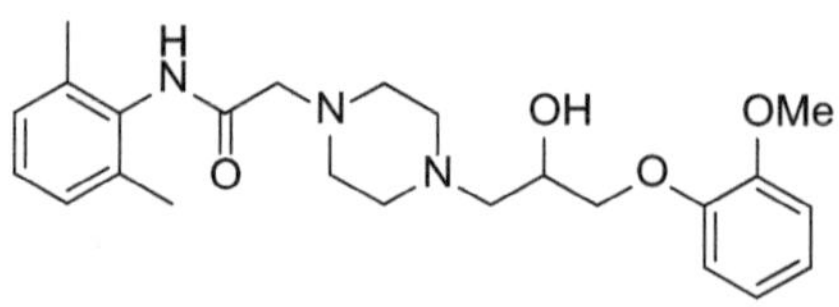

Figure 10.12 Ranolazine.

Bucindolol (Gencaro) is a beta 1 and 2 antagonist being developed by ARCA for the treatment of HF in a phase II/III trial (Figure 10.13). This study is being run in patients with a specific genotype for the beta-1 and/or alpha-2c adrenergic receptor that have shown improved bucindolol benefit in comparison to other beta blockers.[127] While this is an interesting use of precision medicine in the CV field, it may be difficult to gain clinical use when a genetic marker is required.

10.8.7 Funny Channel (I_f) Inhibitor

Multiple currents are responsible for the depolarization of membrane potential that leads to the initiation of the pacemaker potential in the sinoatrial node. One such current is the so-called 'funny' channel (I_f) which is responsible for the inward current of sodium ions which leads to membrane depolarization. Ivabradine is a small molecule inhibitor of the funny channel discovered by Servier shown to decrease heart rate (Figure 10.14).[126] While there are other bradycardic drugs, including beta blockers or calcium channel blockers, ivabradine is unique in its selectivity for the funny channel.[128] As such it is expected that ivabradine would avoid the adverse effects associated with beta blockers and if co-dosed with beta 1 agonist dobutamine, it could limit tachycardia.[129,130]

The SHIFT clinical trial evaluated ivabradine in 6558 patients with symptomatic CHF on top of their optimal medications. Investigators found improvements in death and/or hospitalizations (24% for drug *vs.* 29% for placebo) after a median follow up time of 22 months.[131] In 2012, ivabradine was approved in the EU for the treatment of patients with CHF with elevated heart rate. Amgen and Servier entered into an agreement in 2013 giving Amgen US commercialization rights. In the first quarter of 2014, it was granted fast-track status by the FDA for the treatment of CHF. While the drug has been shown to be well tolerated and have benefits on top of beta

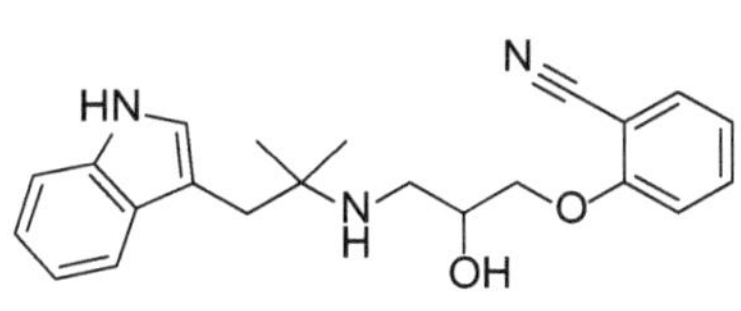

Figure 10.13 Bucindolol.

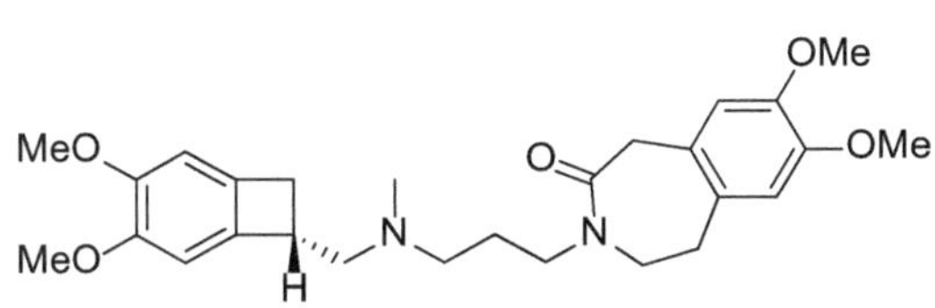

Figure 10.14 Ivabradine.

blockers in CHF patients, the outstanding question is whether or not the FDA will require an outcomes trial for approval.

10.9 Mitochondrial Function

10.9.1 Cardiolipin Stabilizer

Bendavia, an oral peptide (Figure 10.15), has been shown to localize in the mitochondria and have cardioprotective effects in preclinical models of reperfusion injury.[132,133] While originally hypothesized to act as a ROS scavenger, recent studies suggest that Bendavia may be targeting the mitochondrial phospholipid cardiolipin to protect the mitochondrial cristae.[134] It is currently in phase I development for the treatment of HF.

10.9.2 Mitochondrial Permeability Transition Pore (mPTP) Inhibitors

TRO40303 (3,5-seco-4-nor-cholestan-5-one oxime-3-ol) is a small molecule inhibitor of the mitochondrial permeability transition pore (mPTP).[135] It has been shown to delay mPTP opening during reperfusion and reduce the release of apoptosis-inducing factor. Trophos is evaluating IV administration of TRO40303 in a phase II clinical trial in patients undergoing percutaneous coronary intervention after MI.[136] The primary outcome will be measures of infarct size. Cyclosporin A, in addition to its clinical use as an immunosuppressant, has also been reported to act as a mPTP inhibitor[137] and is being evaluated in phase III trials for reperfusion injury.

10.10 Relaxin/Insulin-like Family Peptide Receptor 1 (RXFP1) Agonists

Relaxin is a pregnancy related hormone responsible for increasing vasodilation leading to blood pressure normalization and has also been shown to have anti-fibrotic effects.[138] Novartis is developing serelaxin (RLX030), a

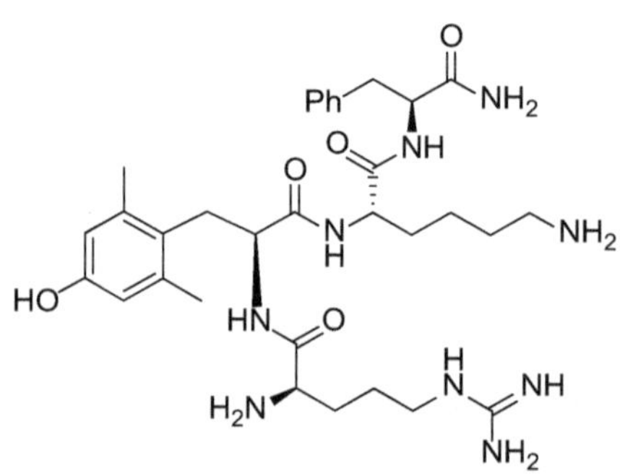

Figure 10.15 Bendavia.

recombinant form of human relaxin-2 with cardiorenal function-enhancing properties. In 2013, they gained breakthrough therapy designation from the FDA based on the results of the phase II/III RELAX-AHF trial. The primary endpoint for this trial was dyspnea, which showed a modest improvement.[139] In early 2014, the FDA advisory committee voted against approval of serelaxin for the treatment of AHF based on RELAX-HF study due to "insufficient evidence to support the proposed indication: to 'improve the symptoms of AHF through reduction of the rate of worsening of HF'".[140] Importantly, researchers saw a 37% reduction in mortality after 180 days, which the FDA referred to as "reassuring". The trial was not powered sufficiently to establish a strong effect and Novartis is following up with a larger phase III trial for AHF. This study is currently recruiting 6375 AHF patients where the primary endpoint will be time to CV death during 180 days and will be complete in 2016. In addition to Novartis, Ambrx/BMS are also in phase I clinical development with a pegylated recombinant relaxin, BMS-986046, for HF.

The promising results from the ACUTE-AHF trial make the case for the discovery of a small molecule very compelling. However, the only published report of a small molecule to date is from the National Institutes of Health (NIH).[138,141] The lead compound (Figure 10.16) identified by screening NIH compound libraries was found to be a potent agonist of RXFP1. There are currently no known RXFP1 small molecules in clinical development.

10.11 P38 Mitogen-activated Protein Kinase (p38 MAPK) Inhibitors

Activation of p38 mitogen-activated protein kinase (p38 MAPK) by cellular stress such as myocardial injury is involved in apoptosis.[142] Inhibition of p38 MAPK has been shown to reduce infarct size, cardiac remodeling and slow the progression of atherosclerosis in preclinical animal models.[143] GSK is developing losmapimod, a p38 MAPK inhibitor for the treatment of acute coronary syndrome (Figure 10.17). Recent phase II results show that drug treatment improved some markers of inflammation at 72 hours but missed the primary end point of inflammation and infarct size at twelve weeks.[143]

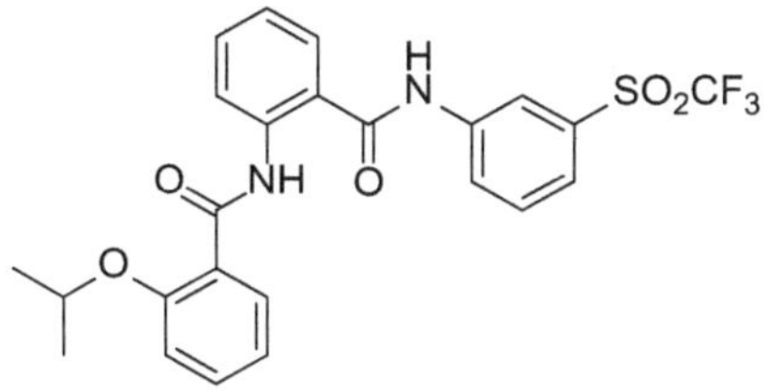

HEK-RXFP1 EC_{50} =94 nM
THP1 EC_{50} =200 nM

Figure 10.16 NIH RXFP1 allosteric agonist.

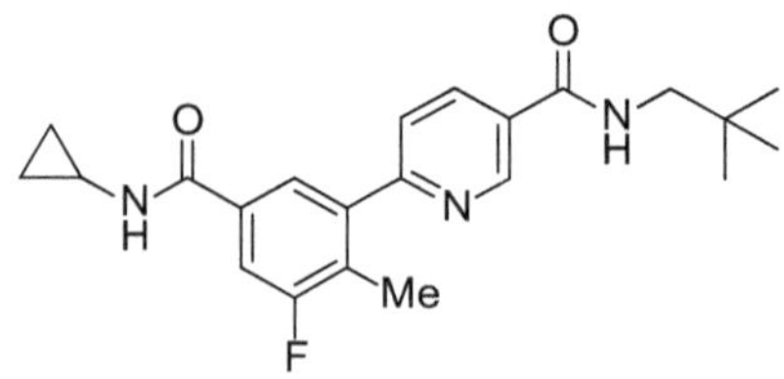

Figure 10.17 Losmapimod.

The authors of the study conclude that the drug might still prove to be therapeutically beneficial and GSK is now recruiting over 25 000 patients for a phase III trial in patients with MI examining time to first CV event over three months. If successful, losmapimod would be the first kinase inhibitor for CVD.[144]

Array BioPharma is also testing their p38 inhibitor ARRY-797 (ARRY-371797) in the clinic.[145] Originally in the clinic for pain, ARRY-797 is now in a small Ph II trial in patients with a rare genetic CVD caused by mutations in the lamin A/C gene that causes dilated cardiomypathy (LMNA-DCM) where the primary endpoint assessed will be improvements in the 6MWT.

10.12 Regenerative Medicine

10.12.1 Stem Cells

Stem cells have been heralded as a potential therapy to reverse effects of HF resulting from, for example, ischemic events as well as chronic states of hypertension.[146] However, to date, most clinical trials with stem cell therapy have not been sufficiently powered to demonstrate improvements in CV end points. In fact, a meta-analysis of stem cell clinical trials has shown that cell-based therapy improved the heart's pumping capacity by only 1.78%, with no improvements in mortality.[147] Some challenges associated with stem cell therapy include the differences in cell source and procurement (allogenic *vs.* autologous), purification and invasive delivery methods. In addition, cell therapy would have to be designed to treat a particular condition, *i.e.* using cells capable of regenerating contractile tissue to treat infarct, as opposed to a single product concept for HF. There are a number of stem cell therapies currently in the clinic for HF. Of particular note is TEVA/Mesoblast's CEP-41750, allogenic bone-marrow-derived mesenchymal cell therapy in phase III development, because they will be looking at major adverse cardiac events over five years in over 1700 patients as opposed to functional cardiac endpoints.

10.12.2 Stromal Cell-Derived Factor-1 (SDF-1)

Juventas is using an alternative strategy to cell therapy for their regenerative medicine platform. They are evaluating JVS-100, a DNA plasmid encoding stromal cell-derived factor-1 (SDF-1, also referred to as CXCL12) for the treatment of HF and critical leg ischemia in a phase II trial. SDF-1 is a cytokine responsible for chemotaxis, adhesion, proliferation and differentiation of cells including progenitor cells to sites of tissue injury.[148] Their phase I results show that endomyocardial injection of JVS-100 showed no adverse events related to JVS 100.[149] The STOP-HF trial is a phase II study evaluating the safety and efficacy of intramyocardial injection of JVS-100 in patients with symptomatic ischemic cardiomyopathy.

10.13 Dyslipidemia

While statins, HMG-CoA reductase inhibitors used to reduce LDL-cholesterol (LDL-c) levels, have been shown to lower the risk for CVD, many patients on statins do not achieve sufficient LDL-c lowering. These patients are often intolerant of a full therapeutic dose due to side effects such as muscle pain and liver enzyme abnormalities.[150] Treatments that lower cholesterol levels *via* novel mechanisms would be valuable to such a patient population and are discussed below.[151] In addition to drugs for large patient groups, a number of new mechanisms are in the clinic looking to serve patients with rare, genetic diseases. These new trials are collecting clinical data which could one day be applicable to HF patients and will be discussed at the end of this section.

10.13.1 Proprotein Convertase Subtilisin/Kexin Type 9 (PCSK9)

One of the most promising new mechanisms for the treatment of dyslipidemia currently in the clinic is neutralization of proprotein convertase subtilisin/kexin type 9 (PCSK9). PCSK9 interacts with the LDL receptor (LDLR) inducing LDLR degradation. Humans with PCSK9 loss of function mutations are reported to have low LDL-c levels and a dramatically reduced risk for coronary heart disease.[152,153] These subjects are also healthy which suggests that pharmacological inhibition of PCSK9 might be a safe and effective therapy.

The most advanced PCSK9-targeting therapies in the clinic are monoclonal antibodies that target the PCSK9-LDLR interaction. These antibodies have shown dramatic effects in lowering LDL-c levels in the clinic (60–75%).[154] However, recent changes in the recommended target LDL-c levels by the American College of Cardiology and American Heart Association, in addition to the fact that this is a new mechanism, necessitate large trials. Regeneron and Sanofi (alirocumab), Amgen (evolocumab) and Pfizer (bococizumab) all have phase III trials looking at LDL-c levels and CV outcomes. In addition to the antibodies in the clinic, BMS is developing BMS-962476 an anti-human PCSK9 adnectin-based protein therapeutic. There are also oligonucleotide approaches including Alnylam's ALN-PCS si-RNA which has completed phase I trials.

A small molecule PCSK9 inhibitor would be competitive with the antibodies; however, there are very few reports of such chemical matter. One such compound, K312, from Kowa Research Institute has completed a phase I trial. This compound is a cholesterol ester transfer protein (CETP) inhibitor that suppresses PCSK9 expression.[155] Kowa also have a number of published patents on chemical matter that reduces PCSK9 mRNA expression.[156–158] Cadila are another company with published patents targeting PCSK9.[159,160]

10.13.2 Cholesterol Ester Transfer Protein (CETP) Inhibitors

Higher HDL-cholesterol (HDL-c) levels are associated with CV benefit[161] but it has not yet been demonstrated that increasing HDL-c will improve CV outcomes. One approach to increase HDL-c levels has been to modulate cholesterol ester transfer proteins (CETPs) which transfer cholesterol esters from HDL-c to pro-atherogenic lipoproteins such as LDL-c. Because CETP deficiency increases HDL-c, it is hypothesized that a CETP inhibitor would be of therapeutic benefit.[151]

For a recent review of CETP chemical matter readers are referred to Mantlo *et al.* and cited references therein.[162] The first CETP inhibitor to be tested in a phase III trial was Pfizer's torcetrapib. A potent CETP inhibitor, torcetrapib was tested in over 15 000 patients and showed efficacy in increasing HDL-c and reducing LDL-c. However, the trial was terminated after a year and a half when investigators saw a significant increase in CV-related events and hospitalizations.[163] Dalcetrapib (Japan Tabacco/Roche) was the second compound to be tested in a large phase III clinical trial but did not show a significant difference in HDL-c or LDL-c and did not meet the primary CV endpoint after 2.6 years.[164] Merck (anacetrapib) and Eli Lilly (evacetrapib) are both running large phase III outcomes trials which will provide insight as to whether the increased mortality seen with torcetrapib was a mechanism-based phenomenon or a chemotype-specific effect (Figure 10.18).

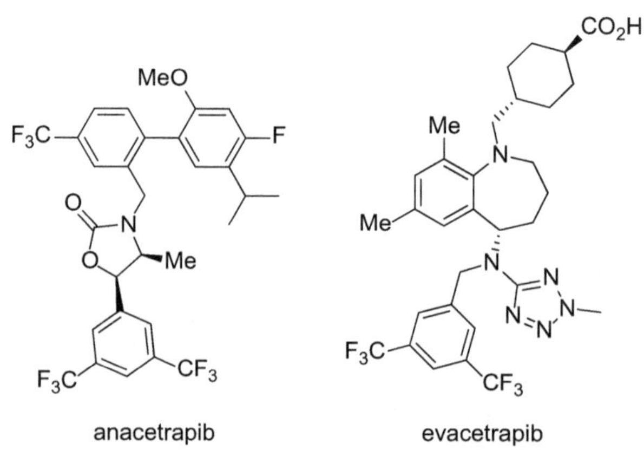

Figure 10.18 CETP inhibitors.

10.13.3 Lipoprotein-associated Phospholipase A2 (Lp-PLA2) Inhibitor

Lipoprotein-associated phospholipase A2 (Lp-PLA2) is a proinflammatory lipid-modifying enzyme associated with atherosclerotic plaque formation and is an identified risk factor for coronary heart disease.[165] Lp-PLA2 hydrolyzes oxidized phospholipids in LDL, generating products that activate inflammatory and pro-atherogenic pathways. It is hypothesized that inhibiting Lp-PLA2 will improve atherosclerosis *via* a novel anti-inflammatory mechanism and lead to reduced CV events.[166] Darapladib is a Lp-PLA2 inhibitor in phase III development by GSK (Figure 10.19). Phase III trial results were recently published and show that the primary endpoints, a reduction in CV death, MI or stroke, were not met.[167] However, since there was a nominal improvement in coronary events, it is unclear what GSK's path forward will be as there might be an opportunity for another trial in a patient population that is statin-intolerant.[168]

10.13.4 Lysosomal Acid Lipase (LIPA or LAL)

Lysosomal acid lipase (LIPA or LAL) is responsible for maintaining lipid homeostasis by breaking down cholesterol esters. Synageva BioPharma is in phase III development with sebelipase alfa (SBC-102) which is a recombinant human LIPA replacement therapy for patients with LIPA deficiency, known as Wolman disease. There are currently no approved therapies for LIPA deficiency and orphan drug status and breakthrough therapy designation have been granted in the US.

10.13.5 Microsomal Triglyceride Transfer Protein (MTTP) Inhibitor

Microsomal triglyceride transfer protein (MTTP) is necessary for the assembly and secretion of VLDL particles.[151] MTTP incorporates lipids into apoB and acts as a chaperone in the proper folding of apolipoprotein B (apoB). Aegerion's lomitapide (Figure 10.20) acts as a microsomal triglyceride transfer protein inhibitor and was recently approved in the US, EU and

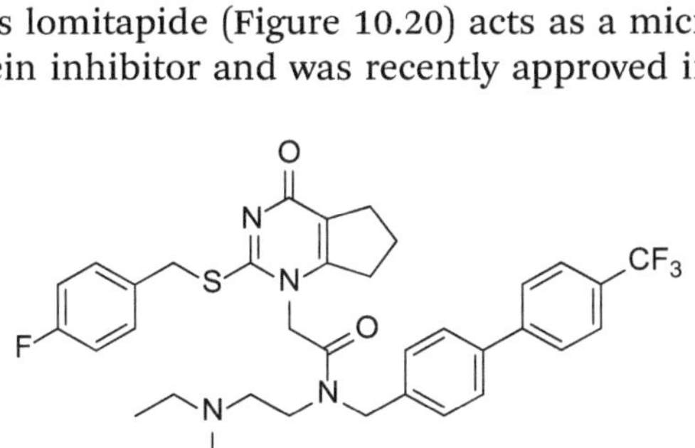

Figure 10.19 Darapladib.

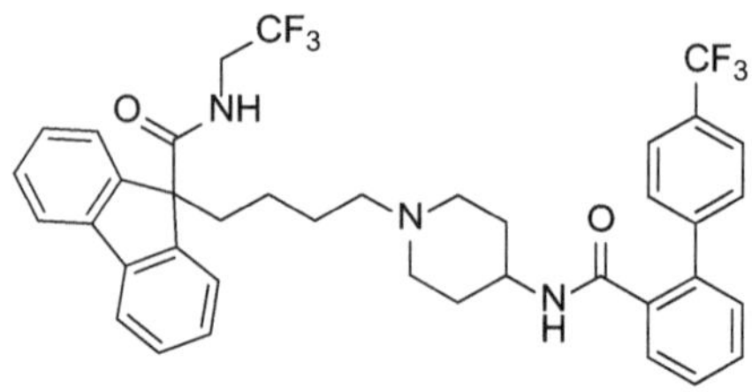

Figure 10.20 Lomitapide.

Canada as a lipid lowering agent in patients with homozygous familial hypercholesterolemia (HoFH) a life-threatening genetic disease.

10.13.6 Apolipoprotein B-100

Apolipoprotein B-100 (apoB-100) is an essential component of atherogenic lipoproteins (*i.e.* VLDL, LDL) and the inhibition of its synthesis reduces circulating lipids.[151] Genzyme is currently marketing mipomersen, an ASO specific for human apoB-100 mRNA for HoFH.[151] Phase III clinical data in HoFH patients on maximum tolerated lipid-lowering drugs showed that LDL-c levels were reduced by 25%.[169] More recent phase III results in patients with high cholesterol and at high risk for CHF showed that mipomersen reduced LDL-c levels by 37%.[170]

10.14 Novel Oral Anticoagulants

Thrombotic disorders are a major complication in CVD and account for a significant portion of associated morbidity and mortality. The anticoagulant warfarin, an oral vitamin K antagonist (VKA) that acts by inhibition of vitamin K epoxide reductase (VKOR), is the most commonly prescribed drug to manage the risk of thrombosis and has proven efficacy for the prevention of stroke.[171] The metabolism and exposure of warfarin are, however, difficult to predict and dependent upon a number of factors, including food intake, alcohol consumption, use of other drugs, and genetic factors – in particular polymorphisms in its major metabolizing enzyme, CYP2C9.[172,173] Furthermore, warfarin is metabolized to a number of hydroxywarfarins which subsequently inhibit CYP2C9 activity.[174] Armetheon is developing tecarfarin, originally discovered by ARYx,[175] also a VKOR inhibitor (Figure 10.21). Tecarfarin, however, is not a substrate for oxidative metabolism, but rather is cleaved to a single inactive acid metabolite *via* carboxyesterase enzymes with minimal risk of drug–drug interaction and not subject to CYP polymorphisms.[176] Tecarfarin has demonstrated good exposure control in patients with atrial fibrillation in comparison to warfarin,[177] and will begin recruitment for a phase III trial which will include patients with prosthetic heart valves. The recently introduced factor Xa (FXa) inhibitors are alternatives to warfarin that also lower the bleeding risk and are showing great

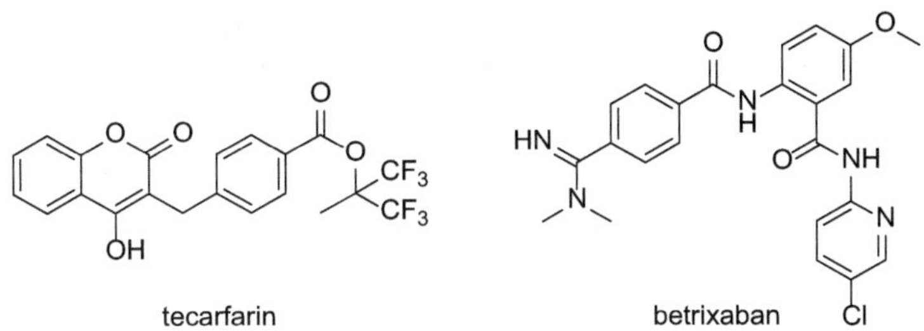

Figure 10.21 Oral anticoagulants.

promise as anticoagulant therapies. These drugs have been reviewed extensively.[178–182]

Like tecarfarin, the novel FXa inhibitor betrixaban (Figure 10.21)[183] has metabolic advantages over other drugs in its class. Betrixaban, currently in phase III trials sponsored by Portola, is unique in that it is not metabolized by CYP3A4 and does not require monitoring or dose titration.[184] In addition, it is cleared *via* biliary excretion with minimal renal clearance, which is predicted to give it an advantage over all other anticoagulants for the treatment of patients with severe renal impairment.[185]

10.15 GLP-1 Analogues

Glucagon-like peptide-1 (GLP-1) is a gut-derived incretin hormone responsible for glucose homeostasis *via* insulin stimulation. In addition to glucoregulatory effects, GLP-1 has also been reported to have CV effects in animal models and in humans.[186] For example, GLP-1 treatment of patients with acute MI showed a significant improvement in ejection fraction. Liraglutide, a GLP-1 analogue with a longer half-life is currently being evaluated in two clinical trials as a treatment for diabetic patients with HF examining cardiac function.

10.16 Conclusions

After years of research, an outstanding question remains regarding the dearth of new HF therapies. Among the reasons cited for the lack of new drugs is the complex etiology of the disease which leads to patient heterogeneity.[3] While there are high costs associated with HF trials which may act as a financial hurdle, there are also considerable questions around clinical study design. For example, what is the optimum timing and duration of intervention? What inclusion criteria should be used for a particular patient population? Background therapy is often varied for a particular patient group and may change over the course of a clinical trial, obfuscating the placebo and treatment group safety and efficacy endpoints. There is also the question of efficacy as there is no consensus on what those endpoints should be for AHF.[3] In addition, regulatory differences between the FDA and

European Medical Agency (EMA) mean variability in study designs. It can mean that a drug may not be able to win approval in both regions from one trial alone. A variety of mechanisms are involved in the pathogenesis of HF and the contribution of each is uncertain. Potentially important mechanisms may remain unknown. As our understanding of these mechanisms grows, so will the precision with which we can target these conditions.

Acknowledgements

We thank Pfizer colleagues Georgios Karamanlidis, Brian Raymer and Kim F. McClure and for helpful discussions during the preparation of this manuscript.

References

1. World Health Organization, 2011, Global status report on noncommunicable diseases 2010, http://www.who.int/nmh/publications/ncd_report_full_en.pdf.
2. D. Lloyd-Jones, R. J. Adams, T. M. Brown, M. Carnethon, S. Dai, G. De Simone, T. B. Ferguson, E. Ford, K. Furie, C. Gillespie, A. Go, K. Greenlund, N. Haase, S. Hailpern, P. M. Ho, V. Howard, B. Kissela, S. Kittner, D. Lackland, L. Lisabeth, A. Marelli, M. M. McDermott, J. Meigs, D. Mozaffarian, M. Mussolino, G. Nichol, V. L. Roger, W. Rosamond, R. Sacco, P. Sorlie, R. Stafford, T. Thom, S. Wasserthiel-Smoller, N. D. Wong and J. Wylie-Rosett, *Circulation*, 2010, **121**, e46.
3. G. M. Felker, P. S. Pang, K. F. Adams, J. G. Cleland, G. Cotter, K. Dickstein, G. S. Filippatos, G. C. Fonarow, B. H. Greenberg, A. F. Hernandez, S. Khan, M. Komajda, M. A. Konstam, P. P. Liu, A. P. Maggioni, B. M. Massie, J. J. McMurray, M. Mehra, M. Metra, J. O'Connell, C. M. O'Connor, I. L. Pina, P. Ponikowski, H. N. Sabbah, J. R. Teerlink, J. E. Udelson, C. W. Yancy, F. Zannad, M. Gheorghiade and on behalf of the International AHFS Working Group, *Circ.: Heart Failure*, 2010, **3**, 314.
4. B. M. Patel and A. A. Mehta, *Eur. J. Pharmacol.*, 2012, **697**, 1.
5. A. Valluri, A. D. Struthers and C. C. Lang, *Curr. Heart Failure Rep.*, 2014, **11**, 31.
6. J. M. Neutel and D. H. Smith, *Cardiovasc. Ther.*, 2013, **31**, 251.
7. H. Makani, S. Bangalore, K. A. Desouza, A. Shah and F. H. Messerli, *Br. Med. J.*, 2013, **346**, f360.
8. R. Corti, J. C. Burnett, Jr., J. L. Rouleau, F. Ruschitzka and T. F. Luscher, *Circulation*, 2001, **104**, 1856.
9. T. G. von Lueder, D. Atar and H. Krum, *Pharmacol. Ther.*, 2014, **144**, 41.
10. Novartis Media Release, 2014, PARADIGM-HF trial of Novartis' LCZ696 for chronic heart failure closes early based on strength of interim

results, http://www.novartis.com/newsroom/media-releases/en/2014/1772754.shtml.

11. F. Zannad, J. J. McMurray, H. Krum, D. J. van Veldhuisen, K. Swedberg, H. Shi, J. Vincent, S. J. Pocock, B. Pitt and Emphasis-HF Study Group, *N. Engl. J. Med.*, 2011, **364**, 11.
12. S. Dhillon, *Drugs*, 2013, **73**, 1451.
13. J. J. V. McMurray, S. Adamopoulos, S. D. Anker, A. Auricchio, M. Boehm, K. Dickstein, V. Falk, G. Filippatos, C. Fonseca, M. A. Gomez-Sanchez, T. Jaarsma, L. Kober, G. Y. H. Lip, A. P. Maggioni, A. Parkhomenko, B. M. Pieske, B. A. Popescu, P. K. Ronnevik, F. H. Rutten, J. Schwitter, P. Seferovic, J. Stepinska, P. T. Trindade, A. A. Voors, F. Zannad, A. Zeiher, J. J. Bax, H. Baumgartner, C. Ceconi, V. Dean, C. Deaton, R. Fagard, C. Funck-Brentano, D. Hasdai, A. Hoes, P. Kirchhof, J. Knuuti, P. Kolh, T. McDonagh, C. Moulin, B. A. Popescu, Z. Reiner, U. Sechtem, P. A. Sirnes, M. Tendera, A. Torbicki, A. Vahanian, S. Windecker, T. McDonagh, U. Sechtem, L. A. Bonet, P. Avraamides, H. A. Ben Lamin, M. Brignole, A. Coca, P. Cowburn, H. Dargie, P. Elliott, F. A. Flachskampf, G. F. Guida, S. Hardman, B. Iung, B. Merkely, C. Mueller, J. N. Nanas, O. W. Nielsen, S. Orn, J. T. Parissis and P. Ponikowski, *Eur. J. Heart Failure*, 2012, **14**, 803.
14. L. A. Allen, S. M. Shetterly, P. N. Peterson, J. H. Gurwitz, D. H. Smith, D. W. Brand, D. L. Fairclough, J. S. Rumsfeld, F. A. Masoudi and D. J. Magid, *Circ.: Heart Failure*, 2014, 7, 43.
15. L. Baerfacker, A. Kuhl, A. Hillisch, R. Grosser, S. Figueroa-Perez, H. Heckroth, A. Nitsche, J.-K. Ergueden, H. Gielen-Haertwig, K.-H. Schlemmer, J. Mittendorf, H. Paulsen, J. Platzek and P. Kolkhof, *ChemMedChem*, 2012, 7, 1385.
16. B. Pitt, L. Kober, P. Ponikowski, M. Gheorghiade, G. Filippatos, H. Krum, C. Nowack, P. Kolkhof, S.-Y. Kim and F. Zannad, *Eur. Heart J.*, 2013, **34**, 2453.
17. Q. Hu, L. Yin and R. W. Hartmann, *J. Med. Chem.*, 2014, **57**, 5011.
18. P. Namsolleck and T. Unger, *Nephrol., Dial., Transplant.*, 2014, **29**, i62.
19. C. D. Schumacher, R. E. Steele and H. R. Brunner, *J. Hypertens.*, 2013, **31**, 2085.
20. E. L. Meredith, G. Ksander, L. G. Monovich, J. P. N. Papillon, Q. Liu, K. Miranda, P. Morris, C. Rao, R. Burgis, M. Capparelli, Q.-Y. Hu, A. Singh, D. F. Rigel, A. Y. Jeng, M. Beil, F. Fu, C.-W. Hu and D. LaSala, *ACS Med. Chem. Lett.*, 2013, **4**, 1203.
21. D. A. Calhoun, W. B. White, H. Krum, W. Guo, G. Bermann, A. Trapani, M. P. Lefkowitz and J. Menard, *Circulation*, 2011, **124**, 1945.
22. L. Yin, Q. Hu, J. Emmerich, M. M.-C. Lo, E. Metzger, A. Ali and R. W. Hartmann, *J. Med. Chem.*, 2014, **57**, 5179.
23. J. D. Violin, A. L. Crombie, D. G. Soergel and M. W. Lark, *Trends Pharmacol. Sci.*, 2014, **35**, 308.

24. J. D. Violin, S. M. DeWire, D. Yamashita, D. H. Rominger, L. Nguyen, K. Schiller, E. J. Whalen, M. Gowen and M. W. Lark, *J. Pharmacol. Exp. Ther.*, 2010, **335**, 572.
25. G. Boerrigter, D. G. Soergel, J. D. Violin, M. W. Lark and J. C. Burnett, Jr., *Circ.: Heart Failure*, 2012, **5**, 627.
26. K.-S. Kim, D. Abraham, B. Williams, J. D. Violin, L. Mao and H. A. Rockman, *Am. J. Physiol.*, 2012, **303**, H1001.
27. D. G. Soergel, R. A. Subach, C. L. Cowan, J. D. Violin and M. W. Lark, *J. Clin. Pharmacol.*, 2013, **53**, 892.
28. E. Kaschina, D. Lauer, P. Schmerler, T. Unger and U. M. Steckelings, *Curr. Hypertens. Rep.*, 2014, **16**, 1.
29. L. Paulis and T. Unger, *Nat. Rev. Cardiol.*, 2010, **7**, 431.
30. J. P. Joseph, A. P. Mecca, R. W. Regenhardt, D. M. Bennion, V. Rodriguez, F. Desland, N. A. Patel, D. J. Pioquinto, T. Unger, M. J. Katovich, U. M. Steckelings and C. Sumners, *Neuropharmacology*, 2014, **81**, 134.
31. J. Gao, I. H. Zucker and L. Gao, *Am. J. Hypertens.*, 2014, **27**, 1248.
32. J. T. Ortiz-Perez, M. Riera, X. Bosch, T. M. De Caralt, R. J. Perea, J. Pascual and M. J. Soler, *PLoS One*, 2013, **8**, e61695.
33. M. Flores-Munoz, B. M. Godinho, A. Almalik and S. A. Nicklin, *PLoS One*, 2012, **7**, e45564.
34. R. A. Santos, A. C. Simoes e Silva, C. Maric, D. M. Silva, R. P. Machado, I. de Buhr, S. Heringer-Walther, S. V. Pinheiro, M. T. Lopes, M. Bader, E. P. Mendes, V. S. Lemos, M. J. Campagnole-Santos, H. P. Schultheiss, R. Speth and T. Walther, *Proc. Natl. Acad. Sci. U. S. A.*, 2003, **100**, 8258.
35. N. Parajuli, T. Ramprasath, V. B. Patel, W. Wang, B. Putko, J. Mori and G. Y. Oudit, *Can. J. Physiol. Pharmacol.*, 2014, **92**, 558.
36. S. Ahmad, J. Varagic, L. Groban, L. Dell'Italia, S. Nagata, N. Kon and C. Ferrario, *Curr. Hypertens. Rep.*, 2014, **16**, 1.
37. S. Ahmad, T. Simmons, J. Varagic, N. Moniwa, M. C. Chappell and C. M. Ferrario, *PLoS One*, 2011, **6**, e28501.
38. G. Nguyen, F. Delarue, C. Burckle, L. Bouzhir, T. Giller and J.-D. Sraer, *J. Clin. Invest.*, 2002, **109**, 1417.
39. H. Mahmud, H. H. Sillje, M. V. Cannon, W. H. van Gilst and R. A. de Boer, *J. Cell. Mol. Med.*, 2012, **16**, 722.
40. N. F. Schroten, C. A. Gaillard, D. J. van Veldhuisen, M. K. Szymanski, H. L. Hillege and R. A. de Boer, *Heart Failure Rev.*, 2012, **17**, 191.
41. D. J. van Veldhuisen, G. C. Linssen, T. Jaarsma, W. H. van Gilst, A. W. Hoes, J. G. Tijssen, W. J. Paulus, A. A. Voors and H. L. Hillege, *J. Am. Coll. Cardiol.*, 2013, **61**, 1498.
42. Y. Rautureau and G. F. Baxter, *Curr. Pharm. Des.*, 2004, **10**, 2477.
43. G. E. Woodard and J. A. Rosado, *Int. Rev. Cell Mol. Biol.*, 2008, **268**, 59.
44. E. Takimoto, *Circ. J.*, 2012, **76**, 1819.
45. L. R. Potter, *Pharmacol. Ther.*, 2011, **130**, 71.
46. H. Jerczynska and Z. Pawlowska, *Postepy Biochem.*, 2008, **54**, 35.

47. C. M. O'Connor, R. C. Starling, A. F. Hernandez, P. W. Armstrong, K. Dickstein, V. Hasselblad, G. M. Heizer, M. Komajda, B. M. Massie, J. J. V. McMurray, M. S. Nieminen, C. J. Reist, J. L. Rouleau, K. Swedberg, K. F. Adams, Jr., S. D. Anker, D. Atar, A. Battler, R. Botero, N. R. Bohidar, J. Butler, N. Clausell, R. Corbalan, M. R. Costanzo, U. Dahlstrom, L. I. Deckelbaum, R. Diaz, M. E. Dunlap, J. A. Ezekowitz, D. Feldman, G. M. Felker, G. C. Fonarow, D. Gennevois, S. S. Gottlieb, J. A. Hill, J. E. Hollander, J. G. Howlett, M. P. Hudson, R. D. Kociol, H. Krum, A. Laucevicius, W. C. Levy, G. F. Mendez, M. Metra, S. Mittal, B. H. Oh, N. L. Pereira, P. Ponikowski, W. H. W. Tang, W. H. Wilson, S. Tanomsup, J. R. Teerlink, F. Triposkiadis, R. W. Troughton, A. A. Voors, D. J. Whellan, F. Zannad and R. M. Califf, *N. Engl. J. Med.*, 2011, **365**, 32.
48. J. D. Sackner-Bernstein, H. A. Skopicki and K. D. Aaronson, *Circulation*, 2005, **111**, 1487.
49. P. M. McKie, S. J. Sangaralingham and J. C. Burnett, Jr., *Curr. Heart Failure Rep.*, 2010, 7, 93.
50. R. A. Rose, *Curr. Opin. Invest. Drugs (BioMed Cent.)*, 2010, **11**, 349.
51. O. Lisy, B. K. Huntley, D. J. McCormick, P. A. Kurlansky and J. C. Burnett, Jr., *J. Am. Coll. Cardiol.*, 2008, **52**, 60.
52. A. H. L. From, *Cardiovasc. Res.*, 2010, **88**, 391.
53. R. Zakeri and J. C. Burnett, *Can. J. Physiol. Pharmacol.*, 2011, **89**, 593.
54. D. M. Dickey and L. R. Potter, *J. Mol. Cell. Cardiol.*, 2011, **51**, 67.
55. F. L. Martin, S. J. Sangaralingham, B. K. Huntley, P. M. McKie, T. Ichiki, H. H. Chen, J. Korinek, G. E. Harders and J. C. Burnett, Jr., *PLoS One*, 2012, 7, e52422.
56. C. Y. W. Lee, H. H. Chen, O. Lisy, S. Swan, C. Cannon, H. D. Lieu and J. C. Burnett, Jr., *J. Clin. Pharmacol.*, 2009, **49**, 668.
57. J. Neutel, W. Rolston, S. Maddock, S. Goldsmith, M. Koren, V. A. Bill, J. Burnett and H. D. Lieu, *J. Am. Coll. Cardiol.*, 2012, **59**, E1037.
58. X. W. Ng, Y. Huang, K. L. Liu, F. Y. C. Boey and S. S. Venkatraman, *J. Pharm. Sci.*, 2014, **103**, 1400.
59. J. C. Burnett, Jr. and C. Y. W. Lee, *Natriuretic polypeptide CU-NP lacking the ability to lower blood pressure for treatment of renal and cardiac dysfunction*, WO2009015011A1, 2009.
60. A. Kilic, V. Rajapurohitam, S. M. Sandberg, A. Zeidan, J. C. Hunter, N. S. Faruq, C. Y. Lee, J. C. Burnett, Jr. and M. Karmazyn, *Cardiovasc. Res.*, 2010, **88**, 434.
61. S. Pan, H. H. Chen, D. M. Dickey, G. Boerrigter, C. Lee, L. S. Kleppe, J. L. Hall, A. Lerman, M. M. Redfield, L. R. Potter, J. C. Burnett, Jr. and R. D. Simari, *Proc. Natl. Acad. Sci. U. S. A.*, 2009, **106**, 11282.
62. E. Buys and P. Sips, *Curr. Opin. Nephrol. Hypertens.*, 2014, **23**, 135.
63. M. Gheorghiade, C. N. Marti, H. N. Sabbah, L. Roessig, S. J. Greene, M. Boehm, J. C. Burnett, U. Campia, J. G. F. Cleland, S. P. Collins, G. C. Fonarow, P. D. Levy, M. Metra, B. Pitt, P. Ponikowski, N. Sato, A. A. Voors, J.-P. Stasch and J. Butler, *Heart Failure Rev.*, 2013, **18**, 123.

64. M. Packer, W. H. Lee, P. D. Kessler, S. S. Gottlieb, N. Medina and M. Yushak, *N. Engl. J. Med.*, 1987, **317**, 799.
65. J.-P. Stasch, P. Pacher and O. V. Evgenov, *Circulation*, 2011, **123**, 2263.
66. M. Follmann, N. Griebenow, M. G. Hahn, I. Hartung, F.-J. Mais, J. Mittendorf, M. Schaefer, H. Schirok, J.-P. Stasch, F. Stoll and A. Straub, *Angew. Chem., Int. Ed.*, 2013, **52**, 9442.
67. C. Gerassimou, A. Kotanidou, Z. Zhou, D. D. C. Simoes, C. Roussos and A. Papapetropoulos, *Br. J. Pharmacol.*, 2007, **150**, 1084.
68. V. Mitrovic, A. Jovanovic and S. Lehinant, *Curr. Heart Failure Rep.*, 2011, **8**, 38.
69. F. B. M. Priviero and R. C. Webb, *J. Cardiovasc. Pharmacol.*, 2010, **56**, 229.
70. G. Boerrigter, L. C. Costello-Boerrigter, A. Cataliotti, H. Lapp, J.-P. Stasch and J. C. Burnett, *Hypertension*, 2007, **49**, 1128.
71. J. Tamargo, J. Duarte, R. Caballero and E. Delpon, *Curr. Opin. Invest. Drugs (BioMed Cent.)*, 2010, **11**, 1039.
72. H.-P. Zhao, H.-M. Jiang and B.-R. Xiang, *Expert Opin. Invest. Drugs*, 2012, **21**, 1449.
73. E. S. Underbakke, A. T. Iavarone and M. A. Marletta, *Proc. Natl. Acad. Sci. U. S. A.*, 2013, **110**, 6777.
74. E. R. Derbyshire and M. A. Marletta, *Annu. Rev. Biochem.*, 2012, **81**, 533.
75. O. Gileadi, *Biochem. Soc. Trans.*, 2014, **42**, 108.
76. A. Kalra, V. Maharaj and S. R. Goldsmith, *Curr. Heart Failure Rep.*, 2014, **11**, 10.
77. S. Nielsen, C. L. Chou, D. Marples, E. I. Christensen, B. K. Kishore and M. A. Knepper, *Proc. Natl. Acad. Sci. U. S. A.*, 1995, **92**, 1013.
78. K. A. Walter, *Am. J. Health-Syst. Pharm.*, 2007, **64**, 1385.
79. K. Kondo and Y. Yamamura, *Case Studies in Modern Drug Discovery and Development*, John Wiley & Sons, Inc., 2012, ch. 13, p. 336.
80. K. Kondo, H. Ogawa, H. Yamashita, H. Miyamoto, M. Tanaka, K. Nakaya, K. Kitano, Y. Yamamura, S. Nakamura, T. Onogawa, T. Mori and M. Tominaga, *Bioorg. Med. Chem.*, 1999, **7**, 1743.
81. Otsuka Pharmaceutical Company, Ltd., 2013, Tolvaptan Product Monograph, http://samsca.ca/assets/doc/monograph_en.pdf.
82. M. A. Konstam, M. Gheorghiade, J. C. Burnett, Jr., L. Grinfeld, A. P. Maggioni, K. Swedberg, J. E. Udelson, F. Zannad, T. Cook, J. Ouyang, C. Zimmer, C. Orlandi and Efficacy of Vasopressin Antagonism in Heart Failure Outcome Study With Tolvaptan Investigators, *JAMA, J. Am. Med. Assoc.*, 2007, **297**, 1319.
83. D. E. Lanfear, H. N. Sabbah, S. R. Goldsmith, S. J. Greene, A. P. Ambrosy, A. J. Fought, M. J. Kwasny, K. Swedberg, C. W. Yancy, M. A. Konstam, A. P. Maggioni, F. Zannad, M. Gheorghiade and Everest Trial Investigators, *Circ.: Heart Failure*, 2013, **6**, 47.
84. P. J. Hauptman, J. Burnett, M. Gheorghiade, L. Grinfeld, M. A. Konstam, D. Kostic, H. B. Krasa, A. Maggioni, J. Ouyang,

K. Swedberg, F. Zannad, C. Zimmer, J. E. Udelson and Everest Investigators, *J. Card. Failure*, 2013, **19**, 390.
85. Y. Matsue, M. Suzuki, M. Seya, R. Iwatsuka, A. Mizukami, W. Nagahori, M. Ohno, A. Matsumura and Y. Hashimoto, *J. Cardiol.*, 2013, **61**, 169.
86. B. T. Bowman and M. H. Rosner, *Core Evidence*, 2013, **8**, 47.
87. G. Liamis, T. D. Filippatos and M. S. Elisaf, *Expert Rev. Clin. Pharmacol.*, 2014, 7, 431.
88. T. Ring, *Kidney Int.*, 2013, **83**, 1205.
89. D. Wile, *Ann. Clin. Biochem.*, 2012, **49**, 419.
90. R. T. Morrison, *Med. Clin. North Am.*, 1997, **81**, 689.
91. V. Hasselblad, W. Gattis Stough, M. R. Shah, Y. Lokhnygina, C. M. O'Connor, R. M. Califf and K. F. Adams, Jr., *Eur. J. Heart Failure*, 2007, **9**, 1064.
92. P. M. McKie, J. A. Schirger, L. C. Costello-Boerrigter, S. L. Benike, L. K. Harstad, K. R. Bailey, D. O. Hodge, M. M. Redfield, R. D. Simari, J. C. Burnett, Jr. and H. H. Chen, *J. Am. Coll. Cardiol.*, 2011, **58**, 2095.
93. W. Mullens and W. H. Tang, *J. Am. Coll. Cardiol.*, 2011, **58**, 2104.
94. M. R. Costanzo, J. T. Heywood, B. M. Massie, J. Iwashita, L. Henderson, M. Mamatsashvili, H. Sisakian, H. Hayrapetyan, P. Sager, D. J. van Veldhuisen and D. Albrecht, *Eur. J. Heart Failure*, 2012, **14**, 922.
95. D. Albrecht, A. D. Strickland and G. M. Grass, *Crosslinked cation-binding polymers for treatment of heart failure*, WO2012108947A1, 2012.
96. F. L. Ruberg and J. L. Berk, *Circulation*, 2012, **126**, 1286.
97. Y. Sekijima, R. L. Wiseman, J. Matteson, P. Hammarstroem, S. R. Miller, A. R. Sawkar, W. E. Balch and J. W. Kelly, *Cell*, 2005, **121**, 73.
98. M. Hanna, *Curr. Heart Failure Rep.*, 2014, **11**, 50.
99. S. Nencetti, A. Rossello and E. Orlandini, *ChemMedChem*, 2013, **8**, 1617.
100. C. E. Bulawa, S. Connelly, M. DeVit, L. Wang, C. Weigel, J. A. Fleming, J. Packman, E. T. Powers, R. L. Wiseman, T. R. Foss, I. A. Wilson, J. W. Kelly and R. Labaudiniere, *Proc. Natl. Acad. Sci. U. S. A.*, 2012, **109**, 9629.
101. X. Jiang, J. N. Buxbaum and J. W. Kelly, *Proc. Natl. Acad. Sci. U. S. A.*, 2001, **98**, 14943.
102. S. C. Penchala, S. Connelly, Y. Wang, M. S. Park, L. Zhao, A. Baranczak, I. Rappley, H. Vogel, M. Liedtke, R. M. Witteles, E. T. Powers, N. Reixach, W. K. Chan, I. A. Wilson, J. W. Kelly, I. A. Graef and M. M. Alhamadsheh, *Proc. Natl. Acad. Sci. U. S. A.*, 2013, **110**, 9992.
103. T. Coelho, D. Adams, A. Silva, P. Lozeron, P. N. Hawkins, T. Mant, J. Perez, J. Chiesa, S. Warrington, E. Tranter, M. Munisamy, R. Falzone, J. Harrop, J. Cehelsky, B. R. Bettencourt, M. Geissler, J. S. Butler, A. Sehgal, R. E. Meyers, Q. Chen, T. Borland, R. M. Hutabarat, V. A. Clausen, R. Alvarez, K. Fitzgerald, C. Gamba-Vitalo, S. V. Nochur, A. K. Vaishnaw, D. W. Y. Sah, J. A. Gollob and O. B. Suhr, *N. Engl. J. Med.*, 2013, **369**, 819.
104. E. J. Ackermann, S. Guo, S. Booten, L. Alvarado, M. Benson, S. Hughes and B. P. Monia, *Amyloid*, 2012, **19**, 43.
105. E. G. Kranias and D. M. Bers, *Subcell. Biochem.*, 2007, **45**, 523.

106. L. Lipskaia, E. R. Chemaly, L. Hadri, A. M. Lompre and R. J. Hajjar, *Expert Opin. Biol. Ther.*, 2010, **10**, 29.
107. M. Jessup, B. Greenberg, D. Mancini, T. Cappola, D. F. Pauly, B. Jaski, A. Yaroshinsky, K. M. Zsebo, H. Dittrich and R. J. Hajjar, *Circulation*, 2011, **124**, 304.
108. K. Zsebo, A. Yaroshinsky, J. J. Rudy, K. Wagner, B. Greenberg, M. Jessup and R. J. Hajjar, *Circ. Res.*, 2014, **114**, 101.
109. B. Greenberg, A. Yaroshinsky, K. M. Zsebo, J. Butler, G. M. Felker, A. A. Voors, J. J. Rudy, K. Wagner and R. J. Hajjar, *JACC: Heart Failure*, 2014, **2**, 84.
110. R. G. Johnson, Jr., *Ann. N. Y. Acad. Sci.*, 1998, **853**, 380.
111. K. M. Zsebo and R. Dahl, *Methods for treating diseases associated with the modulation of Serca*, 2010.
112. R. L. Cornea, S. J. Gruber, E. L. Lockamy, J. M. Muretta, D. Jin, J. Chen, R. Dahl, T. Bartfai, K. M. Zsebo, G. D. Gillispie and D. D. Thomas, *J. Biomol. Screening*, 2013, **18**, 97.
113. C. Tilgmann, P. Pollesello, M. Ovaska, J. Kaivola, J. Pystynen, E. Tiainen, M. Yliperttula, A. Annila and J. Levijoki, *Chem. Biol. Drug Des.*, 2013, **81**, 463.
114. H. Haikala, P. Kaheinen, J. Levijoki, J. Kaivola, M. Ovaska and J. Pystynen, *A method for the prevention and treatment of stunned myocardium using benzopyranones, quinolinones, and other phospholamban inhibitors*, US5968959A, 1999.
115. X. H. Wehrens, S. E. Lehnart, S. R. Reiken, S. X. Deng, J. A. Vest, D. Cervantes, J. Coromilas, D. W. Landry and A. R. Marks, *Science*, 2004, **304**, 292.
116. D. Aronson and H. Krum, *Pharmacol. Ther.*, 2012, **135**, 1.
117. M. George, M. Rajaram, E. Shanmugam and T. M. VijayaKumar, *Eur. J. Clin. Pharmacol.*, 2014, **70**, 765.
118. J. R. Teerlink, *Heart Failure Rev.*, 2009, **14**, 289.
119. B. P. Morgan, A. Muci, P.-P. Lu, X. Qian, T. Tochimoto, W. W. Smith, M. Garard, E. Kraynack, S. Collibee, I. Suehiro, A. Tomasi, S. C. Valdez, W. Wang, H. Jiang, J. Hartman, H. M. Rodriguez, R. Kawas, S. Sylvester, K. A. Elias, G. Godinez, K. Lee, R. Anderson, S. Sueoka, D. Xu, Z. Wang, N. Djordjevic, F. I. Malik and D. J. Morgans, Jr., *ACS Med. Chem. Lett.*, 2010, **1**, 472.
120. C. Pierrakos, D. Velissaris, F. Franchi, L. Muzzi, M. Karanikolas and S. Scolletta, *J. Clin. Med. Res.*, 2014, **6**, 75.
121. J. G. F. Cleland, N. Freemantle, A. P. Coletta and A. L. Clark, *Eur. J. Heart Failure*, 2006, **8**, 105.
122. A. P. Coletta, J. G. F. Cleland, N. Freemantle and A. L. Clark, *Eur. J. Heart Failure*, 2004, **6**, 673.
123. C. Antzelevitch, A. Burashnikov, S. Sicouri and L. Belardinelli, *Heart Rhythm*, 2011, **8**, 1281.
124. 2014, Gilead Announces Phase 1 Data for Investigational Therapy, GS-6615, in Patients with Long QT-3 Syndrome, http://www.gilead.com/

news/press-releases/2014/5/gilead-announces-phase-1-data-for-investigational-therapy-gs6615-in-patients-with-long-qt3-syndrome.
125. W. H. Frishman, *J. Cardiovasc. Pharmacol. Ther.*, 2013, **18**, 310.
126. A. Franco, F. M. Sarullo, Y. Salerno, S. Figliozzi, R. Parrinello, P. Pasquale and G. A. Lanza, *Am. J. Cardiovasc. Drugs*, 2014, **14**, 101.
127. The Beta-Blocker Evaluation of Survival Trial Investigators, *N. Engl. J. Med.*, 2001, **344**, 1659.
128. M. R. Rosen, A. Bucchi and R. B. Robinson, *Eur. Heart J. Suppl.*, 2006, **8**, D3.
129. F. Roubille, B. Lattuca, D. Busseuil, F. Leclercq, J.-M. Davy, E. Rhéaume and J.-C. Tardif, *Med. Hypotheses*, 2013, **81**, 202.
130. G. Riccioni, *Curr. Med. Chem.*, 2013, **20**, 1817.
131. K. Swedberg, M. Komajda, M. Bohm, J. S. Borer, I. Ford, A. Dubost-Brama, G. Lerebours and L. Tavazzi; on behalf of the SHIFT Investigators, *Lancet*, 2010, **376**, 875.
132. R. A. Kloner, S. L. Hale, W. Dai, R. C. Gorman, T. Shuto, K. J. Koomalsingh, J. H. Gorman, III, R. C. Sloan, C. R. Frasier, C. A. Watson, P. A. Bostian, A. P. Kypson and D. A. Brown, *J. Am. Heart Assoc.*, 2012, **1**, 001644.
133. D. A. Brown, S. L. Hale, C. P. Baines, C. L. del Rio, R. L. Hamlin, Y. Yueyama, A. Kijtawornrat, S. T. Yeh, C. R. Frasier, L. M. Stewart, F. Moukdar, S. R. Shaikh, K. H. Fisher-Wellman, P. D. Neufer and R. A. Kloner, *J. Cardiovasc. Pharmacol. Ther.*, 2014, **19**, 121.
134. A. V. Birk, S. Liu, Y. Soong, W. Mills, P. Singh, J. D. Warren, S. V. Seshan, J. D. Pardee and H. H. Szeto, *J. Am. Soc. Nephrol.*, 2013, **24**, 1250.
135. S. Schaller, S. Paradis, G. A. Ngoh, R. Assaly, B. Buisson, C. Drouot, M. A. Ostuni, J.-J. Lacapere, F. Bassissi, T. Bordet, A. Berdeaux, S. P. Jones, D. Morin and R. M. Pruss, *J. Pharmacol. Exp. Ther.*, 2010, **333**, 696.
136. MITOCARE Study Group, *Cardiology*, 2012, **123**, 201.
137. C. Piot, P. Croisille, P. Staat, H. Thibault, G. Rioufol, N. Mewton, R. Elbelghiti, T. T. Cung, E. Bonnefoy, D. Angoulvant, C. Macia, F. Raczka, C. Sportouch, G. Gahide, G. Finet, X. Andre-Fouet, D. Revel, G. Kirkorian, J.-P. Monassier, G. Derumeaux and M. Ovize, *N. Engl. J. Med.*, 2008, **359**, 473.
138. J. Xiao, Z. Huang, C. Z. Chen, I. U. Agoulnik, N. Southall, X. Hu, R. E. Jones, M. Ferrer, W. Zheng, A. I. Agoulnik and J. J. Marugan, *Nat. Commun.*, 2013, **4**, 1953.
139. J. R. Teerlink, G. Cotter, B. A. Davison, G. M. Felker, G. Filippatos, B. H. Greenberg, P. Ponikowski, E. Unemori, A. A. Voors, K. F. Adams, Jr., M. I. Dorobantu, L. R. Grinfeld, G. Jondeau, A. Marmor, J. Masip, P. S. Pang, K. Werdan, S. L. Teichman, A. Trapani, C. A. Bush, R. Saini, C. Schumacher, T. M. Severin and M. Metra; for the RELAXin in Acute Heart Failure (RELAX-AHF) Investigators, *Lancet*, 2013, **381**, 29.
140. U.S. Food and Drug Administration, 2014, FDA briefing document, Cardiovascular and Renal Drugs Advisory Committee Meeting,

http://www.fda.gov/AdvisoryCommittees/CommitteesMeetingMaterials/Drugs/CardiovascularandRenalDrugsAdvisoryCommittee/ucm390442.htm.

141. J. J. Marugan, J. Xiao, M. Ferrer-Alegre, C. Chen, N. Southall, W. Zheng, A. Agoulnik and I. Agoulnik, *Preparation of small molecule modulators of the relaxin receptor 1 for treatment of cardiovascular and other diseases*, WO2013165606A1, 2013.
142. X. L. Ma, S. Kumar, F. Gao, C. S. Louden, B. L. Lopez, T. A. Christopher, C. Wang, J. C. Lee, G. Z. Feuerstein and T.-L. Yue, *Circulation*, 1999, **99**, 1685.
143. L. K. Newby, M. S. Marber, C. Melloni, L. Sarov-Blat, L. H. Aberle, P. E. Aylward, G. Cai, R. J. de Winter, C. W. Hamm, J. F. Heitner, R. Kim, A. Lerman, M. R. Patel, J.-F. Tanguay, J. J. Lepore, H. R. Al-Khalidi, D. L. Sprecher and C. B. Granger, *Lancet*, 2014, **384**, 1187.
144. A. R. Kompa and H. Krum, *Lancet*, 2014, **384**, 1162.
145. D. M. Goldstein, A. Kuglstatter, Y. Lou and M. J. Soth, *J. Med. Chem.*, 2010, **53**, 2345.
146. T. J. Povsic and C. M. O'Connor, *Expert Rev. Cardiovasc. Ther.*, 2010, **8**, 1107.
147. D. M. Clifford, S. A. Fisher, S. J. Brunskill, C. Doree, A. Mathur, S. Watt and E. Martin-Rendon, *The Cochrane Library*, 2012, **2**, CD006536.
148. M. Chatterjee and M. Gawaz, *J. Thromb. Haemostasis*, 2013, **11**, 1954.
149. M. S. Penn, F. O. Mendelsohn, G. L. Schaer, W. Sherman, M. Farr, J. Pastore, D. Rouy, R. Clemens, R. Aras and D. W. Losordo, *Circ. Res.*, 2013, **112**, 816.
150. Z. Reiner, *Nutr., Metab. Cardiovasc. Dis.*, 2014, **24**, 1057.
151. D. Noto, A. B. Cefalu and M. R. Averna, *Curr. Atheroscler. Rep.*, 2014, **16**, 1.
152. J. C. Cohen, E. Boerwinkle, T. H. Mosley, Jr. and H. H. Hobbs, *N. Engl. J. Med.*, 2006, **354**, 1264.
153. Z. Zhao, Y. Tuakli-Wosornu, T. A. Lagace, L. Kinch, N. V. Grishin, J. D. Horton, J. C. Cohen and H. H. Hobbs, *Am. J. Hum. Genet.*, 2006, **79**, 514.
154. A. Mullard, *Nat. Rev. Drug Discovery*, 2012, **11**, 817.
155. K. Miyosawa, Y. Watanabe, T. Murakami, D. Desai, J. Matsumoto, S. Tanabe and M. Aikawa, American Heart Association 2012 Scientific Sessions and Resuscitation Science Symposium, 2012.
156. T. Ohgiya, K. Yamazaki, K. Matsuda, T. Watanabe and T. Maejima, *Pharmaceuticals containing pyrimidines for lowering blood LDL level and treatment of various diseases*, WO2014017569A1, 2014.
157. T. Ohgiya, K. Yamazaki, K. Matsuda, T. Murakami, T. Watanabe and T. Maejima, *Preparation of N,5-disubstituted pyrimidin-2-amine compounds for reducing low-density lipoprotein (LDL) cholesterol level in blood*, WO2013137371A1, 2013.
158. T. Ohgiya, T. Murakami, K. Miyosawa, K. Shibuya, K. Yamazaki and T. Kusakabe, *Preparation of of optically pure (S)-trans-{4-[({2-[({1-[3,5-bis(trifluoromethyl)phenyl] ethyl}{5-[2-(methylsulfonyl)ethoxy]pyrimidin-2-yl}amino)*

methyl]-4-(trifluoromethyl)phenyl}(ethyl)amino)methyl]cyclohexyl}acetic acid, WO2011152508A1, 2011.
159. H. Pingali, V. V. M. S. Kalapatapu, P. Makadia and M. R. Jain, *1,3-Dioxane derivatives as PCSK9 inhibitors and their preparation and use for the treatment of dyslipidemia and related diseases*, WO2011051961A1, 2011.
160. H. Pingali, P. Makadia, V. Pandya and S. V. V. M. Kalapatapu, *Preparation of substituted N-[(tetrahydro-2H-pyran-4-yl)methyl] acetamides as novel compounds for the treatment of dyslipidemia and related diseases*, WO2013132509A1, 2013.
161. P. Barter, *Atheroscler. Suppl.*, 2011, **12**, 267.
162. N. B. Mantlo and A. Escribano, *J. Med. Chem.*, 2014, **57**, 1.
163. P. J. Barter, M. Caulfield, M. Eriksson, S. M. Grundy, J. J. Kastelein, M. Komajda, J. Lopez-Sendon, L. Mosca, J. C. Tardif, D. D. Waters, C. L. Shear, J. H. Revkin, K. A. Buhr, M. R. Fisher, A. R. Tall and B. Brewer; for the ILLUMINATE Investigators, *N. Engl. J. Med.*, 2007, **357**, 2109.
164. G. G. Schwartz, A. G. Olsson, M. Abt, C. M. Ballantyne, P. J. Barter, J. Brumm, B. R. Chaitman, I. M. Holme, D. Kallend, L. A. Leiter, E. Leitersdorf, J. J. McMurray, H. Mundl, S. J. Nicholls, P. K. Shah, J. C. Tardif and R. S. Wright; for the dal-OUTCOMES Investigators, *N. Engl. J. Med.*, 2012, **367**, 2089.
165. A. Thompson, P. Gao, L. Orfei, S. Watson, E. Di Angelantonio, S. Kaptoge, C. Ballantyne, C. Cannon, M. Criqui, M. Cushman, A. Hofman, C. Packard, S. G. Thompson, R. Collins, J. Danesh, J. Willeit, S. Kiechl, C. Wiedermann, B. Psaty, C. Furberg, K.-T. Khaw, M. Sandhu, E. J. Benjamin, R. S. Vasan, R. B. Schnabel, J. Oldgren, G. P. Rossi, M. Cesari, L. Lenzini, M. Zanchett, S. K. James, E. Rimm, I. Hatoum, J. L. Anderson, H. T. May, B. D. Horne, J. F. Carlquist, J. B. Muhlestein, W. Koenig, H. Brenner, D. Rothenbacher, W. Marz, B. Bohm, B. R. Winkelmann, K. Winkler, G. Berglund, M. Persson, V. Roger, Y. Gerber, P. B. Berger, E. S. Brilakis, J. McConnell, C. Meisinger, R. Sacco, M. Elkind, P. J. Talmud, M. O'Donoghue, M. S. Sabatine, D. A. Morrow, M. Caslake, E. Braunwald, C. P. Cannon, R. Bernardo, E. Barrett-Connor, L. B. Daniels, G. A. Laughlin, A. Hofman, I. Kardys, J. C. M. Witteman, J. Nelson, J. P. Corsetti, D. L. Rainwater, A. J. Moss, S. Wassertheil-Smoller, P. Ridker, N. Zariffa, A. Zalewski, M. Walker, A. Wood and P. Perry, *Lancet*, 2010, **375**, 1536.
166. I. F. Charo and R. Taub, *Nat. Rev. Drug Discovery*, 2011, **10**, 365.
167. H. D. White, *N. Engl. J. Med.*, 2014, **370**, 1702.
168. A. Mullard, *Nat. Rev. Drug Discovery*, 2014, **13**, 481.
169. F. J. Raal, R. D. Santos, D. J. Blom, A. D. Marais, M.-J. Charng, W. C. Cromwell, R. H. Lachmann, D. Gaudet, J. L. Tan, S. Chasan-Taber, D. L. Tribble, J. D. Flaim and S. T. Crooke, *Lancet*, 2010, **375**, 998.
170. G. S. Thomas, W. C. Cromwell, S. Ali, W. Chin, J. D. Flaim and M. Davidson, *J. Am. Coll. Cardiol.*, 2013, **62**, 2178.
171. R. G. Hart, L. A. Pearce and M. I. Aguilar, *Ann. Intern. Med.*, 2007, **146**, 857.

172. A. L. Jorgensen, R. J. FitzGerald, J. Oyee, M. Pirmohamed and P. R. Williamson, *PLoS One*, 2012, **7**, e44064.
173. T. Khan, H. Wynne, P. Wood, A. Torrance, C. Hankey, P. Avery, P. Kesteven and F. Kamali, *Br. J. Haematol.*, 2004, **124**, 348.
174. H. Takahashi and H. Echizen, *Clin. Pharmacokinet.*, 2001, **40**, 587.
175. P. Druzgala and C. Becker, *Preparation of 3-benzyl-4-hydroxy-coumarin derivatives as inhibitors of vitamin K epoxide reductase*, WO2005100336A1, 2005.
176. L. M. Bavisotto, D. J. Ellis, P. G. Milner, D. L. Combs, I. Irwin and D. M. Canafax, *J. Clin. Pharmacol.*, 2011, **51**, 561.
177. D. J. Ellis, M. H. Usman, P. G. Milner, D. M. Canafax and M. D. Ezekowitz, *Circulation*, 2009, **120**, 1029.
178. Y. Wang and B. Bajorek, *Am. J. Cardiovasc. Drugs*, 2014, **14**, 175.
179. J. Foley, P. Kirchhof and G. Y. Lip, *Curr. Vasc. Pharmacol.*, 2014, **12**, 373.
180. C. J. Boos, M. Nam and A. J. Camm, *Heart Failure Rev.*, 2014, **19**, 391.
181. L. R. Jackson, 2nd and R. C. Becker, *J. Thromb. Thrombolysis*, 2014, **37**, 380.
182. S. Schulman, *Thromb. Haemostasis*, 2014, **111**, 575.
183. P. Zhang, W. Huang, L. Wang, L. Bao, Z. J. Jia, S. M. Bauer, E. A. Goldman, G. D. Probst, Y. Song, T. Su, J. Fan, Y. Wu, W. Li, J. Woolfrey, U. Sinha, P. W. Wong, S. T. Edwards, A. E. Arfsten, L. A. Clizbe, J. Kanter, A. Pandey, G. Park, A. Hutchaleelaha, J. L. Lambing, S. J. Hollenbach, R. M. Scarborough and B.-Y. Zhu, *Bioorg. Med. Chem. Lett.*, 2009, **19**, 2179.
184. N. C. Chan, J. Hirsh, J. S. Ginsberg and J. W. Eikelboom, *Future Cardiol.*, 2014, **10**, 43.
185. M. Palladino, G. Merli and L. Thomson, *Expert Opin. Invest. Drugs*, 2013, **22**, 1465.
186. A. G. Goodwill, K. J. Mather, A. M. Conteh, D. J. Sassoon, J. N. Noblet and J. D. Tune, *Rev. Endocr. Metab. Disord.*, 2014, **15**, 209.

Part 3
Diabetes

CHAPTER 11

Discovery of Biomarkers of Diabetes Type 1 and Type 2

DILER ASLAN

Department of Medical Biochemistry, Faculty of Medicine, Pamukkale University, Denizli, Turkey
Email: daslan@pau.edu.tr

11.1 Introduction

Biomarkers – genes, proteins, imaging data and other measurable indicators – are essential decision-making tools in medicine as indicators of normal biological processes, pathogenic processes, or pharmacologic response to a therapeutic intervention,[1–3] and can be evaluated according to:

- the pathways (transcriptional, and post-transcriptional, post-translational, and metabolic) in physiological and disease states;
- responses to interventions and environmental exposures;
- their structures (*e.g.* proteins, carbohydrates, nucleic acid-based structures, genes, lipids and products, nutritional molecules (*e.g.* vitamins, minerals), and functions (*e.g.* hormones, enzymes, immune system molecules);
- their usages for prediction of, prevention of, diagnosis of, and interventions for diseases; and
- identification and measurement techniques

They can be classified as Type 0, Type I, and Type II biomarkers.[1,4–6] Type 0 biomarkers measure the natural history of a disease and should

RSC Drug Discovery Series No. 45
Cardiovascular and Metabolic Disease: Scientific Discoveries and New Therapies
Edited by Philip Peplow, James Adams and Tim Young

Published by the Royal Society of Chemistry, www.rsc.org

correlate over time with known clinical indicators. Type I biomarkers indicate the intervention effects, for example of a therapeutic drug, and Type II biomarkers are considered Surrogate Endpoint markers. They are also classified according to their intended usage as: predictive, diagnostic, prognostic, pharmacodynamics (or activity) biomarkers, and surrogate endpoints.[7]

A biomarker must be qualified or validated before use in clinical practice and drug development according to scientific and regulated procedures.[7,8] The whole process is composed of discovery, development and validation processes. Hence, even if a high number of structures can be determined by advanced technologies, their usage as biomarkers needs a long journey.[2,9–12]

Although a single biomarker can be used in decision-making, it is also suggested that a combination of them might give more powerful information. Advanced technology provides the utilization of combinations of genomic biomarkers and metabolic biomarkers.

In this chapter, the biomarkers of type 1 (T1D) and type 2 (T2D) diabetes mellitus are discussed in the context of the types, and relations to the physiological and pathological processes. It also covers clinical biomarkers used as diagnostic tests and in validation of drugs and recommended by diabetes management practice guidelines, and also novel biomarkers associated with transcriptional (gene biomarkers), post-transcriptional (miRNA biomarkers), post-translational (proteins), and metabolic (molecules other than DNA and RNA) regulations in patients with diabetes mellitus and related complications.

11.2 Glucose Homeostasis and Pathogenesis of Diabetes Mellitus

Most of the biomarkers are molecules that play key roles in maintaining normal biological status from transcription to metabolic pathways by modulating the interrelations between cells, tissues, organs and systems of the body. Several key molecules and metabolic pathways which take place in glucose homeostasis, pathogenesis of diabetes and its complications are summarized in this section.

11.2.1 Regulation of Blood Glucose Concentration

Diabetes mellitus is characterized by chronic hyperglycemia due to defects in insulin secretion, insulin action or both. In physiological conditions, blood glucose concentration is regulated by complex multiple pathways which interplay with each other. These pathways and utilization of glucose by tissues are modulated by several hormones. These are insulin and counter regulatory hormones (glucagon, epinephrine, cortisol and growth hormone).[13]

Utilization of glucose by several tissues is regulated by insulin. The main target organs of insulin are liver, skeletal muscle and adipose tissue, and it acts on cells to stimulate glucose, lipid and protein metabolism, as well as RNA and DNA synthesis, by modifying the activity of a variety of enzymes and transport processes.[13,14] Insulin stimulates glucose uptake through GLUT4 in muscle and fat cells, but not liver cells.

Insulin is secreted into blood with proinsulin, and C-peptide. There are antibodies to insulin in the blood of patients who are treated with exogenous insulin, but they are usually present at low concentrations and produce no adverse effects.[13] The molecular mechanism of insulin action has been investigated in order to unravel the pathogenesis of diabetes mellitus.[13–18]

The transport of glucose into cells is regulated by two families of proteins:[13] (1) the sodium/glucose co-transporters in the small bowel and kidney (sodium-dependent glucose transporter-SGLT2), and (2) the facilitative glucose transporters (GLUT1–GLUT7). GLUT4 is an insulin-stimulated transporter and distributed in the skeletal muscle, cardiac muscle and adipose tissue.

MicroRNAs (miRNAs) also have regulatory roles in insulin signalling and glucose homeostasis.[19] miRNAs are non-coding RNAs that are regulators of their mRNA targets in the translation into proteins. Studies show the importance of miRNAs in the maintenance of beta cells that play an initiating roles in the complex mechanism of glucose homeostasis.[20–22] miRNA-375 is the most studied miRNA, is abundantly expressed in the islet cells and exerts functional control over glucose-stimulated insulin release. miRNA-9, miRNA-96, miRNA-124a, miRNA-103 and miR-107 are some of the miRNAs that work in a coordinated manner in the regulation of insulin and glucose homeostasis.[19,22] Recent findings have suggested that miRNAs have hundreds of mRNA targets and have potential as endocrine signalling molecules. Different miRNAs regulate single mRNA translation and turnover, and a single miRNA also regulates more than one mRNA, hence they have the ability to regulate multiple key physiological and pathological processes in a concerted manner. miRNA profiling in glucose-responsive and non-responsive cells and miRNAs in insulin-responsive cells (liver, adipose tissue, skeletal muscle) have been studied extensively to understand glucose homeostasis.

Several transcription factors are also involved in regulating the function of beta cells which play a central role in glucose homeostasis in concert with miRNAs.[19,23] These transcription factors regulate the expression of genes key to various aspects of beta cell function. Forkhead box-containing protein O (FoxO)-1, sterol regulatory element-binding protein (SREBP)-1c, and musculoaponeurotic fibrosarcoma oncogene homolog (Maf) are transcription factors that are involved in glucose-stimulated insulin secretion of pancreatic beta cells.[24,25]

Increased understanding of the roles and fates of these candidate biomarkers that have key roles in the regulation of blood glucose in the mechanisms of hyperglycemia-caused abnormalities summarized in the proceeding paragraphs will aid in the discovery and development of better diagnostic tests and tools for validation of drugs.

11.2.2 Pathogenesis of Diabetes Mellitus

Diabetes mellitus is a highly heterogeneous metabolic disorder characterized by chronic hyperglycemia with disturbances of carbohydrate, fat and protein metabolism due to defects in insulin secretion, insulin action, or both. The chronic hyperglycemia of diabetes is associated with long-term damage, dysfunction, and failure of different organs, especially the eyes, kidneys, nerves, heart, and blood vessels.[13,26,27] As the disease progresses, patients are at risk for the development of specific complications, including retinopathy, nephropathy, neuropathy and atherosclerosis, which cause stroke, gangrene and coronary artery disease.

There are three main types of diabetes mellitus: (1) Type 1 diabetes mellitus (T1D); (2) Type 2 diabetes mellitus (T2D), and (3) Gestational diabetes mellitus (GDM). Other specific types of diabetes mellitus also exist.[27,28] Assigning a type of diabetes mellitus to an individual depends on the circumstances present at the time of diagnosis, and many diabetic individuals do not easily fit a single class.[27] Recent findings also reveal that many patients with diabetes mellitus present with overlapping features of different types of diabetes mellitus.[29]

T1D is usually caused by an autoimmune reaction where the body's defence system attacks the beta cells of pancreas, and it is a chronic autoimmune disorder that is caused in susceptible individuals by environmental factors.[30] The destruction is mediated mostly by T cells. A minority of patients with T1D (type 1B, idiopathic) have no known aetiology and no evidence of autoimmunity. T2D is characterized by insulin resistance and relative insulin deficiency, either or both may be present at the time when diabetes is diagnosed.[27,28,31]

The autoimmune process leading to T1D begins months or years before the clinical presentation. The best characterized antibodies are as follows: autoantibody to islet cell cytoplasm (ICA), hepatocyte nuclear factor (HNF), variable nucleotide tandem repeats (NVTR), insulin autoantibody (IAA), autoantibody to 65 kDa isoform of glutamic acid decarboxylase (GAD65A), autoantibody to insulinoma antigen 2 (IA-2A), autoantibody to insulinoma antigen 2β (IA-2βA), autoantibody to zinc transporter (ZnT8A).[13,31]

Autoantibody biomarkers of immune destruction are present in 85% to 90% of individuals with immune mediated diabetes when fasting hyperglycemia is initially detected. However, there are healthy individuals who have a single autoantibody and are at low risk of developing immune-mediated diabetes. On the other hand, the presence of multiple islet autoantibodies (IAA, GAD65A, IA-2A, and/or IA-2βA) is associated with the risk of immune-mediated diabetes. In this context, multiple biomarkers may be helpful for diagnosis of T1D rather than a single biomarker. More than 40 T1D susceptibility alleles have been identified.[31,32]

The role of inflammatory features in T1D pathogenesis is also growing, and inflammation is now recognized as a factor in beta cell loss in both types of diabetes.[30,33]

T2D results from at least two identifiable pathological defects; one is decreased insulin action, and the other is beta cell dysfunction, which is an inability of the pancreas to produce sufficient insulin to compensate for the insulin resistance.[13,31] Whether a defect in beta cell secretion or peripheral resistance to insulin or both is responsible for T2D has been debated for decades. Despite the lack of consensus, it is clear that T2D is an extremely heterogeneous disease and no single cause is adequate to explain the progression from normal glucose tolerance to diabetes. Progressive loss of beta cell function due to insulin resistance, and genetic and environmental factors, such as diet and exercise, are important determinants in the pathogenesis of T2D. Insulin resistance is found in both obese non-diabetic individuals with the features of the metabolic syndrome and patients with T2D.[13]

Besides the genetic predispositions to both types of diabetes, environmental factors affect both types; and there is also apparent overlap between T1D and T2D in the context of inflammation and immunity.[29,33] This heterogeneity in diabetes mellitus with multiple aetiologies necessitates the discovery of biomarkers and/or biomarker panels specific to the prediction and understanding of every stage of the disease and its complications in the context of subgroups and also for individualized therapy.

11.2.3 Pathogenesis of Microvascular and Macrovascular Complications of Diabetes

The long-term complications of diabetes mellitus caused by hyperglycemia-induced damages are separated into microvascular complications (retinopathy, nephropathy, and neuropathy) and macrovascular complications (coronary artery disease, peripheral artery disease, gangrene, stroke). There is a direct relationship between the degree of plasma glucose control and the risk of retinal, renal and neurologic complications.[31] This has been demonstrated for both T1D and T2D.[31,34] Ketoacidosis and hypoglycaemia are the acute complications of diabetes mellitus, and have life threatening effects.

Several mechanisms and genetic factors are responsible for the pathogenesis of long term complications of diabetes mellitus. These mechanisms include increased polyol pathway flux, protein kinase C (PKC) activation, increased intracellular production of advanced glycation end products (AGEs), increased expression of the receptor for advanced glycation and products (RAGE) and RAGE ligands, increased hexosamine pathway activity, increased oxidative stress, increased abnormalities in energy production, increased abnormalities in posttranslational modification (*e.g.* endoplasmic reticulum stress), increased epigenetic pathways, increased synthesis of growth factors, and cytokines.[34–41]

Oxidative stress with increased superoxide production is the key mechanism which causes the activation of major pathways involved in the pathogenesis of complications.[36,37,42]

11.3 Advanced Technologies for Discovery of Biomarkers in Diabetes Mellitus

Besides the standard laboratory techniques, advanced technologies such as genomics, proteomics,[43–53] metabolomics,[54–60] transcriptomics,[61–63] lipidomics,[64–67] and glycoproteomics[68,69] have been used to identify biomarkers and biomarker combinations as transcriptome, epigenome, proteome, metabolome specific to a disorder, and have potential for providing novel biomarkers of disease progression and drug efficacy.

11.4 Biomarkers of Diabetes Mellitus

The number of candidate biomarkers of diabetes is increasing exponentially with the sharp increase in biomarker research and with the opportunities provided by the advanced technologies.[56,58,70–72] Because they are still in discovery phase, it is challenging to classify the biomarkers as Type 0, Type I and Type II biomarkers, as defined by the Biomarkers and Surrogate Endpoint Working Group (National Institutes of Health),[1] or as predictive, diagnostic, prognostic and pharmacodynamic biomarkers, as required in the "Guidance for Industry and FDA Staff Qualification Process for Drug Development Tools" published by the FDA.[7] In this chapter, the biomarkers of T1D and T2D are presented under the headings of "Biomarkers Assayed for Screening, Diagnosis and Management of diabetes mellitus-Clinical Biomarkers"; "Biomarkers Related to Transcriptional Regulation: genetic biomarkers" (Table 11.1); "Biomarkers Related to Post-Transcriptional Regulation: microRNAs" (Table 11.2); and "Biomarkers Related to Post-Translational regulations: proteome and molecules in the biological processes" (Table 11.3). Although biomarkers are listed in alphabetical order in the tables (except genetic biomarkers), the types of diabetes, the complications of diabetes, the biological samples studied, the technologies used, and relations to the physiological and pathological processes are also displayed in the columns.

11.4.1 Biomarkers Assayed for Screening, Diagnosis and Management of Diabetes Mellitus-clinical Biomarkers

Several important parameters recommended for screening, diagnosis and management of diabetes mellitus are as follows:[13,31]

- Screening:
  - Immunological markers: Autoantibody to islet cell cytoplasm (ICA), autoantibody to insulinoma antigen (IAA), glutamic acid decarboxylase (GAD) antibodies, protein tyrosine phosphatase antibodies (IA-2)
  - Genetic markers: *e.g.* human leukocyte antigen (HLA)
  - Insulin secretions
  - Blood glucose

- Diagnosis (clinical):
  - Blood glucose
  - Oral glucose tolerance test (OGTT)
  - Ketones (urine and blood)
  - Insulin
  - C-peptide
  - Stimulation tests
- Management:
  - Acute: glucose, ketones, pH, bicarbonate, lactate, other abnormalities related to cellular dehydration or therapy (*e.g.* potassium)
  - Chronic: glucose, glycated proteins (HbA1c, fructosamine), urinary protein, evaluations of complications (*e.g.* creatinine, cholesterol, triglycerides), and evaluation of pancreas transplant (C-peptide, insulin).

These parameters should be assayed according to the recommendations in the evidence-based diabetes management guidelines.[27,31,32] Although there are assays for measurement of insulin, C-peptide, proinsulin and glucagon, they are useful primarily for research purposes. Insulin and C-peptide, and insulin alone are being used for classification and prediction of diabetes mellitus. Urine C-peptide is being measured for insulin secretion, but has limited value because of urine variability.[13]

11.4.2 Biomarkers Related to Transcriptional Regulation: Genetic Biomarkers for Determining Susceptibility to Diabetes and its Complications

Findings from several clinical trials, such as the Action to Control Cardiovascular Risk in Diabetes (ACCORD)[73] and the Action in Diabetes and Vascular Disease (ADVANCE),[74] showed that cardiovascular disease was established despite good glycemic control.[34] "Hyperglycemia memory" and glucose-induced effects on molecules which have roles on the posttranslational modifications of histones and other epigenetic pathways can effect gene expression and activation, and may contribute to diabetes complications as suggested.[34]

The diabetes susceptibility loci established by candidate gene, genome-wide linkage and genome-wide association (GWA) studies, and some candidate genes that have diabetogenic single nucleotide polymorphisms (SNPs) can be seen in Table 11.1.

The HLA region is known to have the strongest determinants for T1D.[75] Recent studies have provided more than 50 non-HLA T1D loci, have integrated T1D GWAS data with protein–protein interactions, and determined that eight of the regulated gene SNPs (see Table 11.1) were nominally associated with T1D.

Noble *et al.*[75] pointed out that the genes known to affect T1D susceptibility could be grouped into categories of immune function, insulin expression

Table 11.1 Susceptibility genes for Type 1 and Type 2 Diabetes, and related complications. T1D: Type 1 diabetes; T2D: Type 2 diabetes; DN: Diabetic nephropathy; DR: diabetic retinopathy; MicroV Comp.: microvascular complications; CHD: coronary heart disease; GWA: Genome-wide Association; SNP: single nucleotide polymorphism; Reg: regulation; PM: polymorphism; Treg: regulatory T cells; Cand.: candidate.

Gene Name	Gene/Region	Locus	Types of diabetes/comp.	Identification and or relations
Carnosinase (Carnosine dipeptidase 1)[12,76]	CNDP1	18q22.3	DN	SNP, Protective
Glucose transporter 1[12]	GLUT1	3L		SNP
Aldose Reductase[12,77,78]	AKR1B1	7q22	DR	SNP
Superoxide dismutase-2[79]	SOD2	6q25.3	MicroV. Comp	
Vascular Endothelial Growth Factor[78,80]	VEGF	6p21.1	MicroV. Comp, DR	
Cluster of differentiation 38[81]	CD83	6	T1D	GWA
C-type lectin domain family 16, member A[82]	CLEC16A	16		SNP
Cytotoxic T lymphocyte-associated protein 4 (CTLA-4)[30,83]	CTLA-4	2		NegativeReg, immune response
Chemokine (C-X-C motif) receptor 7 (Atypical chemokine receptor 3)[77]	CXCR7 (ACKR3)	2		GWA
HLA class 2[30,32]	HLA	6p21		DR3/4-DQ2/8 genotype
Human leukocyte antigen[75,83]	HLA	6p21.3		Linkage S.
Interferon-induced helicase 1 (IFIH1)[30]	IFIH1	2		GWA; cytoplasmic sensor for viral infection
Interferon gamma receptor 1[81]	IFNGR1	6	T1D	GWA
Interleukin (IL)-2 receptor-alpha gene[30,75,83]	IL 2RA	10		Treg function
Interleukin 17 receptor D	IL17RD	3		GWA
Interleukin 27 receptor, alpha[81]	IL27RA	19		
Insulin[30,83]	INS	11		Linkage S, Cand. Gene; polymorphism
Myosin IB[81]	MYO1B	2		GWA

Phospholipase C, gamma 2 (phosphatidylinositol-specific)[81]	PLCG2	16		GWA
Lymphoid protein tyrosine phosphatase (LYP)[30,75]	PTPN22	1		signaling pathway
TRAF3 interacting protein 2[81]	TRAF3IP2	6		GWA
ADAM metallopeptidase with thrombospondin type 1 motif, 9[84]	ADAMS9	3p14	T2D	GWA
Adenylate cyclase[84]	ADCY5	3		GWA
Glucose-6-phosphatase, catalytic 2[84]	C6PC2	2q24.3		GWA
Cell division cycle protein 123 homolog/ calcium/calmodulin-dependent protein kinase 1D[84]	CDC123-CAMK1D	10p13-p14		GWA
CDK5 regulatory subunit associated protein 1-like 1[84]	CDKAL1	6p22.2		GWA
Cyclin-dependent kinase inhibitor 2A/B[75,84]	CDKN2A/B	9p21		GWA
Fat mass and obesity associated[84]	FTO	16q12.2		GWA
Glucokinase[84]	GCK	7p15.3-p15.1		GWA
Glucokinase regulator[75,84]	GCKR	2p23		GWA
Hematopoietically expressed homeobox and insulin-degrading enzyme[75,84]	HHEX/IDE	10q23-q25		GWA
Hepatocyte nuclear factor 1a[84]	HNF1A	12q24.2		Cand. gene
HNF1 homeobox B[84]	HNF1B	17q12		Cand. gene
Hepatocyte nuclear factor 4a[84]	HNF4A	20q13.12		Cand. gene
Insulin-like growth factor 2 mRNA binding protein 2[75,84]	IGF2BP2	3q28		GWA
Insulin receptor substrate 1[84]	IRS1	2q36		GWA
Juxtaposed with another zinc finger gene 1[84]	JAZF1	7p15		GWA
Potassium channel, inwardly rectifying subfamily J, member 11[75,84]	KCNJ11	11p15.5		Cand. gene
Potassium channel, voltage-gated, KQT-like subfamily, member 1[84]	KCNQ1	12q21		GWA

Table 11.1 (*Continued*)

Gene Name	Gene/Region	Locus	Types of diabetes/comp.	Identification and or relations
Melatonin receptor 1B[84]	MTNR1B	11q21-q22		GWA
Notch homolog 2, *Drosophila*[84]	NOTCH2	1p12		GWA
Solute carrier family 30, member 8[75,84]	SLC30A8	8q24.1		GWA
Transcription factor 7-like 2[83–86]	TCF7L2	10q25.3		Linkage study; SNP; Wnt signaling pathway
Thyroid adenoma-associated[84]	THADA	2p21		GWA
Tumor necrosis factor-a[84]	TNFa	6p21.3		Cand. gene
Tetraspanin 8 and leucine-rich-repeat-containing G-protein coupled[84]	TSPAN8-LGR5	12q21		GWA
Wolfram syndrome 1 (wolframin)[84]	WFS1	4p16.1		Cand. gene
Angiotensin-converting enzyme 2[87]	ACE2	X	T2D, CHD	G8790A PM
Interleukine-4[88]	IL4	5	T2D, DN	Polymorphism
Peroxisome proliferator–activated receptor g[84,89]	PPARg	3q25	T2D, T2D[a]	Cand. Gene; Lipid metabolism
Angiotensin-I Converting Enzyme[12,78]	ACE	17	T2D, DN	SNP
v-aktmurine thymoma viral oncogene homolog 1[89]	AKT1	14	T2D[a]	Apoptosis
BCL2-associated X protein[89]	BAX	19		Apoptosis
Caspase 3[89]	Casp3	4		Apoptosis
Cyclin-dependent kinase inhibitor 1A[89]	CDKN1A	6		Apoptosis
Glyceraldehyde-3-phosphate dehydrogenase[89]	GAPDH	12		Housekeeping gene
Metalloproteinase 10[89]	MMP10	11		Apoptosis
Metalloproteinase 2[89]	MMP2	16		Apoptosis
NFkB inhibitor interacting Ras-like 2[89]	NKIRAS2	17		Transcription factor
Scavenger receptor class B, member 1[89]	SCARB1	12		Lipid metabolism
Tissue factor pathway inhibitor[89]	TFPI	2		Inflammatory pathway
Vascular endothelial growth factor A[89]	VEGFA	6		Inflammatory pathway

[a]Atherosclerosis biopsy samples.

and beta cell function. Most of them encode products that have roles in the immune response (HLA), the second strongest susceptibility locus is the insulin gene itself, and other candidate loci are involved in beta cell function.

Although environmental factors (*e.g.* diet and weight) play a large part in the development T2D, genetic components also play important roles since T2D often goes undetected for years. Identification of the susceptibility genes to T2D will provide important tools for individualized preventive and therapeutic medicine. GWAS have dramatically increased the number of known T2D susceptibility loci.[83,84,90] Some of them are listed in the Table 11.1, but much more work will be required to obtain a complete picture of how genetic variation alters a protein function leading to diabetes.

11.4.3 Biomarkers Related to Post-transcriptional Regulation: Transcriptome (MicroRNAs)

MicroRNAs as a group in the transcriptome are non-coding RNAs that are regulators of the complementary mRNAs in the translation into proteins.[34] They are involved in normal functioning of all cells and hence their dysregulation is known to contribute to a number of disease mechanisms including diabetes complications.[19–22,34,91–98] Several diabetes-related *m*iRNAs associated with hepatic physiology and/or glucose metabolism, endothelial dysfunction, pathways in the adipose tissue, pathways in skeletal muscle are dysregulated with the progression of diabetic complications and are shown in Table 11.2.

miRNAs have a large number of target mRNAs, and each mRNA also can be regulated by different miRNAs. Therefore, the functional analysis of individual miRNAs is complicated. Diabetes leads to changes in miRNA expressions in many tissues, such as muscle, liver, pancreas, heart and kidney, as seen in Table 11.2. They present a new class of biomarkers for both types of diabetes and their complications. Not only are miRNAs displaying promise as early biomarkers for the disease, but a number of these miRNAs are displaying therapeutic potential in preclinical development.[45,92]

11.4.4 Biomarkers Related to Post-translational Regulation: Proteome and Metabolome (Molecules in the Biological Processes)

Clinical biomarkers recommended by the evidence-based clinical and laboratory medicine practice guidelines are glycated haemoglobin (HbA1c) and glucose for glucose metabolism; HbA1c, triglycerides, HDL-cholesterol, creatinine and albumin for complications of diabetes in both T1D and T2D;[13,31] autoantibodies such as ICA, IAA, GAD, protein tyrosine phosphatase (IA-2), and genetic markers such as HLA for T1D.

Table 11.2 MicroRNA biomarkers associated to the diabetes-related pathways and tissues. T1D: Type 1 diabetes; T2D: Type 2 diabetes; DN: Diabetic nephropathy; DR: Diabetic retinopathy; SMCs: skeletal muscle cells; CVD: cardiovascular disease; AdipoT: adipose tissue; hepatic patho. and glucose metab.: hepatic pathology and glucose metabolism; UpReg: upregulated; ER: endoplasmic reticulum; DownReg: down regulated; endodysfunc: endothelial dysfunction.

miRNA	Types of diabetes and complications	miRNAs associated with diabetes-related pathways and tissues
Let-7f[19,22]	T2D, rat model	SMCs; AdipoT; insulin sensitivity
miRNA-1[22,45,60]	CVD	regulating insulin sensitivity in skeletal muscle and adipose tissue
miRNA-100[22]	T2D rat mod	hepatic patho. and glucose metab.
miRNA-101[96]	T1D	AdipoT
miRNA-103[19,22,45]	T2D rat model	AdipoT, liver; regulating glucose and lipid metabolism in liver; hepatic patho. and glucose metab.
miRN-107[19,45]		AdipoT, liver; regulating glucose and lipid metabolism in liver
miRNA-10a[96]	T1D	
miRNA-122[19,22,45]	T1D mouse model	regulating glucose and lipid metabolism in liver; hepatic patho. and glucose metab.
miRNA-124a[19,45,91,97]	T2D	regulating insulin biosynthesis, insulin secretion and beta-cell survival in pancreatic beta-cells; polymorphism
miRNA-125a[91]		AdipoT
miRNA-125a[22]	T2D rat model	hepatic patho. and glucose metab.
miRNA-125a/b[22]	T2D rat model	AdipoT
miRNA-125b[22]	T2D rat model	hepatic patho. and glucose metab.; endodysfunc.
miRNA-126[22,45,96]	T1D and T2D rat model	hepatic patho. and glucose metab.; SMCs (Circulating miRNAs are promising clinical biomarker for diabetes)
miRNA-1275[96]	T1D	
miRNA-128a		
miRNA-128b[91]		
miRNA-129-3p[22]	T2D mouse model	hepatic patho. and glucose metab.
miRNA-30[19]		AdipoT
miRNA-130a[91]		
miRNA-130a[22]	T2D rat model	SMCs
miRNA-132[92]	UpReg	oxidant stress and ER stress
miRNA-133[22,45,97]	CVD	regulating insulin sensitivity in skeletal muscle and adipose tissue
miRNA-133a[22]	T2D mouse model	hepatic patho. and glucose metab.; AdipoT
miRNA-137	T2D mouse model	hepatic patho. and glucose metab.
miRNA-140[22]	T2D rat model	
miRNA-140-3p[96]	T1D	
miRNA-142-5p[22]	T2D mouse model	hepatic patho. and glucose metab.
miRNA-143[19,22,45,97]	T2D rat model	regulating insulin sensitivity in skeletal muscle and adipose tissue; AdipoT

Table 11.2 (*Continued*)

miRNA	Types of diabetes and complications	miRNAs associated with diabetes-related pathways and tissues
miRNA-144[22]	T2D rat and mouse model	hepatic patho. and glucose metab.; SMCs.; AdipoT
miRNA-145		AdipoT
miRNA-146[22]	T2D rat and mouse model, DR	hepatic patho. and glucose metab.
miRNA-146a[91,92]	DownReg	oxidant stress and ER stress
miRNA-148a[22,96]	T1D, T2D rat model	SMCs
miRNA-150	T2D rat model	SMCs
miRNA-152		
miRNA-155[22]	DR	
miRNA-15b[22,96]	T1D, T2D rat model	hepatic patho. and glucose metab.
miRNA-16[96]	T1D	
miRNA181a[45]		regulating glucose and lipid metabolism in liver
miRNA-183[22]	T2D mouse model	hepatic patho. and glucose metab.
miRNA-186[96]	T1D	
miRNA-18b[96]		
miRNA-191[22]	T2D rat model	hepatic patho. and glucose metab.
miRNA-192[22,92,97]	DN, UpReg	oxidant stress and ER stress
miRNA-193[22]	T2D rat model	hepatic patho. and glucose metab.
miRNA-194[91]		
miRNA-195[22,96]	T1D, T2D rat model	hepatic patho. and glucose metab.
miRNA-199a[22]	T2D rat model	SMCs
miRNA-199a-3p[96]	T1D	
miRNA-19a[22,97]	T1D, T2D mouse model	hepatic patho. and glucose metab.
miRNA-19b[22]	T2D rat model	SMCs
miRNA-200 family	DN, UpReg	oxidant stress and ER stress
miRNA-200a[22]	T2D rat model	hepatic patho. and glucose metab.
miRNA-200b[22]	DR	
miRNA-205[92]	DownReg	oxidant stress and ER stress
miRNA-206[22,45]	CVD	regulating insulin sensitivity in skeletal muscle and adipose tissue
miRNA-207[22]	T2D rat and mouse model	hepatic patho. and glucose metab.; AdipoT
miRNA-20a[96]	T1D	
miRNA-20b[96]	T1D	
miRNA-21[22,45,96]	T1D, DN, CVD, DR	regulating insulin biosynthesis, insulin secretion and beta-cell survival in pancreatic beta-cells
miRNA-210[22]	T2D rat model	hepatic patho. and glucose metab.
miRNA-211	UpReg	oxidant stress and ER stress
miRNA-212[22]	T2D mouse model	hepatic patho. and glucose metab.
miRNA-216a	DN, UpReg	oxidant stress and ER stress
miRNA-22[22]	T2D rat model	hepatic patho. and glucose metab.
miRNA-221[22]		diabendodys
miRNA-222[22]	T2D rat model	hepatic patho. and glucose metab.; endodysfunc.
miRNA-223[19,22]	CVD	muscle; glucose uptake, insulin resistance
miRNA-23a[22]	T2D rat model	hepatic patho. and glucose metab.

Table 11.2 (*Continued*)

miRNA	Types of diabetes and complications	miRNAs associated with diabetes-related pathways and tissues
miRNA-23a/b[22]	T2D rat model	SMCs
miRNA-25[92]	DownReg	oxidant stress and ER stress
miRNA-26b[22,97]	T1D, T2D rat model	hepatic patho. and glucose metab.
miRNA-26p[96]	T1D	
miRNA-27a[91]	T2D	polymorphism; AdipoT
miRNA-27b[19,22,91,96]	T1D, T2D rat model	hepatic patho. and glucose metab.
miRNA-28[92]	UpReg	oxidant stress and ER stress
miRNA-29[19,22]	CVD	muscle, AdipoT, liver; glucose transport
miRNA-29a/b/c[45]		regulating insulin sensitivity in skeletal muscle and adipose tissue
miRNA-290	T2D rat model	hepatic patho. and glucose metab.
miRNA-292		SMCs
miRNA-296		
miRNA-299[22]		
miRNA-29a[22,97]	T2D rat model	hepatic patho. and glucose metab.
miRNA-29a/b/c[22]	T2D rat model,	regulating insulin biosynthesis, insulin secretion and beta-cell survival in pancreatic beta-cells; SMCs, AdipoT
miRNA-29b[22]	T2D rat model	hepatic patho. and glucose metab.
miRNA-29c[22]	T2D rat model	hepatic patho. and glucose metab.
miRNA-30[19,22]	DiabetCompCVD	AdipoT
MiRNA-30d[45]		regulating insulin biosynthesis, insulin secretion and beta-cell survival in pancreatic beta-cells
miRNA-301[22]	T2D rat model	SMCs
miRNA-301a[97]	T1D	
miRNA-30a[22]	T2D rat model	hepatic patho. and glucose metab.
miRNA-30e[22,97]	T1D, T2D rat model	
miRNA-31[95]	T2D	
miRNA-32[97]	T1D	
miRNA-320	CVD	regulating insulin sensitivity in skeletal muscle and adipose tissue; endodysfunc.
miRNA-324-5p[97]	T1D	
miRNA-326	T2D mouse model	hepatic patho. and glucose metab.
miRNA-328		
miRNA-33[22]		
miRNA-33a/b		liver, macrophage; cholesterol, lipid and energy homeostasis
miRNA-335	T1D, T2D mouse and rat model	hepatic patho. and glucose metab.; SMCs, insulin production, fatty acid and triglyceride synthesis
miRNA-338-3p	T1D	regulating insulin biosynthesis, insulin secretion and beta-cell survival in pancreatic beta-cells
miRNA-33a[97]	T1D	
miRNA-34[22]	T2D rat model	SMCs
miRNA-34a		liver, pancreas; lipid metabolism, B cell exocytosis

Table 11.2 (*Continued*)

miRNA	Types of diabetes and complications	miRNAs associated with diabetes-related pathways and tissues
miRNA-340	T1D	
miRNA-342-3p		
miRNA-342-5p[97]		
miRNA-34a[22,91]	Obese mice T1D mouse model MIN6B1	hepatic patho. and glucose metab.
miRNA-34c[22]	T2D mouse model	hepatic patho. and glucose metab.
miRNA-365[22]	T2D rat model	hepatic patho. and glucose metab.
miRNA-375[19,22,45,91,93,97]	MIN6 cell line INS-1E cell line miR-375-knockout mouse model	regulating insulin biosynthesis, insulin secretion and beta-cell survival in pancreatic beta-cells; hepatic patho. and glucose metab. (Circulating miRNAs are promising clinical biomarker for diabetes)
miRNA-377[22,92]	DN, UpReg	oxidant stress and ER stress
miRNA-378[19]		AdipoT
miRNA-379[22]	T2D rat model	SMCs
miRNA-384	T2D mouse model	hepatic patho. and glucose metab.
miRNA-409[22]		
miRNA-422b[22]	T2D rat model	
miRNA-423-5p[97]	T1D	
miRNA-424[97]	T1D, T2D rat model	SMCs
miRNA-434[22]	T2D rat model	SMCs
miRNA-434-3p	T2D mouse model	hepatic patho. and glucose metab.
miRNA-448[22]		
miRNA-450[22]	T2D rat model	SMCs
miRNA-450a[96]	T1D	
miRNA-451[22]	T2D rat model	SMCs
miRNA-454[96]	T1D	
miRNA-503[22]		endodysfunc
miRNA-542-3p	T1D	
miRNA-548c-3p[96]		
miRNA-7[45,96,99]	T1D	regulating insulin biosynthesis, insulin secretion and beta-cell survival in pancreatic beta-cells
miRNA-720	T1D	
miRNA-766		
miRNA-7f		
miRNA-7g[96]		
miRNA-802[45]		regulating glucose and lipid metabolism in liver
miRNA-9[19,45,91,97]		regulating insulin biosynthesis, insulin secretion and beta-cell survival in pancreatic beta-cells
miRNA-93[22]	DN	
miRNA-940	T1D	
miRNA-98		
miRNA-BART12[96]		

Although researchers are racing to find better biomarkers by using advanced technologies, there is no powerful biomarker for diagnosis of diabetes mellitus and prediction of its complications except HbA1c.[35] Recently, adiponectin was confirmed as a robust predictive biomarker of glycemic efficacy in T2D and healthy subjects by the Biomarker Consortium.[100]

Novel biomarkers including the clinical biomarkers studied for understanding the mechanisms of abnormalities caused by hyperglycemia are listed in Table 11.3 (clinical biomarkers are shown in bold characters and by greater number of asterisks according to their clinical relevancies). Although they are listed in alphabetical order, it should be kept in mind that hundreds of molecules can be identified simultaneously as the proteome and metabolome of complex samples (blood, biopsy, urine, *etc.*) by advanced methods and techniques. These methods and techniques are also presented in Table 11.3.

As outlined in Table 11.3, more than one tissue is responsible for the pathogenesis of diabetes and its complications; and metabolites have potential to serve as mediators between tissues or organs and represent the biological output of environmental and genetic influences on a tissue.

Metabolites like mRNA transcripts can also be mapped to genetic regions, providing a powerful tool to establish connections between genetic loci and physiological traits.[25,146] However, gene expression is often not predictive of biological change because of the failure of translation of mRNAs into proteins. In this context, miRNAs that are widely expressed in the body to regulate gene expression post-transcriptionally by cleavage or translational repression of their specific target mRNAs or their profiles, which are specific for cell lineages and tissues play important roles of reflecting pathological changes.

11.5 Emerging Considerations

Advanced technologies have delivered large arrays of data, allowing one to correlate physiological states with patterns of gene expression, protein levels, and metabolite abundance,[25,147] and the metabolic profile can be integrated with genomic, transcriptomic, proteomic profiles and microarray datasets to define gene/metabolite networks.[54] However, a major challenge is the analysis and the interpretation of these large arrays of data in order to validate the candidate biomarkers as single biomarkers and/or panels of biomarkers. Computational methods for analysis of complex datasets have been developed[54,119,148–150] and using biostatistical dimensionality reduction and pattern recognition, a clear separation between "normal" and "treated" or "diseased" panels of biomarkers can be achieved.

These studies and reviews[49,50,60–62,146,151] have demonstrated that with the use of advanced technologies such as genomics, transcriptomics, epigenomics, proteomics, proteoglycomics, metabolomics and lipidomics with bioinformatics, certain biomarker panels will be soon available for early detection, treatment, and understanding molecular mechanisms of

Table 11.3 Novel biomarkers including the clinical biomarkers studied for understanding the mechanisms of abnormalities caused by hyperglycemia.

Biomarker (Functions, Related Processes and/or Type)	Types of diabetes mellitus/ complications[a]	Biological system/ status	Methods/ Technique[b]
1,5 Anhydroglucitol[84,101,102] (gycemic status)	T2D		
1,5-Anhydrosorbitol[57] (carbohydrates)	T2D	P	Mmics
25-Hydroxyvitamin D[84] (oxidative stress and nutrient status)	T2D		
2-Aminoadipic acid[35] (glycative and oxidative stress)			
2-Deoxyerythritol	T2D	P, Inc.	Mmics
2-Hydroxybutanoic acid (a-hydroxybutryrate)[103]	T2D	P	Mmics
2-Hydroxybutyrate[57] (energy metabolism)	T2D	P	Mmics
2-Ketoisocaproic acid	T2D	P, Inc.	Mmics
3,6-anhydrogalactose[103] (a-ketoisocaproate)	T2D		Mmics
3-Deoxy-glucosone hydroimidazolone[35] (glycative and oxidative stress)	T2D	P, Inc.	Mmics
3-Hydroxybutanoic acid (beta-hydroxybutryrate)[103]	T2D	P, Inc.	Mmics
3-Hydroxybutyrate[57] (energy metabolism)	T2D	P	Mmics
3-Nitrotyrosine[35] (glycative and oxidative stress)	DM		
8-Oxo-7,8-dihydroguanosine (8-oxoGuo)[104] (RNA oxidation)	T2D	U	
Acrolein-lysine adducts[105] (oxidative stress)	T1D, T2D	U	
Acyl-alkyl-phosphatidylcholines C34:3, C40:6, C42:5, C44:4, and C44:5[55]	T2D	S, Dec.	T-Mmics
Adiponectin[84,99,100,105,106] (glycative, oxidative stress, inflammation, endodysfunc; predictive for glycemia efficacy)	T2D, DN, CVE	S	
Adiponectin precursor[43]	DM	U, UpReg	Pmics
Advanced glycated end products receptor (RAGE)[84] (oxidative stress and nutrient status)	T2D		
Alanine aminopeptidase[48,105] (defense response)	T1D, T2D, DN, DR	U	
Alanine aminotransferase[84,99,107] (metabolic)	T2D		
Albumin[c,43,105,108] (surrogate endpoint)	T1D, T2D, DN, DR, CVD, RD	U, UpReg	
Alkaline phosphatase[105]	T1D, T2D, DN	U	

Table 11.3 (*Continued*)

Biomarker (Functions, Related Processes and/or Type)	Types of diabetes mellitus/ complications[a]	Biological system/ status	Methods/ Technique[b]
Alpha-1-antitrypsin[43,53,109]	T1D, T2D, RD	U, UpReg	Pmics
Alpha-1-Microglobulin[e,43]	T1D, T2D, DN	U	
Alpha-1-microglobulin/Bikunin Precursor (AMBP)[43]	T1D, T2D, RD	U, DownReg	Pmics
Alpha-2-HS-glycoprotein precursor (Fetuin A)[43]	DM	U, UpReg	
Alpha-hydroxybutyrate (a-HB)[110]	T2D		
Angiotensin-converting enzyme[105] (inflammation and endodysfunc)	T2D, DN	U	
Antithrombin-III[44] (diagnosis)	T1D	U	Pmics
Apolipoprotein A[84] (oxidative stress and nutrient status)	T2D	S	
Apolipoprotein A-I[43]	Diab-RD	U	Pmics
Apolipoprotein B[84,111] (oxidative stress and nutrient status)	T2D		
Apolipoprotein CIII	RD	U	Pmics
Apolipoprotein E[43]	RD	U	Pmics
aPTT[99]			
Arachidonic acid[103]	T2D	P, Dec.	Mmics
Ascorbic acid[84] (oxidative stress and nutrient status)	T2D		
Asymmetric dimethylarginine (ADMA)[99]			
Autoantibodies:[c,13,31,112]	T1D		
Autoantibody to 65-kDa isoform of glutamic acid decarboxylase (GAD65A)			
Autoantibody to insulin (IAA)			
Autoantibody to insulinoma antigen 2 (IA-2A)			
Autoantibody to insulinoma antigen 2b (IA-2bA)			
Autoantibody to islet cell cytoplasm (ICA) Autoantibody to zinc transporter (ZnT8A)			
Autoantigens[113]: ARF6, CAPN11, CDC2, CDK4, CDK6, CDKN1A, DOM3Z, DYRK2, FOXM1, GPR120, GTF2A1, HOXB7, HTR1E, IL13RA2, KCNK13, KLK3, LIMK2, PEA15, POU2F1, SERPINH1, Slc30A8-NC, TBCA, TBRG4, TSPAN31, TTC4, ZFP64	T1D		Pmics
Benzoic acid	T2D	P, Dec.	Mmics
Benzylalcohol[103]	T2D	P, Dec.	Mmics
Beta-2-glycoprotein 1[43]		U, UpReg	Pmics
Beta-2-glycoprotein-1/Apolipoprotein H[105]	T1D, T2D, DN	U	
Beta-2-Microglobulin[43]	T1D, T2D, DN, CVD	U, UpReg	

Table 11.3 (*Continued*)

Biomarker (Functions, Related Processes and/or Type)	Types of diabetes mellitus/ complications[a]	Biological system/ status	Methods/ Technique[b]
Beta-hexosaminidase subunit beta[44]	T1D, DR	U	Pmics
Beta-Ig-H3[105]	T2D, DN, RD	U	
Bifunctional aminoacyl-tRNA synthetase (EPRS)[114]		U	Pmics
Branched-chain and aromatic amino acids[84] (oxidative stress and nutrient status)	T2D		
B-type natriuretic peptide[84] (metabolic)	T2D		
C4d[115]	T1D		Imhistchm
Calgranulin B	RD	U, UpReg	Pmics
Carbonic anhydrase 1[43]	RD	U, UpReg	Pmics
Carboxyethyl lysine[35,116]			
Cathepsin B[105]	T1D, T2D, DN	U	
CD40[99]	Diabetes		
CKD273 classifier[53]	T1D, T2D, DN	U	
Cell adhesion molecule[84] (endothelial and inflammatory markers)	T2D		
Ceruloplasmin[105]	T2D, DN	U	
Clusterin precursor protein (CLU)[114]	Diabetes	U	
Clusterin (apolipoprotein J)[53,117]	T1D, T2D, early RD	U	Pmics
Collagen alpha-1 (I)[43,53,108,109] (predictive for renal damage)	T1D, T2D, RD	U, UpReg	Pmics
Collagen alpha-1 (II)[43,118] (predictive for renal damage)	T1D, T2D, RD	U, UpReg	Pmics
Collagen alpha-1 (III)[43,109,118]	T1D, T2D, RD	U	Pmics
Collagen alpha-1 (IV)	T1D, T2D, RD	U	Pmics
Collagen alpha-1 (V)[43]	T1D, T2D, RD	U	Pmics
Collagen alpha-1 (XV)	Diabetes	U	Pmics
Collagen alpha-1 (XVII)	Diabetes	U	Pmics
Collagen alpha-1 (XVIII)[109]	Diabetes	U	Pmics
Collagen alpha-2 (I)[43,109]	T1D, T2D	U	Pmics
Collagen alpha-2 (VIII)	Diabetes	U	Pmics
Collagen alpha-5(IV)[43]	T1D, T2D, RD	U, UpReg	Pmics
Collagen alpha-6 (IV)[43]	T1D, T2D, RD	U	Pmics
Complement C3[99]			
Complement component C4 A[43]	T1D, T2D, RD	U	Pmics
Complement component C4B3[43]	T1D, T2D, RD	U, UpReg	Pmics
Complement factor B[43]	T1D, T2D, RD	U	Pmics
Complement factor H-related 1[43]	T1D, T2D, RD	U, UpReg	Pmics
Complement factor I light chain[43]	T1D, T2D, RD	U	Pmics
Complement regulatory protein CD59[119]	T2D	P	
Complex-forming glycoprotein HC	T1D, T2D, RD	U, UpReg	Pmics
COOH- terminal proendothelin-1 (CT-proET-1)[120]		P	
Copeptin[121]	T2D, CVE	S	
C-peptide[30]	T1D		

Table 11.3 (*Continued*)

Biomarker (Functions, Related Processes and/or Type)	Types of diabetes mellitus/ complications[a]	Biological system/ status	Methods/ Technique[b]
C-reactive protein[84,99,111,122] (endothelial and inflammatory status, surrogate endpoint)			
Creatine kinase-MB[84] (metabolism)	T2D		
C-Type lectin domain family 3 member B Cubilin[43,108]	T1D, T2D, RD	U, UpReg	Pmics
Cystatin C[122,123]	T1D, DN		
Cystine[103]	T2D	P	Mmics
Decanoyl carnitine[124]			
desphospho-uncarboxylated MGP (dp-ucMGP)[125]	T2D, CVD	S	
Diacylphosphatidylcholines C32:1,C36:1,C38:3, and C40:5[55]	T2D	S, Inc.	T-Mmics
Dipeptidyl aminopeptidase IV[105]	T2D, DN	U	
E-cadherin[108]			
Endothelin-1[105] (inflammation, endodysfunc)	T1D, T2D, DN	U	
Ephrin type-B receptor 4[44]	T1D, DR, DN	U	Pmics
Epidermal growth factor[105]	T1D, T2D, DN, DR	U	
Epithelial-cadherin precursor[43]	T1D, T2D, RD	U, UpReg	Pmics
Ethanolamine[103]	T2D	P, Dec.	Mmics
Factor VII[99]			
FAT tumour suppressor[43,108]	T1D, T2D, RD	U, UpReg	Pmics
Ferritin[84,111]			
Fetuin-A[84,99,126] (metabolism)	T2D, PAD		
Fibrinogen[99]			
Fibrinogen alpha chain		U	Pmics
Fibrinogen beta chain[109]			
Fibrinopeptide A[105] (inflammation, endodysfunc)	T1D, T2D, DN	U	
Fibronectin[84,105] (metabolism)	T1D, T2D, DN	U	
Ficolin 3 precursor[43]	T1D, T2D, RD	U	Pmics
Free testosterone[85] (metabolism)	T2D		
Fructosamine[c,84,127,128] (glycemia)			
Fructose[103]	T2D	P, Inc.	Mmics
Fructosyl-lysine[35] (glycative and oxidative stress)			
Gamma -Glutamyl-transferase (GGT)[84,99,105,107,129] (metabolic)	T1D, T2D, DN, DR	U	
Ganglioside GM2 activator[44]	T1D, DR	U	Pmics
Gelsolin[129] (diagnosis)	T1D	U	Pmics
Gluconic acid[103]	T2D	P, Inc.	Mmics
Glucosamine[57] (carbohydrates)	T2D	P	Mmics
Glucose[c,57,84,103,107]	T2D	P, Inc.	Mmics
Glucose-1-phosphate (additional: Glucose)[57] (carbohydrates)	T2D	P	Mmics
Glucuronic acid[103]	T2D	P, Inc.	Mmics
Glutathione peroxidase precursor[43] (oxidative stress)	T1D, T2D, RD	U	Pmics

Table 11.3 (*Continued*)

Biomarker (Functions, Related Processes and/or Type)	Types of diabetes mellitus/ complications[a]	Biological system/ status	Methods/ Technique[b]
Glycated albumin[d,84,127,128] (glycemia)	T2D		
Glycated hemoglobin (HbA1c)[c,84,107,127,128] (glycemia, surrogate endpoint)	T1D, T2D		
Glycated insulin[84] (glycemia)			
Glycated LDL[84] (glycemia)			
Glyceraldehyde-derived advanced glycation end products (AGE)[130] (vascular inflammation)	T2D		
Glycerol-3-phosphate (glycerol-a-phosphate)[103]	T2D	P, Dec.	Mmics
Glycine[55,103]	T2D	S, P, Dec.	T-Mmics, Mmics
Glycosaminoglycan[105]	T1D, T2D, DN, DR	U	
Glycosylated amylin[84] (glycemia)	T2D		
Glyoxal hydroimidazolone[35] (glycative and oxidative stress)			
Glyoxylate[57] (miscellaneous)	T2D	P	Mmics
Haptoglobin precursor, and Haptoglobin[43,108,131] (predictive for early renal functional decline; Prognostic of DN)	T1D, T2D, RD,	U	Pmics
Haptoglobin-related protein precursor[43,108]	T1D, T2D, RD	U	Pmics
HDL-Cholesterol[107]			
Heart-fatty acid binding protein (H-FABP)[132]	T2D, RD	U	
Hemopexin[43]	T1D, T2D, RD	U, UpReg	Pmics
Hemopexin precursor[43]	T1D, T2D, RD	U	Pmics
Hepatocyte growth factor (HGF)[133,134] (fibrosis)	DN		
Hepatocyte nuclear factor (HNF)[13,31]	T1D		
Heptadecanoic acid[103]	T2D	P, Inc.	Mmicss
Hexose[55]	T2D	S, Inc.	T-Mmicss
High-molecular-weight adiponectin[99]			
Histidine[103]	T2D	P, Inc.	Mmicss
histidine triad nucleotide-binding protein 1 (HINT1)[114]	T2D	U	Pmics
Histidine-rich glycoprotein[134]	RD	U	Pmics
Homocysteine[122] (oxidative stress and nutrient) status	T2D, DN	S	
ICAM			
ICAM-1[99]			
Ig Heavy chain	T1D, T2D, RD	U, UpReg	Pmics
Ig K chain C region			
Ig K chain V-II region cum			
Ig K chain V-III region SIE[43]			

Table 11.3 (*Continued*)

Biomarker (Functions, Related Processes and/or Type)	Types of diabetes mellitus/ complications[a]	Biological system/ status	Methods/ Technique[b]
Immunoglobulin G[105]	T1D, T2D, DN, DR, RD	U	Pmics
Immunoglobulin-free light chains[84]	T1D, T2D, DN, DR	U	Pmics
Inositol pentakisphosphate 2-Kinase[43]	T1D, T2D, RD	U, UpReg	Pmics
Insulin[c]	T2D		
Insulin-like growth factor I Insulin receptor[84] (metabolism)			
Interleukin-1 receptor antagonist	T2D		
Interleukin-18	T2D		
Interleukin-2 receptor antagonist[84,99] (endothelial and inflammatory status)	T2D		
Interleukin-6[84,99,105,122] (inflammation, endodysfunc)	T1D, T2D, DN	U	
Interleukine-1beta[135]	T1D, T2D		
Inulobiose[103]	T2D	P, Inc.	Mmics
Isoprostanes[105] (oxidative stress)	T1D, T2D, DN	U	
Kallikrein-3[43]	T1D, T2D, RD	U	Pmics
Kidney injury molecule-1 (kim-1)[12]	DN	U	
Kininogen precursor[43]	T1D, T2D, RD	U, UpReg	Pmics
Lactate[57,99] (energy metabolism)	T2D	P	Mmics
Laminin[105]	T1D, T2D, DN, DR	U	
LDL[99]			
Leptin[84,99] (oxidative stress and nutrient status)	T2D		
Leucine[103]	T2D	P, Inc.	Mmics
Leukotriene E4 (U-LTE4)[136]	T2D	U	
Linoleoyl- glycerophosphocholine (L-GPC)[110]	T2D		
Lipocalin-Type prostaglandin D synthase[105]	T2D, DN, CVD/ MVD	U	
Liver-type fatty acid–binding protein[137]	T2D	U	
LpaPA2 activity			
LpaPA2 mass[99]			
Lysine[103]	T2D	P, Inc.	Mmics
lysoPC (C14:0)[124]			
Lysophosphatidylcholine C18:2[30,55]	T1D, T2D	S, Dec.	T-Mmics, Mmics
Magnesium[99]			
Malondialdehyde[105] (oxidative stress)	T1D, T2D	U	
Mannosamine	T2D	P	Mmics
Mannose[57] (carbohydrates)			
MASP-2-related protein[43]	T1D, T2D, RD	U	Pmics

Table 11.3 (*Continued*)

Biomarker (Functions, Related Processes and/or Type)	Types of diabetes mellitus/ complications[a]	Biological system/ status	Methods/ Technique[b]
Matrix metalloproteinase-9 (MMP-9)[138]			
Megalin[43,139]	T1D, T2D, DN, RD	U, UpReg	Pmics
Membrane-associated progesterone receptor component 1[109]		U	Pmics
Methionine sulfoxide[35] (glycative and oxidative stress)			
Methylglyoxal hydroimidazolones[116] (glycative and oxidative stress)	T1D, DN	P	Pmics
Monocyte chemoattractant protein-1 (MCP-1)[99,105] (inflammation, endodysfunc)	T1D, T2D, DN	U	
Monomeric a-2-macroglobulin[140]	T1D, T2D	S	
MR-Pro atrial natriuretic peptide[84](metabolism)	T2D		
N-Acetyl-*B*-D-glucosaminidase (NAG)[d,105]	T1D, T2D, DN,DR; CVD	U	Pmics
Neopterin[84] (endothelial and inflammatory status)	T2D		
Nephrin[12]	T1D, DN	U	
Neutrophil gelatinase-associated lipocaline (NGAL)[141]	T2D, DN	S, U	Pmics
Nitric oxide[105] (inflammation, endodysfunc)	T1D, T2D, DN	U	
Nonesterified free fatty acids[99]			
N-terminal pro brain natriuretic peptide (NT-proBNP)[84,142] (metabolism)	T2D, CV	S	
Oleic acid[103]	T2D	P, Inc.	Mmics
Orosomucoid (alpha-1-Acid glycoprotein)[e,43,99,105,117] (inflammation and endodysfunc)	T1D, T2D, RD, CVD/MVD	U, UpReg	Pmics
Osteoprotegerin[99,137]			
Palmitic acid	T2D	P, Inc.	Mmics
Palmitoleic acid[103]			
Pancreatic polypeptide[84] (metabolism)	T2D		
Pentosidin[105] (oxidative stress)	T1D, T2D, DN	U	
Phenylalanine[55]	T2D	S, Inc.	T-Mmics
Pigment epithelium-derived factor[43]		U, UpReg	Pmics
Plasminogen activator inhibitor-1[84] (endothelial and inflammatory status)	T2D		
Polyclonal serum immunoglobulin free light chains (FLCs)[143]	T2D, CVD		
Proapo-A-I protein[43]	T1D, T2D, RD	U	Pmics
Progranulin[117]	T1D	U	Pmics
Prostaglandin-H2-isomerase precursor	T1D, T2D, RD	U, UpReg	Pmics

Table 11.3 (*Continued*)

Biomarker (Functions, Related Processes and/or Type)	Types of diabetes mellitus/ complications[a]	Biological system/ status	Methods/ Technique[b]
Prostatic acid phosphatase precursor[43]	T1D, T2D, RD	U	Pmics
P-selectin[99]			
Pyrraline[105] (oxidative stress)	T1D, T2D, DN	U	
Relaxin-like factor INSL3	T1D, T2D, RD	U	Pmics
Retinol binding protein-4[43,99]			
Retinol-Binding Proteins[43,105,108]	T1D, T2D, DN, DR, CVD/MVD	U	
Retinol-Binding Protein Precursor[43]	T1D, T2D, RD	U, UpReg	Pmics
Ribonuclease 2[43]	T1D, T2D, RD	U	Pmics
Sex Hormone-Binding Globulin	T1D, T2D, RD	U	Pmics
Sialic acid[99]			
Sphingomyelin C16:1[55]	T2D	S, Dec.	T-Mmics
Stearic acid[103]	T2D	P, Inc.	Mmics
Tamm – Horsfall Protein/ Uromodulin[105,117]	T1D, T2D, DN, RD, early RD	U	Pmics
Thromboxane[e,105] (inflammation, endodysfunc)	T2D, DN, DR		
Transforming growth factor-beta (TGF-beta)[e,105] (inflammation, endodysfunc)	T1D, T2D, DN	U	
TGF-beta inducible gene h3 (beta-h3)[12]	T2D, DN	U	
Tissue plasminogen activator antigen[84] (endothelial and inflammatory status)	T2D		
Tumour necrosis factor-α (TNF-alfa)[e,105,135] (inflammation, endodysfunc)	T1D, T2D, DN, DR	U	
Transferrin[43,105]	T1D, T2D, DN, DR, CVD/MVD, RD	U	Pmics
Transthyretin[43]		U	Pmics
Transthyretin Precursor[43]	T1D, T2D, RD	U, UpReg	Pmics
Transthyretin Precursor, Tenascin-X[43]	T1D, T2D, RD	U	Pmics
Triglycerides[107]			
Thrombomodulin	T1D, T2D, DR	U	
Thromboxane[105] (inflammation, endodysfunc)	T1D, T2D, DR	U	
Tumor necrosis factor receptor 1 (TNFR1)[144]	T2D, DN	P	
Tumor necrosis factor receptor 2 (TNFR2)[99]			
Type IV Collagen[d,105]	T1D, T2D, DN, DR; CVD	U	
Ubiquitin fusion protein (UbA52)[43]	T1D, T2D, RD	U	Pmics
Uridine[103]	T2D	P, Inc.	Mmics

Table 11.3 (*Continued*)

Biomarker (Functions, Related Processes and/or Type)	Types of diabetes mellitus/ complications[a]	Biological system/ status	Methods/ Technique[b]
Urinary 8-Hydroxydeoxyguanosine (8- OHdG)[e,137,145] (intracellular oxidative stress, DNA damage)			
Urine isoprostanes[99]			
Uromodulin, Fragment[43,53]	T1D, T2D, DN, RD, DownReg	U	Pmics
Vascular endothelial growth factor (VEGF)[e,105] (inflammation, endodysfunc)	T1D, T2D, DN, DR	U	
Variable nucleotide tandem repeats (NVTR)[13,31]	T1D		
Vitamin D-Binding Protein[43]		U	Pmics
Vitamin K-dependent protein Z[44]	T1D, DR, DN	U	Pmics
von Willebrand factor (vWF)[84,99] (endothelial and inflammatory status)			
Xylose[103]	T2D	P, Inc.	Mmics
Zinc alpha-2-Macroglobulin[43]	T1D, T2D, RD	U	Pmics
Zinc-alpha-2-Glycoprotein 1[43]	T1D, T2D, RD	U, UpReg	Pmics

[a]The empty cells belong to the studies that the type of diabetes mellitus was not presented.
[b]only "mics" technology is presented. T1D: type 1 diabetes; T2D: type 2 diabetes; DN: diabetic nephropathy; DR: diabetic retinopathy; RD: renal damage; PAD: Peripheral arterial disease; CVE: cardiovascular events; CVD: cardivascular disease; MVD: macrovascular disease; P: Plasma; S: Serum; U: Urine; Inc.: increased; Dec. decreased; UpReg: Upregulated; DownReg: Down regulated; Mmics: metabolomics; Pmics: proteomics; T-Mics: Targeted metabolomics; Imhistchm: Immunohistochemical.
[c]Clinical relevant.
[d]highly clinical relevant, research outcomes reproducible across studies.
[e]Clinical relevant, might need further studies.[105]

diabetes, and diabetes-associated pathophysiological changes, and also assessment of disease progression and efficacy of therapy.

Biomarker discovery studies are composed of five main phases[9,10,152] of preclinical exploratory studies; establishment of an assay with a clear intended clinical use and validation of the analytical performance of the test; validation of diagnostic accuracy studies composed of large cohort studies or intervention studies; evaluation of the clinical sensitivity and specificity of the test on a prospective cohort; and evaluation of the overall benefits and risks. Hence, multidisciplinary teams composed of all stakeholders related to diabetes, such as researchers from medicine, laboratory medicine, health sciences, biostatistics, bioinformatics, mathematics, engineering, *etc.*, and clinicians, including other healthcare providers and related areas, should be established with high information technology resources.

References

1. Biomarker Definitions Working Group, *Clin. Pharmacol. Ther.*, 2002, **69**, 89.
2. R. Mayeux, *Neurotherapeutics*, 2004, **1**, 182.
3. S. Naylor, *Expert Rev. Mol. Diagn.*, 2003, **3**, 525.
4. G. S. Firestein, *Nat. Clin. Pract. Rheumatol.*, 2006, **2**, 635.
5. R. Frank and R. Hargreaves, *Nat. Rev. Drug Discovery*, 2003, **2**, 566.
6. S. Naylor, *Drug Discovery World*, 2005, 21.
7. Guidance for Industry and FDA Staff Qualification Process for Drug Development Tools 2014, available at http://www.fda.gov/Drugs/GuidanceComplianceRegulatoryInformation/Guidances/ucm335850.htm.
8. European Medicines Agency. Biomarkers. available at http://www.ema.europa.eu/ema/index.jsp?curl = pages/special_topics/general/general_content_000349.jsp.
9. M. Kumar and S. K. Sarin, *Curr. Trends Life Sci.*, 2002, 403.
10. A. R. Horvath, S. J. Lord, A. StJohn, S. Sandberg, C. M. Cobbaert, S. Lorenz, P. J. Monaghan, W. D. Verhagen-Kamerbeek, C. Ebert and P. M. Bossuyt, *Clin. Chim. Acta*, 2014, **427**, 49.
11. N. Rifai, M. A. Gillette and S. A. Carr, *Nat. Biotechnol.*, 2006, **24**, 971.
12. G. Tesch, S. Amur, J. T. Schousboe, J. N. Siegel, L. J. Lesko and J. P. Bai, *AAPS J.*, 2010, **12**, 243.
13. D. B. Sacks, in *Tietz Textbook of Clinical Chemistry and Molecular Diagnostics*, ed. C. A. Burtis, E. R. Ashwood and D. E. Bruns. Elsevier Inc, US, 2006, pp. 846–863.
14. C. R. Kahn and M. F. White, *J. Clin. Invest.*, 1988, **82**, 1151.
15. P. Gual, Y. Le Marchand-Brustel and J. F. Tanti, *Biochimie*, 2005, **87**, 99.
16. A. M. Johnston, L. Pirola and E. Van Obberghen, *FEBS Lett.*, 2003, **546**, 32.
17. A. R. Saltiel and C. R. Kahn, *Nature*, 2001, **414**, 799.
18. A. R. Saltiel and J. E. Pessin, *Trends Cell Biol.*, 2002, **12**, 65.
19. V. Rottiers and A. M. Naar, *Nat. Rev. Mol. Cell Biol.*, 2012, **13**, 239.
20. C. Guay and R. Regazzi, *Nat. Rev. Endocrinol.*, 2013, **9**, 513.
21. C. Guay, E. Roggli, V. Nesca, C. Jacovetti and R. Regazzi, *Transl. Res.*, 2011, **157**, 253.
22. D. S. Karolina, A. Armugam and S. Sepramaniam, *Expert Rev. Endocrinol. Metab.*, 2012, 7, 281.
23. S. Shao, Z. Fang, X. Yu and M. Zhang, *Biochem. Biophys. Res. Commun.*, 2009, **384**, 401.
24. C. Talchai, S. Xuan, H. V. Lin, L. Sussel and D. Accili, *Cell*, 2012, **150**, 1223.
25. C. T. Ferrara, *Metabolic Pathways of Type 2 Diabetes Intersection of Genetics, Transcriptomics, and Metabolite Profiling*. Department of Pharmacology and Cancer Biology, Duke University, 2008.

26. World Health Organization. Definition, Diagnosis and Classification of Diabetes Mellitus and its Complications Part 1: Diagnosis and Classification of of Diabetes Mellitus 1999. available at http://whqlibdoc.who.int/hq/1999/who_ncd_ncs_99.2.pdf.
27. American Diabetes Association (ADA), *Diabetes Care*, 2014, **37**, S81.
28. World Health Organization/International Diabetes Federation. Definition And Diagnosis of Diabetes Mellitus And Intermediate Hyperglycemia 2006, 11, available at http://www.who.int/diabetes/publications/Definition and diagnosis of diabetes_new.pdf.
29. T. Tuomi, N. Santoro, S. Caprio, M. Cai, J. Weng and L. Groop, *Lancet*, 2014, **383**, 1084.
30. T. L. van Belle, K. T. Coppieters and M. G. von Herrath, *Physiol. Rev.*, 2011, **91**, 79.
31. D. B. Sacks, M. Arnold, G. L. Bakris, D. E. Bruns, A. R. Horvath, M. S. Kirkman, A. Lernmark, B. E. Metzger and D. M. Nathan, *Clin. Chem.*, 2011, **57**, e1.
32. Australian Diabetes Society. National Evidence-Based Clinical Care Guidelines for Type 1 Diabetes in Children, Adolescents and Adults, 2011, available at http://www.apeg.org.au/portals/0/guidelines1.pdf.
33. Advances and Emerging Opportunities in Diabetes Research: A Strategic Planning Report of the Diabetes Mellitus Interagency Coordinating Committee. 2011, available at http://www.niddk.nih.gov/about-niddk/strategic-plans-reports/Pages/advances-emerging-opportunities-in-diabetes-research.aspx.
34. J. M. Forbes and M. E. Cooper, *Physiol. Rev.*, 2013, **93**, 137.
35. P. J. Beisswenger, *Amino acids*, 2012, **42**, 1171.
36. M. Brownlee, *Nat. Biotechnol.*, 2001, **414**.
37. M. Brownlee, *Diabetes*, 2005, **54**, 1615.
38. M. J. Fowler, *Clin. Diabetes*, 2008, **26**.
39. M. Kitada, Z. Zhang, A. Mima and G. L. King, *J. Diabetes Invest.*, 2010, **1**, 77.
40. C. Ling and L. Groop, *Diabetes*, 2009, **58**, 2718.
41. T. J. Lyons and A. Basu, *Transl. Res.*, 2012, **159**, 303.
42. F. Giacco and M. Brownlee, *Circ. Res.*, 2010, **107**, 1058.
43. R. Ben Ameur, L. Molina, C. Bolvin, C. Kifagi, F. Jarraya, H. Ayadi, F. Molina and C. Granier, *Nephrol., Dial., Transplant.*, 2010, **25**, 2866.
44. A. Caseiro, A. Barros, R. Ferreira, A. Padrao, M. Aroso, C. Quintaneiro, A. Pereira, R. Marinheiro, R. Vitorino and F. Amado, *Transl. Res.*, 2014, **163**, 188.
45. Y. Mao, R. Mohan, S. Zhang and X. Tang, *Pharmacol. Res.*, 2013, **75**, 37.
46. M. L. Merchant, B. A. Perkins, G. M. Boratyn, L. H. Ficociello, D. W. Wilkey, M. T. Barati, C. C. Bertram, G. P. Page, B. H. Rovin, J. H. Warram, A. S. Krolewski and J. B. Klein, *J. Am. Soc. Nephrol.*, 2009, **20**, 2065.
47. H. H. Otu, H. Can, D. Spentzos, R. G. Nelson, R. L. Hanson, H. C. Looker, W. C. Knowler, M. Monroy, T. A. Libermann, S. A. Karumanchi and R. Thadhani, *Diabetes Care*, 2007, **30**, 638.

48. P. V. Rao, X. Lu, M. Standley, P. Pattee, G. Neelima, G. Girisesh, K. V. Dakshinamurthy, C. T. Roberts, Jr. and S. R. Nagalla, *Diabetes Care*, 2007, **30**, 629.
49. K. Rossing, H. Mischak, M. Dakna, P. Zurbig, J. Novak, B. A. Julian, D. M. Good, J. J. Coon, L. Tarnow and P. Rossing, *J. Am. Soc. Nephrol.*, 2008, **19**, 1283.
50. K. Rossing, H. Mischak, P. Rossing, J. P. Schanstra, A. Wiseman and D. M. Maahs, *Proteomics: Clin. Appl.*, 2008, **2**, 997.
51. E. M. Scott, A. M. Carter and J. B. Findlay, *Diab. Vasc. Dis. Res.*, 2005, **2**, 54.
52. T. Sundsten, M. Eberhardson, M. Goransson and P. Bergsten, *Proteome Sci.*, 2006, **4**, 22.
53. P. Zurbig, G. Jerums, P. Hovind, R. J. Macisaac, H. Mischak, S. E. Nielsen, S. Panagiotopoulos, F. Persson and P. Rossing, *Diabetes*, 2012, **61**, 3304.
54. J. R. Bain, R. D. Stevens, B. R. Wenner, O. Ilkayeva, D. M. Muoio and C. B. Newgard, *Diabetes*, 2009, **58**, 2429.
55. A. Floegel, N. Stefan, Z. Yu, K. Muhlenbruch, D. Drogan, H. G. Joost, A. Fritsche, H. U. Haring, M. Hrabe de Angelis, A. Peters, M. Roden, C. Prehn, R. Wang-Sattler, T. Illig, M. B. Schulze, J. Adamski, H. Boeing and T. Pischon, *Diabetes*, 2013, **62**, 639.
56. N. Friedrich, *J. Endocrinol.*, 2012, **215**, 29.
57. I. Padberg, E. Peter, S. Gonzalez-Maldonado, H. Witt, M. Mueller, T. Weis, B. Bethan, V. Liebenberg, J. Wiemer, H. A. Katus, D. Rein and P. Schatz, *PLoS One*, 2014, **9**, e85082.
58. A. Preet, T. M. Karve, N. Rizk and A. K. Cheema, *J. Diabetes Metab.*, 2013, **01**, 1.
59. R. Wang-Sattler, Z. Yu, C. Herder, A. C. Messias, A. Floegel, Y. He, K. Heim, M. Campillos, C. Holzapfel, B. Thorand, H. Grallert, T. Xu, E. Bader, C. Huth, K. Mittelstrass, A. Doring, C. Meisinger, C. Gieger, C. Prehn, W. Roemisch-Margl, M. Carstensen, L. Xie, H. Yamanaka-Okumura, G. Xing, U. Ceglarek, J. Thiery, G. Giani, H. Lickert, X. Lin, Y. Li, H. Boeing, H. G. Joost, M. H. de Angelis, W. Rathmann, K. Suhre, H. Prokisch, A. Peters, T. Meitinger, M. Roden, H. E. Wichmann, T. Pischon, J. Adamski and T. Illig, *Mol. Syst. Biol.*, 2012, **8**, 615.
60. A. H. Zhang, S. Qiu, H. Y. Xu, H. Sun and X. J. Wang, *Clin. Chim. Acta*, 2014, **429**, 106.
61. D. L. Eizirik, M. Sammeth, T. Bouckenooghe, G. Bottu, G. Sisino, M. Igoillo-Esteve, F. Ortis, I. Santin, M. L. Colli, J. Barthson, L. Bouwens, L. Hughes, L. Gregory, G. Lunter, L. Marselli, P. Marchetti, M. I. McCarthy and M. Cnop, *PLoS Genet.*, 2012, **8**, e1002552.
62. Y.-H. Lee, K. Joshipura, J. L. Vergara and D. T. Wong, *Genomic Med., Biomarkers, Health Sci.*, 2012, **4**, 7.
63. K. I. Woroniecka, A. S. Park, D. Mohtat, D. B. Thomas, J. M. Pullman and K. Susztak, *Diabetes*, 2011, **60**, 2354.

64. P. J. Meikle, G. Wong, D. Tsorotes, C. K. Barlow, J. M. Weir, M. J. Christopher, G. L. MacIntosh, B. Goudey, L. Stern, A. Kowalczyk, I. Haviv, A. J. White, A. M. Dart, S. J. Duffy, G. L. Jennings and B. A. Kingwell, *Arterioscler., Thromb., Vasc. Biol.*, 2011, **31**, 2723.
65. Y. Shi, *J. Biomed. Res.*, 2010, **24**, 6.
66. C. M. Sorensen, J. Ding, Q. Zhang, T. Alquier, R. Zhao, P. W. Mueller, R. D. Smith and T. O. Metz, *Clin. Biochem.*, 2010, **43**, 948.
67. T. Zhao, J. Zhao and W. Yang, *Diabetes/Metab. Res. Rev.*, 2010, **26**, 357.
68. M. Ly, T. N. Laremore and R. J. Linhardt, *OMICS*, 2010, **14**, 389.
69. F. Priego-Capote, A. Scherl, M. Muller, P. Waridel, F. Lisacek and J. C. Sanchez, *Mol. Cell. Proteomics*, 2010, **9**, 579.
70. D. Aslan, *J. Med. Biochem.*, 2011, **30**, 207.
71. E. Rodriguez-Suarez, J. Siwy, P. Zurbig and H. Mischak, *Biochim. Biophys. Acta*, 2014, **1844**, 884.
72. *Biomarkers: In Medicine, Drug Discovery, and Environmental Health*, ed. V. S. Vaidya and J. V. Bonventre. John Wiley & Sons, Inc, New Jersey, 2010, Part 1, pp. 1–88.
73. F. Ismail-Beigi, *Clin. Chem.*, 2011, **57**, 261.
74. S. R. Heller and ADVANCE Collaborative Group, *Diabetes Care*, 2009, **32**(Suppl 2), S357.
75. J. A. Noble and H. A. Erlich, *Cold Spring Harbor Perspect. Med.*, 2012, **2**, a007732.
76. B. Janssen, D. Hohenadel, P. Brinkkoetter, V. Peters, N. Rind, C. Fischer, I. Rychlik, M. Cerna, M. Romzova, E. de Heer, H. Baelde, S. J. Bakker, M. Zirie, E. Rondeau, P. Mathieson, M. A. Saleem, J. Meyer, H. Koppel, S. Sauerhoefer, C. R. Bartram, P. Nawroth, H. P. Hammes, B. A. Yard, J. Zschocke and F. J. van der Woude, *Diabetes*, 2005, **54**, 2320.
77. S. Abhary, K. P. Burdon, K. J. Laurie, S. Thorpe, J. Landers, L. Goold, S. Lake, N. Petrovsky and J. E. Craig, *Diabetes Care*, 2010, **33**, 1834.
78. D. P. Ng, *J. Ophthalmol.*, 2010, 172593.
79. C. Tian, S. Fang, X. Du and C. Jia, *Diabetologia*, 2011, **54**, 803.
80. J. Y. Gong and Y. H. Sun, *PloS One*, 2013, **8**, e84069.
81. R. Bergholdt, C. Brorsson, A. Palleja, L. A. Berchtold, T. Floyel, C. H. Bang-Berthelsen, K. S. Frederiksen, L. J. Jensen, J. Storling and F. Pociot, *Diabetes*, 2012, **61**, 954.
82. S. A. Soleimanpour, A. Gupta, M. Bakay, A. M. Ferrari, D. N. Groff, J. Fadista, L. A. Spruce, J. A. Kushner, L. Groop, S. H. Seeholzer, B. A. Kaufman, H. Hakonarson and D. A. Stoffers, *Cell*, 2014, **157**, 1577.
83. L. Groop and F. Pociot, *Mol. Cell. Endocrinol.*, 2014, **382**, 726.
84. A. M. McKillop and P. R. Flatt, *Diabetes Care*, 2011, **34**, 2624.
85. M. Buraczynska, A. Swatowski, D. Markowska-Gosik, A. Kuczmaszewska and A. Ksiazek, *Diabetes Res. Clin. Pract.*, 2011, **93**, 390.
86. I. Danquah, T. Othmer, L. K. Frank, G. Bedu-Addo, M. B. Schulze and F. P. Mockenhaupt, *BMC Med. Genet.*, 2013, **14**, 96.

87. J. Chaoxin, S. Daili, H. Yanxin, G. Ruwei, W. Chenlong and T. Yaobin, *Eur. Rev. Med. Pharmacol. Sci.*, 2013, **17**, 2654.
88. M. Kazemi Arababadi, *Iran J. Kidney Dis.*, 2010, **4**, 302.
89. M. R. Bernal-Lopez, L. Garrido-Sanchez, V. Gomez-Carrillo, J. L. Gallego-Perales, V. Llorente-Cortes, F. Calleja, R. Gomez-Huelgas, L. Badimon and F. J. Tinahones, *Diabetes Care*, 2013, **36**, 1006.
90. E. Wheeler and I. Barroso, *Briefings Funct. Genomics*, 2011, **10**, 52.
91. C. Ciccacci, D. Di Fusco, L. Cacciotti, R. Morganti, C. D'Amato, C. Greco, S. Rufini, G. Novelli, F. Sangiuolo, V. Spallone and P. Borgiani, *Acta Diabetol.*, 2013, **50**, 867.
92. M. Kato, N. E. Castro and R. Natarajan, *Free Radical Biol. Med.*, 2013, **64**, 85.
93. X. Li, *Gene*, 2014, **533**, 1.
94. S. Rome, *Clin. Biochem.*, 2013, **46**, 937.
95. G. Sebastiani, L. Nigi, I. Spagnuolo, E. Morganti, C. Fondelli and F. Dotta, *J. Biomed. Sci. Eng.*, 2013, **06**, 58.
96. P. Takahashi, D. J. Xavier, A. F. Evangelista, F. S. Manoel-Caetano, C. Macedo, C. V. Collares, M. C. Foss-Freitas, M. C. Foss, D. M. Rassi, E. A. Donadi, G. A. Passos and E. T. Sakamoto-Hojo, *Gene*, 2014, **539**, 213.
97. X. Tang, G. Tang and S. Ozcan, *Biochim. Biophys. Acta*, 2008, **1779**, 697.
98. M. van de Bunt, K. J. Gaulton, L. Parts, I. Moran, P. R. Johnson, C. M. Lindgren, J. Ferrer, A. L. Gloyn and M. I. McCarthy, *PloS One*, 2013, **8**, e55272.
99. J. B. Echouffo-Tcheugui, S. D. Dieffenbach and A. P. Kengne, *Diabetes Res. Clin. Pract.*, 2013, **101**, 255.
100. J. A. Wagner, E. C. Wright, M. M. Ennis, M. Prince, J. Kochan, D. J. Nunez, B. Schneider, M. D. Wang, Y. Chen, S. Ghosh, B. J. Musser and M. T. Vassileva, *Clin. Pharmacol. Ther.*, 2009, **86**, 619.
101. F. F. Parl, M. F. O'Leary, A. B. Kaiser, J. M. Paulett, K. Statnikova and E. K. Shultz, *Clin. Chem.*, 2010, **56**, 417.
102. M. W. True, *J. Diabetes Sci. Technol.*, 2009, **3**, 743.
103. O. Fiehn, W. T. Garvey, J. W. Newman, K. H. Lok, C. L. Hoppel and S. H. Adams, *PloS One*, 2010, **5**, e15234.
104. K. Broedbaek, V. Siersma, T. Henriksen, A. Weimann, M. Petersen, J. T. Andersen, E. Jimenez-Solem, L. J. Hansen, J. E. Henriksen, S. J. Bonnema, N. de Fine Olivarius and H. E. Poulsen, *Diabetes Care*, 2013, **36**, 669.
105. A. Matheson, M. D. Willcox, J. Flanagan and B. J. Walsh, *Diabetes/Metab. Res. Rev.*, 2010, **26**, 150.
106. B. Schottker, C. Herder, D. Rothenbacher, M. Roden, H. Kolb, H. Muller and H. Brenner, *Diabetes Care*, 2013, **36**, 1703.
107. M. B. Schulze, C. Weikert, T. Pischon, M. M. Bergmann, H. Al-Hasani, E. Schleicher, A. Fritsche, H. U. Haring, H. Boeing and H. G. Joost, *Diabetes Care*, 2009, **32**, 2116.

108. S. Riaz, S. S. Alam, S. K. Srai, V. Skinner, A. Riaz and M. W. Akhtar, *Diabetes Technol. Ther.*, 2010, **12**, 979.
109. D. M. Maahs, J. Siwy, A. Argiles, M. Cerna, C. Delles, A. F. Dominiczak, N. Gayrard, A. Iphofer, L. Jansch, G. Jerums, K. Medek, H. Mischak, G. J. Navis, J. M. Roob, K. Rossing, P. Rossing, I. Rychlik, E. Schiffer, R. E. Schmieder, T. C. Wascher, B. M. Winklhofer-Roob, L. U. Zimmerli, P. Zurbig and J. K. Snell-Bergeon, *PloS One*, 2010, **5**, e13051.
110. E. Ferrannini, A. Natali, S. Camastra, M. Nannipieri, A. Mari, K. P. Adam, M. V. Milburn, G. Kastenmuller, J. Adamski, T. Tuomi, V. Lyssenko, L. Groop and W. E. Gall, *Diabetes*, 2013, **62**, 1730.
111. V. Salomaa, A. Havulinna, O. Saarela, T. Zeller, P. Jousilahti, A. Jula, T. Muenzel, A. Aromaa, A. Evans, K. Kuulasmaa and S. Blankenberg, *PloS One*, 2010, **5**, e10100.
112. J. Lebastchi and K. C. Herold, *Cold Spring Harbor Perspect. Med.*, 2012, **2**, a007708.
113. S. Miersch, X. Bian, G. Wallstrom, S. Sibani, T. Logvinenko, C. H. Wasserfall, D. Schatz, M. Atkinson, J. Qiu and J. LaBaer, *J. Proteomics*, 2013, **94**, 486.
114. L. Chu, G. Fu, Q. Meng, H. Zhou and M. Zhang, *Diabetes Res. Clin. Pract.*, 2013, **101**, 187.
115. P. Rowe, C. Wasserfall, B. Croker, M. Campbell-Thompson, A. Pugliese, M. Atkinson and D. Schatz, *Diabetes Care*, 2013, **36**, 3815.
116. P. J. Beisswenger, S. K. Howell, G. B. Russell, M. E. Miller, S. S. Rich and M. Mauer, *Diabetes Care*, 2013, **36**, 3234.
117. D. Schlatzer, D. M. Maahs, M. R. Chance, J. E. Dazard, X. Li, F. Hazlett, M. Rewers and J. K. Snell-Bergeon, *Diabetes Care*, 2012, **35**, 549.
118. A. Alkhalaf, P. Zurbig, S. J. Bakker, H. J. Bilo, M. Cerna, C. Fischer, S. Fuchs, B. Janssen, K. Medek, H. Mischak, J. M. Roob, K. Rossing, P. Rossing, I. Rychlik, H. Sourij, B. Tiran, B. M. Winklhofer-Roob, G. J. Navis and P. Group, *PloS One*, 2010, **5**, e13421.
119. P. Ghosh, R. Sahoo, A. Vaidya, S. Cantel, A. Kavishwar, A. Goldfine, N. Herring, L. Bry, M. Chorev and J. A. Halperin, *Am. J. Hematol.*, 2013, **88**, 670.
120. I. Drion, N. Kleefstra, G. W. Landman, A. Alkhalaf, J. Struck, K. H. Groenier, S. J. Bakker and H. J. Bilo, *Diabetes Care*, 2012, **35**, 2354.
121. J. Rutishauser, *Ther. Umsch.*, 2009, **66**, 731.
122. T. Wang, Q. Wang, Z. Wang, Z. Xiao and L. Liu, *ISRN Endocrinol.*, 2013, **2013**, 1.
123. J. Y. Oh, *Korean Diabetes J.*, 2010, **34**, 84.
124. C. Y. Ha, J. Y. Kim, J. K. Paik, O. Y. Kim, Y. H. Paik, E. J. Lee and J. H. Lee, *Clin. Endocrinol.*, 2012, **76**, 674.
125. G. W. Dalmeijer, Y. T. van der Schouw, E. J. Magdeleyns, C. Vermeer, W. M. Verschuren, J. M. Boer and J. W. Beulens, *Diabetes Care*, 2013, **36**, 3766.

126. L. H. Eraso, N. Ginwala, A. N. Qasim, N. N. Mehta, R. Dlugash, S. Kapoor, S. Schwartz, M. Schutta, N. Iqbal, E. R. Mohler, 3rd and M. P. Reilly, *Diabetes Care*, 2010, **33**, 408.
127. S. P. Juraschek, M. W. Steffes and E. Selvin, *Clin. Chem.*, 2012, **58**, 1648.
128. E. Selvin, L. M. Francis, C. M. Ballantyne, R. C. Hoogeveen, J. Coresh, F. L. Brancati and M. W. Steffes, *Diabetes Care*, 2011, **34**, 960.
129. E. K. Sims and C. Evans-Molina, *Transl. Res.*, 2014, **163**, 183.
130. N. Tahara, S. Yamagishi, M. Takeuchi, A. Honda, A. Tahara, Y. Nitta, N. Kodama, M. Mizoguchi, H. Kaida, M. Ishibashi, N. Hayabuchi, T. Matsui and T. Imaizumi, *Diabetes Care*, 2012, **35**, 2618.
131. N. M. Bhensdadia, K. J. Hunt, M. F. Lopes-Virella, J. Michael Tucker, M. R. Mataria, J. L. Alge, B. A. Neely, M. G. Janech, J. M. Arthur and Veterans Affairs Diabetes Trial (VADT) study group, *Kidney Int.*, 2013, **83**, 1136.
132. F. L. Nauta, W. E. Boertien, S. J. Bakker, H. van Goor, W. van Oeveren, P. E. de Jong, H. Bilo and R. T. Gansevoort, *Diabetes Care*, 2011, **34**, 975.
133. T. Nakamura, K. Sakai, T. Nakamura and K. Matsumoto, *J. Gastroenterol. Hepatol.*, 2011, **26**(Suppl 1), 188.
134. K. Satani, H. Konya, T. Hamaguchi, A. Umehara, T. Katsuno, T. Ishikawa, K. Kohri, Y. Hasegawa, A. Suehiro, E. Kakishita and M. Namba, *Diabetic Med.*, 2006, **23**, 617.
135. J. I. Odegaard and A. Chawla, *Cold Spring Harbor Perspect. Med.*, 2012, **2**, a007724.
136. A. Rafnsson and M. Back, *Dis. Markers*, 2013, **35**, 475.
137. S. Araki, M. Haneda, D. Koya, T. Sugaya, K. Isshiki, S. Kume, A. Kashiwagi, T. Uzu and H. Maegawa, *Diabetes Care*, 2013, **36**, 1248.
138. N. J. van der Zijl, R. Hanemaaijer, M. E. Tushuizen, R. K. Schindhelm, J. Boerop, C. Rustemeijer, H. J. Bilo, J. H. Verheijen and M. Diamant, *Clin. Biochem.*, 2010, **43**, 635.
139. S. Ogasawara, M. Hosojima, R. Kaseda, H. Kabasawa, K. Yamamoto-Kabasawa, H. Kurosawa, H. Sato, N. Iino, T. Takeda, Y. Suzuki, I. Narita, K. Yamagata, Y. Tomino, F. Gejyo, Y. Hirayama, S. Sekine and A. Saito, *Diabetes Care*, 2012, **35**, 1112.
140. T. Takada, Y. Kodera, M. Matsubara, Y. Kawashima, T. Maeda, Y. Fujita and M. Shichiri, *Atherosclerosis*, 2013, **228**, 270.
141. D. Bolignano, A. Lacquaniti, G. Coppolino, V. Donato, M. R. Fazio, G. Nicocia and M. Buemi, *Kidney Blood Pressure Res.*, 2009, **32**, 91.
142. G. Bruno, A. Landi, F. Barutta, G. Ghezzo, C. Baldin, L. Spadafora, A. Schimmenti, T. Prinzis, P. Cavallo Perin and G. Gruden, *Diabetes Care*, 2013, **36**, 2677.
143. S. Bellary, J. M. Faint, L. K. Assi, C. A. Hutchison, S. J. Harding, N. T. Raymond and A. H. Barnett, *Diabetes Care*, 2014, **37**, 2028.
144. P. J. Saulnier, E. Gand, S. Ragot, G. Ducrocq, J. M. Halimi, C. Hulin-Delmotte, P. Llaty, D. Montaigne, V. Rigalleau, R. Roussel, G. Velho, P. Sosner, P. Zaoui and S. Hadjadj, *Diabetes Care*, 2014, **37**, 1425.

145. T. Nishikawa, T. Sasahara, S. Kiritoshi, K. Sonoda, T. Senokuchi, T. Matsuo, D. Kukidome, N. Wake, T. Matsumura, N. Miyamura, M. Sakakida, H. Kishikawa and E. Araki, *Diabetes Care*, 2003, **26**, 1507.
146. I. Riedmaier and M. W. Pfaffl, *Methods*, 2013, **59**, 3.
147. M. Bugliani, R. Liechti, H. Cheon, M. Suleiman, L. Marselli, C. Kirkpatrick, F. Filipponi, U. Boggi, I. Xenarios, F. Syed, L. Ladriere, C. Wollheim, M. S. Lee and P. Marchetti, *Mol. Cell. Endocrinol.*, 2013, **367**, 1.
148. M. Huang, Q. Liang, P. Li, J. Xia, Y. Wang, P. Hu, Z. Jiang, Y. He, L. Pang, L. Han, Y. Wang and G. Luo, *Mol. Biosyst.*, 2013, **9**, 2134.
149. I. N. Sarkar, *J. Transl. Med.*, 2010, **8**, 22.
150. J. Taneera, S. Lang, A. Sharma, J. Fadista, Y. Zhou, E. Ahlqvist, A. Jonsson, V. Lyssenko, P. Vikman, O. Hansson, H. Parikh, O. Korsgren, A. Soni, U. Krus, E. Zhang, X. J. Jing, J. L. Esguerra, C. B. Wollheim, A. Salehi, A. Rosengren, E. Renstrom and L. Groop, *Cell Metab.*, 2012, **16**, 122.
151. O. Massa, M. Alessio, L. Russo, G. Nardo, V. Bonetto, F. Bertuzzi, A. Paladini, D. Iafusco, P. Patera, G. Federici, T. Not, C. Tiberti, R. Bonfanti and F. Barbetti, *J. Proteomics*, 2013, **82**, 263.
152. N. Rifai, M. Gillette and S. Carr, *Nat. Biotechnol.*, 2006, **24**, 971.

CHAPTER 12

Risk Factors for Diabetes Type 1 and Type 2

KIRSTEN J. COPPELL*[a] AND RACHAEL M. McLEAN[b,c]

[a] Edgar Diabetes and Obesity Research, Department of Medicine, Dunedin School of Medicine, University of Otago, Dunedin, New Zealand; [b] Department of Preventive and Social Medicine, Dunedin School of Medicine, University of Otago, Dunedin, New Zealand; [c] Department of Human Nutrition, University of Otago, Dunedin, New Zealand
*Email: kirsten.coppell@otago.ac.nz

12.1 Introduction

Diabetes is characterised by higher than normal blood glucose levels (hyperglycaemia), which can cause complications including cardiovascular disease, renal failure, visual impairment, peripheral neuropathy, peripheral vascular disease and lower limb amputation. Most of those with diabetes have either type 1 diabetes (T1DM) or type 2 diabetes (T2DM), with approximately 10% having T1DM and approximately 85–90% having T2DM. It is estimated that worldwide about 400 million people have diabetes. The incidence and prevalence of both forms of diabetes are increasing, particularly for T2DM. Although genetic predisposition establishes susceptibility, rapid change in environmental factors is the most probable explanation for the increasing incidence rates of both T1DM and T2DM.

The diagnosis of diabetes is established according to blood glucose levels or the more recent haemoglobin (HbA1c) criteria (Table 12.1).

RSC Drug Discovery Series No. 45
Cardiovascular and Metabolic Disease: Scientific Discoveries and New Therapies
Edited by Philip Peplow, James Adams and Tim Young

Published by the Royal Society of Chemistry, www.rsc.org

Table 12.1 The American Diabetes Association and World Health Organisation criteria for the diagnosis of diabetes.

	WHO[1]	ADA 2010[2]
Diabetes		
Fasting plasma glucose	$\geq$7.0 mmol L^{-1} (126 mg dL^{-1})	$\geq$126 mg dL^{-1} (7.0 mmol L^{-1})
Or		
2 hour plasma glucose (in the 75 g OGTT)	$\geq$11.1 mmol L^{-1} (200 mg dL^{-1})	$\geq$200 mg dL^{-1} (11.1 mmol L^{-1})
Or		
Random plasma glucose	$\geq$11.1 mmol L^{-1} (200 mg dL^{-1})	$\geq$200 mg dL^{-1} (11.1 mmol L^{-1})
Or		
HbA1c	$\geq$6.5% (48 mmol mol^{-1})	$\geq$6.5% (48 mmol mol^{-1})
Intermediate Hyperglycaemia		
Fasting plasma glucose	6.1–6.9 mmol L^{-1} (110–125 mg dL^{-1})	100–125 mg dL^{-1} (5.6–6.9 mmol L^{-1})
Or	Or	
2-hour plasma glucose (in the 75 g OGTT)	$\geq$7.8 and <11.1 mmol L^{-1} ($\geq$140 mg dL^{-1} and <200 mg dL^{-1})	140–199 mg dL^{-1} (7.8–11.0 mmol L^{-1})
Or HbA1c		5.7–6.4% (39–46 mmol mol^{-1})

12.2 Type 1 Diabetes: T1DM

Type 1 diabetes (T1DM), previously termed insulin dependent diabetes, is an autoimmune disease characterised by destruction of pancreatic insulin secreting β-cells and subsequent insulin deficiency. The immune-mediated destruction of the β-cells necessitates the lifelong need for exogenous insulin. Although the onset of T1DM can occur during adulthood, it is typically diagnosed during childhood and teenage years, and is one of the most common chronic conditions of childhood. Overall T1DM is slightly more common among males than in females,[3] and the incidence varies seasonally and geographically. New cases of T1DM are more likely to present in autumn and winter than in spring and summer.[3,4] Geographically, the lowest reported age-standardised incidence for children aged <15 years was 0.1/100 000 per year in China and Venezuela and the highest was 40.9/100 000 per year in Finland, for the 1990–1999 period.[5]

The incidence of T1DM has steadily increased worldwide, with the greatest increase being observed amongst children aged <4 years. This increase has been most apparent since the early 1990s, and the overall the rate of increase has been about 3% per annum. Rates of increase as high as 5.4% per annum have been observed among 0–4 year old children in European countries.[6] In Finland, which has the highest incidence of T1DM, the average age-standardised annual incidence rate doubled from 31.4/100 000 in 1980 to 64.2/100 000 in 2005, with an accelerated increase since the early 1990s.[7] However, there are suggestions that the incidence of T1DM may have levelled since the early 2000s in Finland,[8] as well as in Sweden,[9] Norway[10] and Sardinia,[11] and specifically in children aged 0–14 years. Observations over a longer time period will determine whether the incidence of T1DM is truly levelling off or beginning to decrease, or is still increasing overall.[12]

12.2.1 Aetiology – Genetics and the Environment

The aetiology of T1DM is not clear, despite widespread investigations. However, it is agreed that the cause of T1DM is multifactorial with both genetics and environmental determinants involved, but the identification of specific environmental factors and the interactions between susceptibility genes and these factors is not known. The influence of an environmental factor(s) is universally considered important as the rise in the incidence of T1DM has been more rapid than can be accounted for by genes alone.

The autoimmune process develops prior to the clinical onset of T1DM. Although the details of the process are uncertain, the asymptomatic period is highly variable, usually lasting months among infants and young children, yet more than 10 years in older individuals.[13,14] Models of the pathogenesis of T1DM suggest that amongst genetically susceptible individuals, an event initiates β-cell destruction, which may or may not progress to clinical disease.[13,15] Seroconversion to one or more of the five known diabetes-associated autoantibodies is used as a biomarker to indicate the duration of the asymptomatic preclinical phase.[16] It is thought that progression is dependent on a driving exogenous antigen, similar to the role of gluten in coeliac disease. That is, progression to clinical diabetes requires a combination of genetic disease susceptibility, a critically timed trigger and high subsequent exposure to a driving antigen. Also, in T1DM it is likely that a number of other environmental factors modify the β-cell destruction process, either positively or negatively. That is, it is thought that some unknown environmental factors may be protective, while others may be detrimental.

12.2.1.1 Genes

Observed familial aggregation of T1DM strongly suggested the involvement of genetic risk factors.[17] Among those with a first-degree family member with T1DM, the risk of T1DM is estimated to be 15-fold compared with the general population.[18] Twin studies provide further evidence for the importance of genetic risk factors. Among monozygotic twins the disease concordance rate is up to 65%[19] compared with 6–10% for dizygotic twins.[20,21] With advances in genetic techniques, there is now compelling evidence that genetic factors determine susceptibility for the development of T1DM, with more than 40 susceptibility loci having been identified in genetic studies.[22] The HLA gene on chromosome 6 provides approximately one-half of the genetic susceptibility that increases the risk of T1DM. Genes outside the HLA region are also associated with T1DM risk, but their effects are relatively small.[22]

While genetic factors are necessary, interactions with environmental factor(s) are complex and not yet understood. Further unexplained observations include observations that less than 10% of those genetically susceptible to T1DM go on to develop the disease, of new T1DM cases 85% do not have a family member with T1DM, and children of fathers with T1DM have a higher risk of developing T1DM (7%) compared with children of mothers with T1DM (2%).[23] (Refer to Chapter 13 for a detailed description of the genetic factors involved in the aetiology of T1DM.)

12.2.1.2 Environmental Risk Factors

It is widely accepted that non-genetic factors have an important role in the aetiology of T1DM, and that environmental factors have a role in both the development and rate of progression of T1DM.[23] Many potential environmental risk factors have been investigated, and although associations have been observed, no one single risk factor has been convincingly shown to be causally implicated. Although many studies have tried to isolate a single environmental causative factor, it is likely that more than one factor is involved.

While the timing of the pathogenic process is uncertain, prospective studies from birth suggest islet cell autoimmunity can begin very early in life, possibly even prenatally. In the BABYDIAB study, which prospectively followed offspring of parents with T1DM from birth, 1.4% of the offspring had autoantibodies by 9 months of age, and 3.6% by 2 years of age.[24] Further, among 1005 genetically susceptible children from the general population in the Finnish Diabetes Prediction and Prevention (DIPP) study, 6.3% tested positive for at least one autoantibody by 2 years of age.[25]

Long term follow-up of participants in the BABYDIAB and DIPP studies have both shown that the first appearance of autoantibodies can be as young as 6 months, but most appear between 6 months and 3 years of age in susceptible children. Furthermore, most children who presented with clinical T1DM before adolescence in these studies developed autoimmunity at a young age, but the rate of progression to T1DM was variable.[26] These studies have and will continue to provide valuable information about the natural history of T1DM, but are unlikely to provide explanations about disease causality for the majority of people with T1DM who were at low risk of developing the disease (either because they had no family history of T1DM or were not considered genetically susceptible). Further, these two longitudinal studies were not specifically designed to identify causal environmental factors, whereas the Environmental Determinants of Diabetes in the Young (TEDDY) study, aims to identify environmental factors that trigger, or possibly protect against, islet autoimmunity and T1DM in genetically at risk children.[27]

12.2.2 Prenatal Environmental Factors – The *In Utero* Environment

Pancreatic β-cells begin to function in the growing foetus by mid gestation and are therefore potentially vulnerable to an environmental factor(s).

Intrauterine infection, nutrition and toxins are all considered important factors that may influence β-cell maturation.

Intrauterine infection is known to be associated with an increased risk of T1DM. Rubella infection of a non-immune mother during pregnancy causes foetal viraemia and subsequent congenital rubella syndrome. Although features of the syndrome vary according to the timing and severity of the maternal infection, up to 20% are reported to have T1DM.[28] The isolation of rubella virus from the pancreas of infected children supports this observed association. Congenital rubella is a vaccine preventable disease, and is now rare in most parts of the world. The evidence linking rubella and T1DM is inconclusive, but is considered sufficient to support the concept that intrauterine infection can lead to T1DM.[29] Maternal enterovirus infection, in particular coxsackie virus, has long been considered a possible environmental risk for T1DM. Findings to date have been inconsistent mainly due to small study size and variable immunoassay techniques.[29] However, a recent somewhat larger study does suggest maternal enterovirus infection may trigger β-cell autoimmunity and increase the risk of T1DM in susceptible children.[30]

Good maternal nutrition is important for the growing foetus, and different aspects of the diet have been considered to influence the risk of T1DM. Several studies have examined different foods and nutrients consumed by women during pregnancy, but to date no particular foods or nutrients have been identified that may significantly increase or decrease the risk of T1DM in offspring,[31–33] except possibly vitamin D. Low maternal vitamin D status is an hypothesised risk factor because it is thought to be involved with the regulation of the immune system and inflammatory responses, and thus may have direct effects on β-cells. Observational studies examining the association between maternal vitamin D intake and T1DM in their offspring are few, and results are inconsistent,[29] possibly due to limitations of such studies such as recall bias. A meta-analysis of observational studies concluded there was insufficient evidence to support an association between maternal intake of vitamin D and risk of T1DM in offspring.[34]

In contrast, while toxic and a health hazard for many reasons, maternal smoking during pregnancy has been associated with a small reduced risk of T1DM in their offspring.[29] Not all studies have reported an association, but where an inverse association between maternal smoking and T1DM in offspring has been observed, the reported relative risks have consistently been between about 0.5 to 0.7.[29,35,36] The underlying mechanisms by which maternal smoking during pregnancy possibly reduces the risk of T1DM in offspring is not known but may relate to influences on immune function or epigenetic modifications such as DNA methylation, or both.[37,38]

12.2.3 Prenatal Environmental Factors – Intrapartum Factors

Factors related to birthing including the type of delivery, the age of the mother at delivery and birth order are associated with an increased risk of T1DM.

A number of studies since the 1960s have examined the effect of maternal age at delivery on the risk of T1DM, and most have suggested an increased risk among infants of mothers who were older when they gave birth. A meta-analysis confirmed an association between older maternal age at delivery and increased risk of T1DM in offspring.[39] Overall, children whose mothers were aged over 35 years had a 10% increase in T1DM compared with children whose mothers were aged 25–30 years. The specific reason for this observation is unknown but it has been speculated that maternal age may be an indicator of accumulated maternal exposure to a key factor(s).[29]

Delivery of infants by caesarean section is associated with a 20% increase in the risk of T1DM compared with normal vaginal delivery.[40] The type of microbiota that an infant is first exposed to and the length of exposure during the two different delivery methods are not the same. This leads to infants delivered vaginally developing a microbiome reflective of vaginal flora while those delivered by caesarean section initially have microbiota dominated by skin commensals. It has been speculated that the altered infant gut microbiota following caesarean delivery may negatively influence the development, and subsequent function of the immune system.[41] As the gut microbiota is closely linked with the immune system, it is thought that alterations in gut microbiota may influence the role and function of this system, thereby increasing the risk of autoimmune disorders including T1DM.[42] However, the gut is a complex environment which is also influenced by other factors such as dietary exposures, and it is unclear whether observed correlations are causative or not.

A high birth weight is also associated with a small increase in the risk of T1DM, even after adjustment for factors such as maternal diabetes, gestational age, maternal age and birth order. Children whose birth weight is 4 kg or more, have on average a 10% increase in their risk of T1DM compared with those who weigh 3.0–3.5 kg at birth.[43] A high birth weight is thought to be associated with more active insulin-secreting β-cells, leading to higher insulin secretion and possibly greater susceptibility of the cells to damage.[44]

Increasing birth order is associated with a small reduction in T1DM risk. A pooled analysis found that second or later born children had an approximately 10% lower risk of developing T1DM compared with first born children.[45] The most popular explanation for this observation is the 'hygiene hypothesis', which suggests first born children are exposed to fewer infections than later born children in the early neonatal period, but little evidence supports this. Moreover, this hypothesis is contradictory to the suggestion that infection, particularly some viral infections, increase risk. Other possible explanations for increasing birth order reducing risk have been suggested and include prenatal mechanisms such as maternal immunity.[29] The observation that inter-birth interval is associated with T1DM risk, with children born after a short interval (less than 3 years) having a lower risk than those born after a longer interval, may be a confounding factor.[46]

12.2.4 Postnatal Environmental Factors

Several different postnatal environmental risk factors have been associated with T1DM including rate of growth, infection and nutrition during infancy.

12.2.4.1 Postnatal Growth

Excess weight gain in childhood was first considered as a risk factor for T1DM in the 1970s,[47] and was revisited in the 1990s when a simultaneous increase in both the incidence of T1DM and the incidence of child obesity occurred worldwide. This led to the development of the 'accelerator' hypothesis, which proposes that 'type I and type II diabetes are the same disorder of insulin resistance set against different genetic backgrounds'.[48] The rate (tempo) of loss of β-cells is what distinguishes the two types of diabetes, with an accelerated decline in those with T1DM compared to a slower decline in those with T2DM. As has been suggested for high birth weight, increased childhood growth and weight gain is thought to stress β-cells, thereby making them more vulnerable to attack by an environmental trigger factor.[49]

Whether the obesity epidemic is a driving factor for the increased incidence of T1DM, particularly in children, is not clear. When country specific data were examined there was no correlation between the incidence of T1DM in childhood and the prevalence of child overweight and obesity,[50] and the prevalence of β-cell autoantibodies in overweight non-diabetic school children was similar (1.9%) to that in normal weight school children (4.4%).[51] However, a systematic review, and meta-analysis concluded that there was an association between higher than normal body mass index (BMI) levels and subsequent increased risk of child onset T1DM.[50] Further, not only weight gain but accelerated height velocity may be an important initiating factor in genetically susceptible individuals. In a case control study in Finland, children who developed T1DM were not only heavier but also taller in infancy and throughout childhood.[52]

12.2.4.2 Viruses

Infection with a virus or viruses in infancy or childhood has long been thought to have a role in the aetiology of T1DM, but the role of viruses, if any, is unclear.[53] Viruses, particularly persistent viral infections, may have a role in triggering islet autoimmunity, or in the progression from autoimmunity to clinical T1DM.[16,53] The list of viruses that have been considered to be associated with T1DM includes mumps, rubella, cytomegalovirus, ljunganvirus, retroviruses, rotavirus and enteroviruses. Enteroviruses are considered to be the most promising viral candidate in the aetiology of T1DM. However, results to date from the four prospective cohort studies for the identification of an environmental trigger are not consistent.[54] Only the Finnish DIPP study observed higher proportions of enterovirus in sera of children with islet autoimmunity compared with age-matched controls,[55] while no differences were observed in both the US based Diabetes Autoimmunity Study in the Young (DAISY)[56] and German BABYDIET study.[57] Similarly, the Norwegian MIDIA cohort study in high risk children did not support a causal relationship between enterovirus infection and T1DM risk.[58] Other viruses such as rotavirus have been considered as potentially having an aetiological role, but evidence is limited.[59,60]

12.2.4.3 Nutritional Factors

12.2.4.3.1 Breastfeeding. Whether breastfeeding protects against T1DM is still a controversial issue.[16,54] In the BABYDIAB, DAISY and DIPP studies, the risk of developing autoantibodies was not increased with reduced total or exclusive breastfeeding.[61–63] Other studies have found short duration of breastfeeding is a risk factor for the appearance of autoantibodies, but the results of a recent pooled analysis suggested exclusive breastfeeding for the first 2 weeks of life could marginally reduce the risk of T1DM, whereas longer exclusive or non-exclusive breastfeeding did not appear to provide any protection.[64] However, there was marked variation in the observed associations between the individual studies as well as methodological limitations such as definitions of exclusive breast feeding and recall bias in the questionnaire-based case control studies.[64] Further, differences in breastfeeding practices within and between countries were not usually addressed in these studies.[16]

12.2.4.3.2 Complementary Feeding. Exposure to cow's milk is a possible trigger of T1DM, particularly as cross-reactivity has been found between bovine insulin and human insulin. The age at which infant milk formula is introduced may also influence risk. However, observations from studies of cow's milk exposure in relation to T1DM are inconsistent.[16] This may in part be due to differences in infant formula feeding practices between countries and cultures including the timing and type of other foods introduced to the infant's diet. The use of partly or extensively hydrolysed infant formulas also varies between countries and was considered a potential confounding factor, but hydrolysed milk was recently found not to be associated with T1DM.[65]

12.2.4.3.3 Infant Feeding. Several different foods including gluten-containing foods, root vegetables, fruit and berries are associated with the development of islet autoimmunity.[66–69] Of these foods, the most likely dietary candidate is gluten. This protein is the driving antigen in coeliac disease, an autoimmune disease associated with T1DM. Results from studies in both animals and humans suggest early exposure to dietary gluten is associated with an increased risk of islet cell autoimmunity.

As gluten proteins are partly resistant to enzymatic processing, the intestinal lumen is likely to accumulate non-degradable gluten peptides.[70]

These peptides may directly activate an intestinal immune response[71] or may be processed by intestinal bacteria, leading to the formation of immunogenic gluten peptides.[72] In experimental animal models, the highest incidence of T1DM was in those animals fed a wheat-based diet, whereas in humans, gluten has not been shown to significantly alter autoantibody levels, but can decrease the insulin response.[66] However, the introduction of gluten-containing food before 3 months or after 6 months of age has been shown to be associated with an increased risk of β-cell autoantibodies.[61,63]

Observational studies have suggested that insufficient vitamin D in infancy increased the risk of T1DM, leading to studies where infants were given vitamin D supplementation. A meta-analysis of five of these intervention studies concluded vitamin D supplementation in early childhood was associated with a reduced risk of T1DM.[73] However, in Finland and other northern countries where T1DM incidence is very high, about 80% of parents give recommended daily vitamin D to their infants until at least 12 months of age.[74] Also, the prospective DAISY study has recently observed that neither vitamin D intake nor 25(OH)D levels during childhood were associated with an increased risk of T1DM.[75]

12.2.5 Summary

The incidence of T1DM has increased worldwide over the last 30–40 years, but may be levelling in some parts of the world. It is an immune-mediated disease with an asymptomatic preclinical period, that is triggered by an agent in genetically susceptible individuals. Following the initiation, progression of β-cell destruction to clinical disease is variable and most likely driven by an exogenous antigen. While numerous environmental candidate risk factors have been considered, an aetiological agent(s) has not yet been identified. It is possible that more than one factor may be involved in the causal pathway.

12.3 Type 2 Diabetes: T2DM

Type 2 diabetes (T2DM) is a chronic progressive disorder characterised by insulin resistance and elevated blood glucose concentrations. The onset of T2DM is typically insidious and is diagnosed either by an elevated blood glucose concentration, or more recently by glycated haemoglobin (HbA1C): a measure which reflects blood glucose levels over the previous 10–12 weeks. The early stages of T2DM are usually dominated by insulin resistance and a compensatory increased production of insulin by the pancreas. As the function of the β-cells declines linearly with time, a decline in insulin production occurs and insulin deficiency predominates, and after 10 years following diagnosis, about 50% of T2DM patients require insulin treatment to maintain good glycaemic control.[76]

T2DM is often preceded by a period of impaired glucose metabolism. This period (often referred to as 'pre-diabetes' or 'intermediate glycaemia') may consist of impaired glucose tolerance (IGT) or impaired fasting glucose (IFG) or both.[77] Between 20–45% of people with pre-diabetes will progress to diabetes within 2 to 5 years,[78,79] so people with pre-diabetes are an ideal population with which to conduct intervention studies examining risk factors for diabetes. It is likely however that some environmental influences operate much earlier in life, and these are much more difficult to study due to the long time frame over which diabetes develops. For risk factors early in life, we rely on longitudinal data from cohort studies and life course epidemiological studies, which are more prone to confounding and other methodological uncertainties.

There are few data that specifically examine T2DM prevalence, as most surveys report total diabetes only. The typical gradual onset of T2DM, and the variability in the course of the disease when people are diagnosed with T2DM, precludes good estimates of incidence. Further, as T2DM tends to be asymptomatic during the initial stages of the disease, asymptomatic cases are often undiagnosed. Thus, estimates of diabetes prevalence which rely on self-reporting only will underestimate diabetes prevalence. Nevertheless, there is little doubt that the prevalence of T2DM has increased worldwide in recent years. The estimated worldwide prevalence of total diabetes was 2.8% in 2000,[80] and was estimated to be 6.4% in 2010, and is expected to increase to 7.7% by 2030.[81] Prevalence of diabetes is especially high among indigenous populations, particularly Pacific groups[82] and Asian populations.[83,84] Typically the onset of T2DM is in adulthood. However, the increasing prevalence of obesity in all age groups has been accompanied by an increased incidence of T2DM in adolescence, and even pre-adolescence.[85]

12.3.1 Genetic Basis of Type 2 Diabetes

There is good evidence that a genetic basis for T2DM exists (see Chapter 13). A small number of rare forms of T2DM have strong familial associations and are linked with particular genes and heritability.[86] However for the vast majority of cases, the relationship is more complicated. Although T2DM commonly occurs within families, identification of strong genetic associations has proved elusive. Genome wide association (GWAS) studies have to date identified more than 40 candidate genes which predispose to the development of T2DM. However, there is only a modest effect size of each particular genetic variant, and it is clear that there is an important interaction between genetic, environmental and lifestyle factors.[86,87]

12.3.2 Environmental Risk Factors for T2DM

Of the environmental and lifestyle risk factors for T2DM, the strongest evidence is for obesity (particularly central obesity) being the most important.[88] Central or abdominal adiposity is associated with relatively more insulin resistance, as well as elevated triglycerides and inflammation, all of which contribute to an increased risk of T2DM and its complications. There is strong ecological evidence of this association, with a rising prevalence of both obesity and T2DM in many high income countries,[89,90] which is supported by cohort studies in different populations.[88] Controlled intervention studies of bariatric surgery show that subsequent dramatic weight loss is associated with remission of T2DM in a substantial proportion of patients (around two-thirds of patients in one study).[91,92] Weight loss through a short term 8 week very low calorie diet (600 kcal per day) can also achieve a similar result with reversal of the abnormalities underlying T2DM.[93]

Several large randomised trials of lifestyle interventions in people with increased risk of developing T2DM (such as pre-diabetes or IGT) have shown that interventions involving dietary change and increased physical activity resulting in weight loss can delay or prevent the onset of T2DM.[94–97] These and other studies show that a healthy dietary pattern, and regular physical activity are protective.

12.3.3 Diet

While there is clear evidence that overall dietary pattern is a risk factor for T2DM, the extent to which individual nutrients contribute to risk remains less clear. Excess energy intake, associated with weight gain is clearly associated with increased risk. Diet quality, rather than overall macronutrient composition seems to be relevant, with the type of dietary fat and carbohydrate particularly important. The adoption of 'Western Style' dietary patterns are likely to be contributing to increasing prevalence in middle income countries such as India and China. Data from national and other surveys have shown increased intake of animal products and fat in China and India have paralleled increases in prevalence of T2DM.[87] Diets with a high proportion of saturated to total fat intake appear to increase insulin resistance, and therefore increase risk of T2DM,[98] as have diets high in *trans* fats. Recently there has been interest in the potential beneficial effects of consumption of dairy products such as milk, cheese or yoghurt on cardiometabolic risk. The Dietary Approaches to Stop Hypertension (DASH) diet was high in low-fat dairy, and was associated with blood pressure lowering.[99] A recent meta-analysis showed that higher total dairy intake is associated with a significantly lower risk of T2DM, with particular beneficial effects shown for low-fat dairy products such as low-fat milk, yoghurt and cheese.[100] Exactly which component of dairy products is implicated with this beneficial effect is unclear, but the EPIC-Interact cohort study recently showed that a relatively high intake of odd-chain saturated fats 15 : 0 (pentadecanoic acid) and 17 : 0 (heptadecanoic acid) such as those found in dairy products was associated with decreased risk of incident T2DM.[101]

A high intake of highly processed carbohydrates, including those with a relatively high glycaemic load, is generally considered to be associated with increased risk of T2DM, although the evidence supporting this association independent of obesity is less clear. A high intake of white or polished rice has been shown to be associated with an increased risk of T2DM, especially in Chinese and Japanese populations.[102] An increasing body of evidence relates intake of sugar and sugar-sweetened beverages (SSBs) to increased risk of diabetes. A clear association has been demonstrated between increased sugar and SSB intake and obesity in adults and children.[103,104] This is largely due to the high energy density and relatively low satiety associated with the consumption of sugar rich foods and beverages. Some studies also suggest an association between a high intake of sugar and SSBs, with an increased incidence of T2DM independent of obesity and weight gain. Other potential mechanisms include a rapid increase in blood glucose and insulin concentrations associated with sugar and other rapidly digested carbohydrates, which can result in increased insulin resistance and inflammation.[105]

On the other hand, diets high in dietary fibre, including those rich in whole-grains and cereals appear to be protective.[106] A recent meta-analysis showed a high intake of whole grain and total grain intake reduces risk of T2DM by 32 and 17%, respectively, per 3 servings per day. It is recommended therefore that people choose brown rice, wholegrain breads and cereals instead of the more refined and processed carbohydrate alternatives. A high fruit and vegetable intake is also often included in dietary recommendations; however a recent meta-analysis found that only a high intake of green leafy vegetables was significantly associated with a decreased risk of T2DM.[107]

12.3.4 Alcohol

Cohort studies have shown a protective effect associated with light to moderate alcohol consumption of 1–2 drinks per day (compared to heavy drinkers or non-drinkers) on T2DM incidence, similar to that demonstrated between alcohol and coronary heart disease. The extent to which this is due to residual confounding remains unclear, as light to moderate alcohol consumption is highly correlated with other healthy behaviours.[108] Heavy alcohol intake (greater than 50 grams or 5 standard drinks per day for women or 60 grams or 6 standard drinks per day for men) is associated with increased risk.[109,110]

12.3.5 Physical Activity

Many studies have shown the beneficial effect of physical activity and the detrimental effect of sedentary behaviour on T2DM risk. Randomised controlled trials of lifestyle intervention show beneficial effects of increased physical activity alone, as well as in conjunction with weight loss and improved diet quality.[94,96] Conversely, sedentary behaviours such as television watching are associated with an increased risk of diabetes in observational studies.[87]

12.3.6 Other Risk Factors

It is possible that very early exposures influence diabetes risk. The observation that low birth weight babies who experience rapid catch up growth may be at increased risk of T2DM led to the development of the Foetal Programming Hypothesis[111] and an interest in optimising maternal nutrition as a means of reducing the incidence of T2DM. Obesity in childhood has also been implicated as a risk factor.[112] A history of gestational diabetes (glucose intolerance in pregnancy) increases the risk of T2DM in the mother, but there is no obvious effect of diabetes risk on the infant.[113]

12.3.6.1 Smoking

Many studies have shown an association between cigarette smoking and incidence of T2DM. Recent meta-analyses indicate an increased risk of T2DM of around 40% for current and former smokers,[114] with a dose–response relationship shown.[115,116] Although smoking is often associated with other unhealthy behaviours (such as lower physical activity and poor diet quality), smokers also tend to have lower body weight than non-smokers. Confounding may still be a factor in observational studies, however a recent large well controlled cohort study has shown a similar strength of association. Several plausible biological mechanisms have been proposed: smoking is associated with increased central or abdominal adiposity in both men and women, β-cell dysfunction and impaired insulin sensitivity, as well as chronic inflammation.[115]

12.3.7 Summary

The prevalence of T2DM has reached epidemic proportions worldwide, particularly over the last 20–30 years, with no suggestion of it decreasing. This increase has occurred in parallel with the widespread adoption of a 'Western lifestyle' and increasing prevalence of obesity. While there is a genetic basis for T2DM, environmental factors are key contributors to disease causation. Obesity is a key risk factor, and while excess energy intake is associated with weight gain, whether individual nutrients or other non-dietary factors contribute to risk is not clear. However, being physically active appears to be protective and lifestyle programmes which aim to improve dietary patterns and increase physical activity have been shown to prevent the onset and progression of T2DM, with long term benefits observed.

12.4 Conclusions

T1DM and T2DM are both complex disorders with different clinical presentations, although in some situations there is overlap of the two clinical entities. This is becoming increasingly apparent with the increasing prevalence of obesity and the increasing number of pre-adolescents and adolescents developing T2DM. However, the aetiology of both clinical entities involves a complex interplay between genes and environmental factors, and there is likely to be gene–gene, gene–environment and environment–environment interactions in the development of the diseases.

There is ample evidence that environmental factors are involved in the aetiology of both types of diabetes, but the nature and timing of the interactions are poorly understood. While many different possible risk factors have been studied over a number of years, and progress made with our understanding of the pathogenesis of both T1DM and T2DM, identification of definitive risk factors for T1DM still alludes us, with the implication being efforts to effectively prevent this multisystem disorder are hampered. In contrast, while the pathogenesis of T2DM is not fully understood sufficient evidence exists to enable the prevention or a slowing of the progression of the disease through environment and lifestyle changes. With the worldwide increase in the prevalence and number of people with diabetes, predominantly T2DM, until further evidence is available, the implementation of effective long term lifestyle programmes is critical if the ongoing diabetes epidemic is to be abated.

References

1. I. World Health Organization and II. International Diabetes Federation, *Definition and Diagnosis of Diabetes Mellitus and Intermediate Hyperglycaemia. Report of a WHO/IDF Consultation* WHO Document Production Services, Geneva, Switzerland, 2006.
2. American Diabetes Association, *Diabetes Care*, 2010, **33**(Suppl 1), S62.
3. J. Ostman, G. Lonnberg, H. J. Arnqvist, G. Blohme, J. Bolinder, A. Ekbom Schnell, J. W. Eriksson, S. Gudbjornsdottir, G. Sundkvist and L. Nystrom, *J. Intern. Med.*, 2008, **263**, 386.
4. C. Levy-Marchal, C. Patterson and A. Green, *Diabetologia*, 1995, **38**, 823.
5. DIAMOND Project Group, *Diabet. Med.*, 2006, **23**, 857.
6. C. C. Patterson, G. G. Dahlquist, E. Gyurus, A. Green and G. Soltesz, *Lancet*, 2009, **373**, 2027.
7. V. Harjutsalo, L. Sjoberg and J. Tuomilehto, *Lancet*, 2008, **371**, 1777.
8. V. Harjutsalo, R. Sund, M. Knip and P. H. Groop, *J. Am. Med. Assoc.*, 2013, **310**, 427.
9. Y. Berhan, I. Waernbaum, T. Lind, A. Mollsten and G. Dahlquist, *Diabetes*, 2011, **60**, 577.
10. T. Skrivarhaug, L. C. Stene, A. K. Drivvoll, H. Strom and G. Joner, *Diabetologia*, 2014, **57**, 57.
11. G. Bruno, M. Maule, A. Biggeri, A. Ledda, C. Mannu, F. Merletti and M. Songini, *Diabetes*, 2013, **62**, 3542.
12. J. Tuomilehto, *Curr. Diabetes Rep.*, 2013, **13**, 795.
13. M. A. Atkinson, *Cold Spring Harbor Perspect. Med.*, 2012, **2**, a007641.
14. M. Knip, *Horm. Res.*, 2002, **1**, 6.

15. M. Knip, R. Veijola, S. M. Virtanen, H. Hyoty, O. Vaarala and H. K. Akerblom, *Diabetes*, 2005, **54**, S125.
16. M. Knip and O. Simell, *Cold Spring Harbor Perspect. Med.*, 2012, **2**, a007690.
17. K. Hemminki, X. Li, J. Sundquist and K. Sundquist, *Diabetologia*, 2009, **52**, 1820.
18. R. S. Spielman, L. Baker and C. M. Zmijewski, *Ann. Hum. Genet.*, 1980, **44**, 135.
19. M. J. Redondo, J. Jeffrey, P. R. Fain, G. S. Eisenbarth and T. Orban, *N. Engl. J. Med.*, 2008, **359**, 2849.
20. V. Hyttinen, J. Kaprio, L. Kinnunen, M. Koskenvuo and J. Tuomilehto, *Diabetes*, 2003, **52**, 1052.
21. K. O. Kyvik, A. Green and H. Beck-Nielsen, *Br. Med. J.*, 1995, **311**, 913.
22. P. Concannon, S. S. Rich and G. T. Nepom, *N. Engl. J. Med.*, 2009, **360**, 1646.
23. M. J. Redondo, L. Yu, M. Hawa, T. Mackenzie, D. A. Pyke, G. S. Eisenbarth and R. D. Leslie, *Diabetologia*, 2001, **44**, 354.
24. M. Hummel, E. Bonifacio, S. Schmid, M. Walter, A. Knopff and A. G. Ziegler, *Ann. Intern. Med.*, 2004, **140**, 882.
25. T. Kimpimaki, P. Kulmala, K. Savola, A. Kupila, S. Korhonen, T. Simell, J. Ilonen, O. Simell and M. Knip, *J. Clin. Endocrinol. Metab.*, 2002, **87**, 4572.
26. A. J. Williams and P. J. Bingley, *Diabetologia*, 2012, **55**, 1873.
27. TEDDY Study Group, *Ann. N. Y. Acad. Sci.*, 2008, **1150**, 1.
28. J. M. Forrest, M. A. Menser and J. A. Burgess, *Lancet*, 1971, **2**, 332.
29. L. C. Stene and E. A. Gale, *Diabetologia*, 2013, **56**, 1888.
30. H. Viskari, M. Knip, S. Tauriainen, H. Huhtala, R. Veijola, J. Ilonen, O. Simell, H. M. Surcel and H. Hyoty, *Diabetes Care*, 2012, **35**, 1328.
31. M. M. Lamb, M. A. Myers, K. Barriga, P. Z. Zimmet, M. Rewers and J. M. Norris, *Pediatr. Diabetes*, 2008, **9**, 135.
32. S. Niinisto, H. M. Takkinen, L. Uusitalo, J. Rautanen, J. Nevalainen, M. G. Kenward, M. Lumia, O. Simell, R. Veijola, J. Ilonen, M. Knip and S. M. Virtanen, *Br. J. Nutr.*, 2014, **111**, 895.
33. S. M. Virtanen, L. Uusitalo, M. G. Kenward, J. Nevalainen, U. Uusitalo, C. Kronberg-Kippila, M. L. Ovaskainen, T. Arkkola, S. Niinisto, T. Hakulinen, S. Ahonen, O. Simell, J. Ilonen, R. Veijola and M. Knip, *Pediatr. Diabetes*, 2011, **12**, 95.
34. J. Y. Dong, W. G. Zhang, J. J. Chen, Z. L. Zhang, S. F. Han and L. Q. Qin, *Nutrients*, 2013, **5**, 3551.
35. A. Haynes, M. N. Cooper, C. Bower, T. W. Jones and E. A. Davis, *Diabetologia*, 2014, **57**, 469.
36. A. L. Marshall, A. Chetwynd, J. A. Morris, M. Placzek, C. Smith, A. Olabi and D. Thistlethwaite, *Diabet. Med.*, 2004, **21**, 1035.
37. B. R. Joubert, S. E. Haberg, R. M. Nilsen, X. Wang, S. E. Vollset, S. K. Murphy, Z. Huang, C. Hoyo, O. Midttun, L. A. Cupul-Uicab, P. M. Ueland, M. C. Wu, W. Nystad, D. A. Bell, S. D. Peddada and S. J. London, *Environ. Health Perspect.*, 2012, **120**, 1425.

38. S. P. Ng and J. T. Zelikoff, *Reprod. Toxicol.*, 2007, **23**, 428.
39. C. R. Cardwell, L. C. Stene, G. Joner, M. K. Bulsara, O. Cinek, J. Rosenbauer, J. Ludvigsson, M. Jane, J. Svensson, M. J. Goldacre, T. Waldhoer, P. Jarosz-Chobot, S. G. Gimeno, L. M. Chuang, R. C. Parslow, E. J. Wadsworth, A. Chetwynd, P. Pozzilli, G. Brigis, B. Urbonaite, S. Sipetic, E. Schober, G. Devoti, C. Ionescu-Tirgoviste, C. E. de Beaufort, D. Stoyanov, K. Buschard and C. C. Patterson, *Diabetes*, 2010, **59**, 486.
40. C. R. Cardwell, L. C. Stene, G. Joner, O. Cinek, J. Svensson, M. J. Goldacre, R. C. Parslow, P. Pozzilli, G. Brigis, D. Stoyanov, B. Urbonaite, S. Sipetic, E. Schober, C. Ionescu-Tirgoviste, G. Devoti, C. E. de Beaufort, K. Buschard and C. C. Patterson, *Diabetologia*, 2008, **51**, 726.
41. K. Vehik and D. Dabelea, *Diabetes*, 2012, **61**, 36.
42. M. A. Atkinson and A. Chervonsky, *Diabetologia*, 2012, **55**, 2868.
43. C. R. Cardwell, L. C. Stene, G. Joner, E. A. Davis, O. Cinek, J. Rosenbauer, J. Ludvigsson, C. Castell, J. Svensson, M. J. Goldacre, T. Waldhoer, J. Polanska, S. G. Gimeno, L. M. Chuang, R. C. Parslow, E. J. Wadsworth, A. Chetwynd, P. Pozzilli, G. Brigis, B. Urbonaite, S. Sipetic, E. Schober, C. Ionescu-Tirgoviste, C. E. de Beaufort, D. Stoyanov, K. Buschard and C. C. Patterson, *Diabetologia*, 2010, **53**, 641.
44. R. J. Brown and K. I. Rother, *Pediatr. Diabetes*, 2008, **9**, 14.
45. C. R. Cardwell, L. C. Stene, G. Joner, M. K. Bulsara, O. Cinek, J. Rosenbauer, J. Ludvigsson, J. Svensson, M. J. Goldacre, T. Waldhoer, P. Jarosz-Chobot, S. G. Gimeno, L. M. Chuang, C. L. Roberts, R. C. Parslow, E. J. Wadsworth, A. Chetwynd, G. Brigis, B. Urbonaite, S. Sipetic, E. Schober, G. Devoti, C. Ionescu-Tirgoviste, C. E. de Beaufort, D. Stoyanov, K. Buschard, K. Radon, C. Glatthaar and C. C. Patterson, *Int. J. Epidemiol.*, 2011, **40**, 363.
46. C. R. Cardwell, J. Svensson, T. Waldhoer, J. Ludvigsson, V. Sadauskaite-Kuehne, C. L. Roberts, R. C. Parslow, E. J. Wadsworth, G. Brigis, B. Urbonaite, E. Schober, G. Devoti, C. Ionescu-Tirgoviste, C. E. de Beaufort, G. Soltesz and C. C. Patterson, *Diabetes*, 2012, **61**, 702.
47. J. D. Baum, M. Ounsted and M. A. Smith, *Lancet*, 1975, **2**, 866.
48. T. J. Wilkin, *Int. J. Obes.*, 2009, **33**, 716.
49. J. P. Palmer, S. Helqvist, G. A. Spinas, J. Molvig, T. Mandrup-Poulsen, H. U. Andersen and J. Nerup, *Diabetes*, 1989, **38**, 1211.
50. K. C. Verbeeten, C. E. Elks, D. Daneman and K. K. Ong, *Diabet. Med.*, 2011, **28**, 10.
51. I. M. Libman, E. Barinas-Mitchell, S. Marcovina, F. Bacha, T. Hannon, H. Tfayli, S. J. Lee, S. Bansal, R. Robertson and S. Arslanian, *Pediatr. Diabetes*, 2011, **12**, 207.
52. E. Hypponen, S. M. Virtanen, M. G. Kenward, M. Knip and H. K. Akerblom, *Diabetes Care*, 2000, **23**, 1755.
53. E. A. Gale, *Clin. Exp. Immunol.*, 2012, **168**, 1.
54. K. S. Ronningen, *Arch. Immunol. Ther. Exp.*, 2013, **61**, 459.

55. S. Oikarinen, M. Martiskainen, S. Tauriainen, H. Huhtala, J. Ilonen, R. Veijola, O. Simell, M. Knip and H. Hyoty, *Diabetes*, 2011, **60**, 276.
56. P. M. Graves, H. A. Rotbart, W. A. Nix, M. A. Pallansch, H. A. Erlich, J. M. Norris, M. Hoffman, G. S. Eisenbarth and M. Rewers, *Diabetes Res. Clin. Pract.*, 2003, **59**, 51.
57. M. L. Simonen-Tikka, M. Pflueger, P. Klemola, C. Savolainen-Kopra, T. Smura, S. Hummel, S. Kaijalainen, K. Nuutila, O. Natri, M. Roivainen and A. G. Ziegler, *Diabetologia*, 2011, **54**, 2995.
58. O. Cinek, L. C. Stene, L. Kramna, G. Tapia, S. Oikarinen, E. Witso, T. Rasmussen, P. A. Torjesen, H. Hyoty and K. S. Ronningen, *Diabetologia*, 2014, **57**, 2193.
59. M. Blomqvist, S. Juhela, S. Erkkila, S. Korhonen, T. Simell, A. Kupila, O. Vaarala, O. Simell, M. Knip and J. Ilonen, *Clin. Exp. Immunol.*, 2002, **128**, 511.
60. M. C. Honeyman, B. S. Coulson, N. L. Stone, S. A. Gellert, P. N. Goldwater, C. E. Steele, J. J. Couper, B. D. Tait, P. G. Colman and L. C. Harrison, *Diabetes*, 2000, **49**, 1319.
61. J. M. Norris, K. Barriga, G. Klingensmith, M. Hoffman, G. S. Eisenbarth, H. A. Erlich and M. Rewers, *J. Am. Med. Assoc.*, 2003, **290**, 1713.
62. S. M. Virtanen, M. G. Kenward, M. Erkkola, S. Kautiainen, C. Kronberg-Kippila, T. Hakulinen, S. Ahonen, L. Uusitalo, S. Niinisto, R. Veijola, O. Simell, J. Ilonen and M. Knip, *Diabetologia*, 2006, **49**, 1512.
63. A. G. Ziegler, S. Schmid, D. Huber, M. Hummel and E. Bonifacio, *J. Am. Med. Assoc.*, 2003, **290**, 1721.
64. C. R. Cardwell, L. C. Stene, J. Ludvigsson, J. Rosenbauer, O. Cinek, J. Svensson, F. Perez-Bravo, A. Memon, S. G. Gimeno, E. J. Wadsworth, E. S. Strotmeyer, M. J. Goldacre, K. Radon, L. M. Chuang, R. C. Parslow, A. Chetwynd, K. Karavanaki, G. Brigis, P. Pozzilli, B. Urbonaite, E. Schober, G. Devoti, S. Sipetic, G. Joner, C. Ionescu-Tirgoviste, C. E. de Beaufort, K. Harrild, V. Benson, E. Savilahti, A. L. Ponsonby, M. Salem, S. Rabiei and C. C. Patterson, *Diabetes Care*, 2012, **35**, 2215.
65. M. Knip, H. K. Akerblom, D. Becker, H. M. Dosch, J. Dupre, W. Fraser, N. Howard, J. Ilonen, J. P. Krischer, O. Kordonouri, M. L. Lawson, J. P. Palmer, E. Savilahti, O. Vaarala and S. M. Virtanen, *J. Am. Med. Assoc.*, 2014, **311**, 2279.
66. J. C. Antvorskov, K. Josefsen, K. Engkilde, D. P. Funda and K. Buschard, *Diabetologia*, 2014, **57**, 1770.
67. T. L. Crume, J. Crandell, J. M. Norris, D. Dabelea, M. T. Fangman, D. J. Pettitt, L. Dolan, B. L. Rodriguez, R. O'Connor and E. J. Mayer-Davis, *Eur. J. Clin. Nutr.*, 2014, **13**, 159.
68. B. Frederiksen, M. Kroehl, M. M. Lamb, J. Seifert, K. Barriga, G. S. Eisenbarth, M. Rewers and J. M. Norris, *J. Am. Med. Assoc. Pediatr.*, 2013, **167**, 808.
69. S. M. Virtanen, H. M. Takkinen, J. Nevalainen, C. Kronberg-Kippila, M. Salmenhaara, L. Uusitalo, M. G. Kenward, M. Erkkola, R. Veijola, O. Simell, J. Ilonen and M. Knip, *Diabet. Med.*, 2011, **28**, 965.

70. J. L. Piper, G. M. Gray and C. Khosla, *J. Pharmacol. Exp. Ther.*, 2004, **311**, 213.
71. R. Stepankova, H. Tlaskalova-Hogenova, J. Sinkora, J. Jodl and P. Fric, *Scand. J. Gastroenterol.*, 1996, **31**, 551.
72. E. J. Helmerhorst, M. Zamakhchari, D. Schuppan and F. G. Oppenheim, *PLoS One*, 2010, **5**, e13264.
73. C. S. Zipitis and A. K. Akobeng, *Arch. Dis. Child.*, 2008, **93**, 512.
74. M. Rasanen, C. Kronberg-Kippila, S. Ahonen, L. Uusitalo, S. Kautiainen, M. Erkkola, R. Veijola, M. Knip, M. Kaila and S. M. Virtanen, *Eur. J. Clin. Nutr.*, 2006, **60**, 1317.
75. M. Simpson, H. Brady, X. Yin, J. Seifert, K. Barriga, M. Hoffman, T. Bugawan, A. E. Baron, R. J. Sokol, G. Eisenbarth, H. Erlich, M. Rewers and J. M. Norris, *Diabetologia*, 2011, **54**, 2779.
76. U. K. Prospective Diabetes Study Group, *Diabetes*, 1995, **44**, 1249.
77. L. Perreault and K. Faerch, *J. Diabetes Complications*, 2014, **28**, 226.
78. R. T. Ackermann, Y. J. Cheng, D. F. Williamson and E. W. Gregg, *Am. J. Prev. Med.*, 2011, **40**, 11.
79. E. Selvin, M. W. Steffes, H. Zhu, K. Matsushita, L. Wagenknecht, J. Pankow, J. Coresh and F. L. Brancati, *N. Engl. J. Med.*, 2010, **362**, 800.
80. S. Wild, G. Roglic, A. Green, R. Sicree and H. King, *Diabetes Care*, 2004, **27**, 1047.
81. J. E. Shaw, R. A. Sicree and P. Z. Zimmet, *Diabetes Res. Clin. Pract.*, 2010, **87**, 4.
82. C. H. Yu and B. Zinman, *Diabetes Res. Clin. Pract.*, 2007, **78**, 159.
83. J. W. Lee, F. L. Brancati and H. C. Yeh, *Diabetes Care*, 2011, **34**, 353.
84. M. Shih, Y. Du, A. S. Lightstone, P. A. Simon and M. C. Wang, *Prev. Med.*, 2014, **63**, 90.
85. D. Dabelea, E. J. Mayer-Davis, S. Saydah, G. Imperatore, B. Linder, J. Divers, R. Bell, A. Badaru, J. W. Talton, T. Crume, A. D. Liese, A. T. Merchant, J. M. Lawrence, K. Reynolds, L. Dolan, L. L. Liu and R. F. Hamman, *J. Am. Med. Assoc.*, 2014, **311**, 1778.
86. M. I. McCarthy, *N. Engl. J. Med.*, 2010, **363**, 2339.
87. F. B. Hu, *Diabetes Care*, 2011, **34**, 1249.
88. J. P. Despres, *Ann. Med.*, 2001, **33**, 534.
89. K. J. Coppell, J. I. Mann, S. M. Williams, E. Jo, P. L. Drury, J. C. Miller and W. R. Parnell, *N. Z. Med. J.*, 2013, **126**, 23.
90. J. O. Hill, J. M. Galloway, A. Goley, D. G. Marrero, R. Minners, B. Montgomery, G. E. Peterson, R. E. Ratner, E. Sanchez and V. R. Aroda, *Diabetes Care*, 2013, **36**, 2430.
91. D. E. Arterburn, A. Bogart, N. E. Sherwood, S. Sidney, K. J. Coleman, S. Haneuse, P. J. O'Connor, M. K. Theis, G. M. Campos, D. McCulloch and J. Selby, *Obes. Surg.*, 2013, **23**, 93.
92. L. Sjostrom, *J. Intern. Med.*, 2013, **273**, 219.
93. E. L. Lim, K. G. Hollingsworth, B. S. Aribisala, M. J. Chen, J. C. Mathers and R. Taylor, *Diabetologia*, 2011, **54**, 2506.

94. C. L. Gillies, K. R. Abrams, P. C. Lambert, N. J. Cooper, A. J. Sutton, R. T. Hsu and K. Khunti, *Br. Med. J.*, 2007, **334**, 19.
95. W. C. Knowler, E. Barrett-Connor, S. E. Fowler, R. F. Hamman, J. M. Lachin, E. A. Walker and D. M. Nathan, *N. Engl. J. Med.*, 2002, **346**, 393.
96. G. Li, P. Zhang, J. Wang, E. W. Gregg, W. Yang, Q. Gong, H. Li, H. Li, Y. Jiang, Y. An, Y. Shuai, B. Zhang, J. Zhang, T. J. Thompson, R. B. Gerzoff, G. Roglic, Y. Hu and P. H. Bennett, *Lancet*, 2008, **371**, 1783.
97. J. Tuomilehto, J. Lindstrom, J. G. Eriksson, T. T. Valle, H. Hamalainen, P. Ilanne-Parikka, S. Keinanen-Kiukaanniemi, M. Laakso, A. Louheranta, M. Rastas, V. Salminen and M. Uusitupa, *N. Engl. J. Med.*, 2001, **344**, 1343.
98. WHO/FAO, *Diet, nutrition and the prevention of chronic diseases. Report of a Joint WHO/FAO Expert Consultation. World Health Organisation Technical Report Series 916* 0512-3054 (Print) 0512-3054 (Linking), World Health Organisation, Geneva, 2003.
99. F. M. Sacks, L. P. Svetkey, W. M. Vollmer, L. J. Appel, G. A. Bray, D. Harsha, E. Obarzanek, P. R. Conlin, E. R. Miller, 3rd, D. G. Simons-Morton, N. Karanja and P. H. Lin, *N. Engl. J. Med.*, 2001, **344**, 3.
100. D. Gao, N. Ning, C. Wang, Y. Wang, Q. Li, Z. Meng, Y. Liu and Q. Li, *PLoS One*, 2013, **8**, e73965.
101. N. G. Forouhi, A. Koulman, S. J. Sharp, F. Imamura, J. Kroger, M. B. Schulze, F. L. Crowe, J. M. Huerta, M. Guevara, J. W. Beulens, G. J. van Woudenbergh, L. Wang, K. Summerhill, J. L. Griffin, E. J. Feskens, P. Amiano, H. Boeing, F. Clavel-Chapelon, L. Dartois, G. Fagherazzi, P. W. Franks, C. Gonzalez, M. U. Jakobsen, R. Kaaks, T. J. Key, K. T. Khaw, T. Kuhn, A. Mattiello, P. M. Nilsson, K. Overvad, V. Pala, D. Palli, J. R. Quiros, O. Rolandsson, N. Roswall, C. Sacerdote, M. J. Sanchez, N. Slimani, A. M. Spijkerman, A. Tjonneland, M. J. Tormo, R. Tumino, A. D. van der, Y. T. van der Schouw, C. Langenberg, E. Riboli and N. J. Wareham, *Lancet Diabetes Endocrinol.*, 2014, **2**, 810.
102. E. A. Hu, A. Pan, V. Malik and Q. Sun, *Br. Med. J.*, 2012, **344**, e1454.
103. V. S. Malik, A. Pan, W. C. Willett and F. B. Hu, *Am. J. Clin. Nutr.*, 2013, **98**, 1084.
104. L. Te Morenga, S. Mallard and J. Mann, *Br. Med. J.*, 2012, **346**, e7492.
105. V. S. Malik, B. M. Popkin, G. A. Bray, J. P. Despres and F. B. Hu, *Circulation*, 2010, **121**, 1356.
106. J. S. de Munter, F. B. Hu, D. Spiegelman, M. Franz and R. M. van Dam, *PLoS Med.*, 2007, **4**, e261.
107. P. Carter, L. J. Gray, J. Troughton, K. Khunti and M. J. Davies, *Br. Med. J.*, 2010, **341**, c4229.
108. T. S. Naimi, D. W. Brown, R. D. Brewer, W. H. Giles, G. Mensah, M. K. Serdula, A. H. Mokdad, D. W. Hungerford, J. Lando, S. Naimi and D. F. Stroup, *Am. J. Prev. Med.*, 2005, **28**, 369.

109. D. O. Baliunas, B. J. Taylor, H. Irving, M. Roerecke, J. Patra, S. Mohapatra and J. Rehm, *Diabetes Care*, 2009, **32**, 2123.
110. L. L. Koppes, J. M. Dekker, H. F. Hendriks, L. M. Bouter and R. J. Heine, *Diabetes Care*, 2005, **28**, 719.
111. H. A. de Boo and J. E. Harding, *Aust. N. Z. J. Obstet. Gynaecol.*, 2006, **46**, 4.
112. E. H. Yeung, C. Zhang, G. M. Louis, W. C. Willett and F. B. Hu, *Diabetes Care*, 2010, **33**, 1364.
113. L. Bellamy, J. P. Casas, A. D. Hingorani and D. Williams, *Lancet*, 2009, **373**, 1773.
114. US Department of Health and Human Services, *The Health Consequences of Smoking—50 Years of Progress: A Report of the Surgeon General.*, Centers for Disease Control and Prevention (US), Atlanta (GA), 2014.
115. A. M. Spijkerman, A. D. van der, P. M. Nilsson, E. Ardanaz, D. Gavrila, A. Agudo, L. Arriola, B. Balkau, J. W. Beulens, H. Boeing, B. de Lauzon-Guillain, G. Fagherazzi, E. J. Feskens, P. W. Franks, S. Grioni, J. M. Huerta, R. Kaaks, T. J. Key, K. Overvad, D. Palli, S. Panico, M. L. Redondo, O. Rolandsson, N. Roswall, C. Sacerdote, M. J. Sanchez, M. B. Schulze, N. Slimani, B. Teucher, A. Tjonneland, R. Tumino, Y. T. van der Schouw, C. Langenberg, S. J. Sharp, N. G. Forouhi, E. Riboli and N. J. Wareham, *Diabetes Care*, 2014, **37**, 3164.
116. C. Willi, P. Bodenmann, W. A. Ghali, P. D. Faris and J. Cornuz, *J. Am. Med. Assoc.*, 2007, **298**, 2654.

CHAPTER 13

Gene–Environment Interactions in Diabetes

RINKI MURPHY

Department of Medicine, Faculty of Medical and Health Sciences, University of Auckland, Auckland, New Zealand
Email: R.Murphy@auckland.ac.nz

13.1 Introduction

Despite considerable progress in the understanding of its pathophysiology and risk factors, diabetes continues to increase in prevalence. Type 2 diabetes accounts for approximately 90% of diabetes, while type 1 diabetes accounts for approximately 5% of diabetes, with other rare monogenic causes accounting for approximately 2% of cases.[1] The global health crisis in diabetes however, is largely due to an increase in type 2 diabetes. Given the rapid time-course of the epidemic of type 2 diabetes, this cannot be due to population level genetic changes, and is almost certainly driven by recent environmental changes. Nonetheless, diabetes disproportionately affects certain populations suggesting the possibility of gene–environment interactions, whereby the effects of pre-existing diabetes susceptibility genes may be magnified by exposure to recent changes in non-genetic factors.

Several environmental and genetic factors have consistently been shown to increase the risk of type 1 and type 2 diabetes, while genetic factors themselves are sufficient to cause rarer forms of monogenic diabetes (see Figure 13.1). Defining the complex interplay of both genetic and environmental factors in diabetes is important in guiding public health measures and clinical management of diabetes to (1) optimize targeting of preventive

RSC Drug Discovery Series No. 45
Cardiovascular and Metabolic Disease: Scientific Discoveries and New Therapies
Edited by Philip Peplow, James Adams and Tim Young

Published by the Royal Society of Chemistry, www.rsc.org

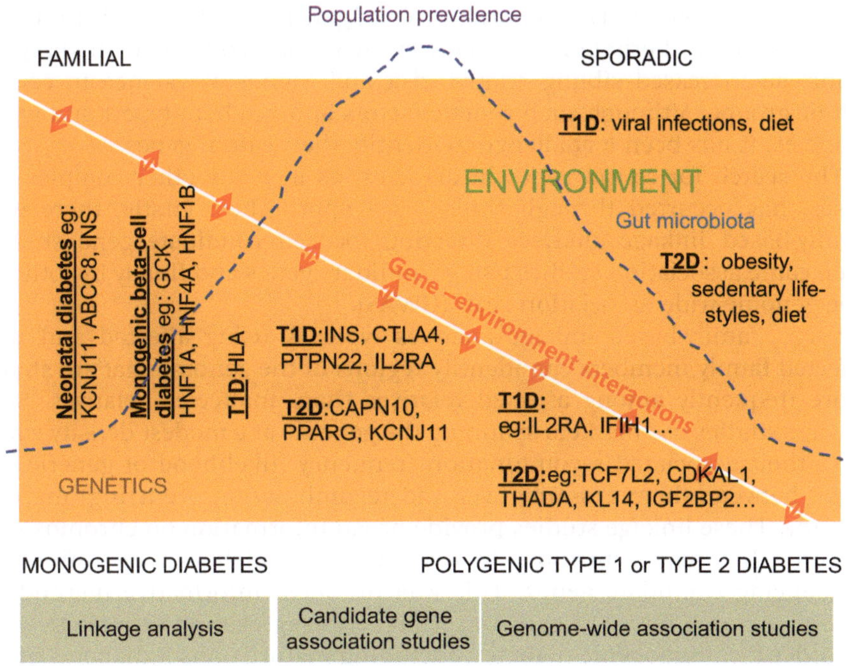

Figure 13.1 The contribution of genetic factors within the left, lower segment and environment factors on the right, upper segment, to the population prevalence of diabetes (indicated in the dotted line) is rarely entirely genetic (left of bell-shaped dotted curve) as seen in neonatal diabetes and monogenic beta cell forms of diabetes, and rarely entirely environmental (right of bell-shaped dotted curve), but largely a combination of genetic and environmental factors across the spectrum (indicated by small red arrows). Examples of genetic loci for diabetes and how these were identified are indicated in boxes below.

and therapeutic diabetes interventions to those most likely to respond well to them, (2) improve cost-effectiveness of existing diabetes preventive and therapeutic strategies, (3) reduce adverse consequences of diabetes management, (4) identify novel interventions that are beneficial only in a defined genetic subgroup of the population.

This chapter will summarize the genetic contribution to type 1 and type 2 diabetes, and discuss progress in evaluating gene–environment and gene–therapy interactions. Finally, limitations of the current data on gene–environment and gene–gene interactions are presented and future potential for further studies is highlighted.

13.2 Genetic Contribution to Diabetes

Given the same environmental risk factors, including dietary and lifestyle factors, certain people are more prone to diabetes because of different genetic backgrounds. The important role of genetics in diabetes is indicated by the stronger heritability of diabetes in monozygotic than in dizygotic twins, an increased sibling relative risk and higher prevalence in certain ethnic groups. Although such genetic factors apply to both type 1 and type 2 diabetes, it has been a challenge to identify their causal genes.

The search for causal genes in diabetes, as in many other complex diseases, has occurred through three main approaches: Firstly, there were family-based linkage analyses directing focused candidate–gene studies, then came association studies using candidate genes and finally hypothesis-free genome-wide association scan (GWAS).

Large, family-based linkage analyses typically using affected and non-affected family members can identify regions of the genome that are shared more frequently among affected relatives than unaffected relatives. Such linkage studies use markers spanning the genome at a modest density which have their associated recombination frequency (likelihood of genetic material between them crossing over and recombination occurring) mapped already. These linkage studies provide broad information on chromosomal regions that may contribute to disease risk, which can be refined by sequencing candidate genes. This was the most informative method for several extreme forms of early-onset diabetes segregating as a single-gene disorder *i.e.*: monogenic diabetes, including maturity onset diabetes of the young (MODY), neonatal diabetes and maternally inherited diabetes and deafness. Such discoveries led to improved diagnosis, prognosis and therapies for many of these specific types of monogenic diabetes.[2] The most important linkage discovery for type 1 diabetes was in the HLA region on chromosome 6p21, which remains the strongest genetic risk factor to date.

In contrast to family based-linkage studies, case-control association studies are able to detect genes with much more modest effects on risk, as long as those alleles are relatively common. Such genetic association studies used specifically selected markers in candidate genes of interest that were genotyped in patients with the disease and in unaffected controls. In type 2 diabetes, several new loci were identified for the first time, such as *PPARG* and *KCNJ11*.

Since the advent of GWAS, which was enabled by the international HapMap project (which catalogued the majority of common human genetic variants), and larger sample sizes, more and more highly likely causal genetic loci have been discovered for both type 1 and type 2 diabetes but with lower and lower effect sizes. It has been predicted that there are many rare variants, possibly with larger effects contributing to diabetes that cannot be detected using current genotyping platforms. To uncover the remaining "missing heritability" in complex diseases such as type 1 and type 2 diabetes, investigators are working on large, high-throughput sequencing efforts involving thousands of DNA samples from affected subjects and a similar number of controls. Such powerful, next-generation DNA sequencing is still very expensive and time consuming, but is likely to reduce in cost over time and to become the dominant method for identifying new causal genes.

One of the main challenges that remain is to determine how the identified genetic sequence variants linked with type 1 and type 2 diabetes affect the expression and function of the gene products relating to diabetes onset in order to be translated into meaningful benefits for patient care.

13.2.1 Genetics of Type 1 Diabetes

Type 1 diabetes is an immune-mediated form of diabetes resulting from pancreatic beta cell destruction, which usually leads to absolute insulin deficiency. This was previously also known as insulin dependent diabetes, or juvenile-onset diabetes due to the more frequent and relatively straightforward diagnosis in children. Currently, the majority of patients with type 1 diabetes are adults and the diagnosis is generally made on clinical, catabolic symptoms suggestive of insulin deficiency such as polyuria, polydipsia, weight loss, or marked hyperglycemia that is unresponsive to oral agents or presentation with diabetic ketoacidosis. In some cases, additional diagnostic measurements such as low or undetectable plasma C-peptide, presence of autoantibodies to insulin (IAA), glutamic acid decarboxylase (GAD), islet cell (ICA), protein tyrosine phosphatase (IA2A) or zinc transporter protein (ZnT8) are sought. While these antibodies are useful in classifying diabetes as type 1 or determining progression to type 1 diabetes, they are not thought to be directly involved in its pathogenesis, which is thought to be mainly driven by T cells.

Although the majority of patients (>85%) with type 1 diabetes lack a positive family history, there is greater familial clustering. The mean prevalence of type 1 diabetes is 6% in siblings compared to 0.4% in the general US white population, so the sibling risk ratio is 15 (6/0.4). Therefore, the risk for type 1 diabetes in siblings of patients is 15 fold higher than the risk in the general population without an affected sibling.[3] The risk for offspring of fathers with type 1 diabetes is approximately 12% compared to 6% for offspring of affected mothers[4] suggesting complex inheritance. A higher concordance of type 1 diabetes among monozygotic twins than dizygotic twins is seen (probandwise concordance rate of 23% among monozygotic twins and 5% among dizygotic twins).[5] This risk is dependent on the age of onset, so if the first monozygotic twin develops type 1 diabetes before the age of 6 years, the risk of the second twin developing this also is at least 60%, but if the first monozygotic twin develops type 1 diabetes after age of 25 years then the risk is much lower.[6,7]

The worldwide incidence of type 1 diabetes is highest in Europeans, ranging from 0.1 in China to 40 per 100 000 person year in Finland.[8] The lower rates of type 1 diabetes incidence persisting in second generation immigrants and among foreign born children who were adopted by Swedish parents, supports the contribution of ethnicity related genetic heterogeneity in type 1 diabetes risk.[9] That is, in the presence of the same environmental risk factors, some people are more prone to diabetes than others because of different genetic backgrounds.

The search for the susceptibility genes underlying type 1 diabetes began with linkage studies showing the most important and consistent linkage with the human leukocyte antigen (HLA) region on chromosome 6p21.[10] The HLA genes encode highly polymorphic proteins, which are essential in immune recognition of self *versus* non-self, and the presentation of antigens by T cells. The HLA cluster of cell surface proteins is divided into class 1 (A, B, C) and class 2 (DP, DQ, DR). Class 1 molecules are expressed widely and present intracellular antigen to $CD8^+$ T cells for destruction. Class 2 molecules are expressed mainly on certain antigen presenting cells, such as dendritic cells, macrophages, B cells and thymus epithelium. These class 2 molecules are composed of A and B chains and present antigens to $CD4^+$ T cells, which produce a variety of effects to eliminate the pathogen. Approximately half of the genetic risk for type 1 diabetes is conferred by the genetic region containing the HLA class 2 genes: HLA-DRB1, -DQA1 and -DQB1 (odds ratio >6.5). Specifically the alleles with the highest associated risk are DRB1*04/DQB1*03:02 (OR of 11) and DRB1*03:01/DQB*02:01 (OR 3.6).[11] The haplotype of DRB1*03:01/DQB*02:01 forms a haplotype encompassing the complement gene region in which a human endogenous retrovirus of class K occurs within the C4 gene [HERV-K(C4)]. Recently, low copy number of HERV-K(C4) has been identified as a strong marker of type 1 diabetes risk which captures the disease association of the DRB1*03:01/DQB*02:01 and DRB1*15:01/DQB1*06:02 HLA haplotypes.[12]

Linkage studies were followed by candidate association studies, which identified 4 additional well replicated risk loci, including *INS*,[13] *CTLA4*,[14] *PTPN22*,[15] and IL2RA.[16] Of these, *INS* and *PTPN22* have the highest odds ratios for type 1 diabetes at around 2 and 1.8, respectively. The HapMap project showed that the genome contains discrete linkage disequilibrium (LD) blocks, with limited variation within each of these blocks. Therefore, genotyping a smaller set of single nucleotide polymorphisms (500 000 out of 3 billion) is sufficient to detect almost all the common haplotypes present, thus improving genotyping accuracy and reducing cost. This was the basis of GWAS, which rapidly increased the discovery of genes associated with type 1 diabetes and has since identified almost 60 additional loci,[17–19] each with odds ratios below 2. Further meta-analyses of type 1 diabetes GWAS datasets have been used to increase the statistical power and discover novel loci. While the type 1 diabetes loci can be broadly classified according to their function of either non-immune or immune,[20] the exact mechanisms by which these loci confer susceptibility to the disease is not known.

13.2.2 Genetics of Type 2 Diabetes

Type 2 diabetes accounts for 90–95% of those with diabetes and is characterized by insulin resistance and relative insulin deficiency (in contrast to absolute insulin deficiency seen in type 1 diabetes). This was previously referred to as adult-onset diabetes or non-insulin-dependent diabetes due to at least initially not requiring insulin treatment. Most patients with this form of diabetes are obese, while those who are not obese by standard height and weight criteria have increased visceral fat, typically with increased abdominal girth. Type 2 diabetes is distinguished from type 1 diabetes on the basis of clinical features such as central obesity, strong family history, previous history of gestational diabetes or polycystic ovarian syndrome in women, gradual onset of osmotic symptoms and the presence of associated metabolic disorders such as hypertension or dyslipidemia. Most cases of newly diagnosed type 2 diabetes respond to non-insulin glucose lowering agents such as metformin or sulfonylureas.

The worldwide prevalence of type 2 diabetes has undergone dynamic increase in recent years, particularly in Asian countries where approximately 10–20% of adults living in urban China and India are affected. Compared with Europeans, Asians develop diabetes at younger ages, and at lower levels of obesity, suggesting genetic susceptibility and gene–environment interactions. A family history of type 2 diabetes is increasingly common due to the increased prevalence of the condition. Family history together with impaired fasting glucose and overweight is associated with 16-fold increased risk for future type 2 diabetes. The relative risk for a sibling with type 2 diabetes is about three times the general population.[21] First degree relatives of patients with type 2 diabetes who do not have diabetes themselves, still show early defects in insulin secretion and action. This could be due to shared environmental habits such as excess food intake and poor physical activity, which are important risk factors for type 2 diabetes. However, a genetic susceptibility is suggested by the stronger heritability of type 2 diabetes in monozygotic than in dizygotic twins, where the probandwise concordance rate has been reported at 34% among monozygotic twins and 16% among dizygotic twins.[5]

The discovery of underlying causal genes among families with frequent early onset diabetes occurred through linkage analysis. This approach identified relatively rare beta cell monogenic defects manifesting as MODY and neonatal diabetes. However this was not very successful for identifying loci for type 2 diabetes. Early association studies with candidate gene analysis identified loci such as calpain 10 (*CAPN10*),[22] peroxisome proliferator-activated receptor gamma (*PPARG*),[23] and potassium inwardly rectifying channel subfamily J member 11 (*KCNJ11*).[24] The last two genes encode proteins, which act as targets for classes of drugs used in diabetes treatment. However, the most successful approach for discovering type 2 diabetes genes has been the systematic, large scale, hypothesis-free surveys of association between common DNA sequence variants and type 2 diabetes. This type of GWAS successfully identified the *TCF7L2* locus (encoding transcription factor 7-like 2), which had not previously been identified as a biological candidate.[25] *TCF7L2* shows the strongest and most widely replicated association with type 2 diabetes (40% increased risk per risk allele); however, recently TBCD14 has been identified among the Greenlandic population as being associated with a much greater 10 fold increased risk.[26] Over 60 other loci harboring genes such as *CDKAL1*, *SLC30A8*, *NOTCH2*, *THADA*, *KL14*, *HHEX/IDE*, *IGF2BP2*, *CDKN2A/2B* with lower degrees of increased risk have been found (reviewed in ref. 27).[27]

GWAS of diabetes related traits such as fasting glucose and/or homeostatic model of assessment of beta-cell function (HOMA-B) index have found associations with distinct loci such as *FADS1*, *MDD*, *SLC2A2*, *CRY2*, *GLIS3*, while those associated with 2 hour glucose after an oral glucose tolerance test include other loci such as *ADCY5*, *VSP13C*, *GIPR*, while those associated with HbA1c include *ANK1*, *TMPRSS6*, *HFE*, *FN3K* (reviewed in ref. 27 and 28). In contrast to the numerous diabetes susceptibility loci related to impaired beta cell function, only a few have been found to be associated with insulin resistance or fasting insulin, such as *IGF1*, *IRS1*, *GCKR*. This implies that mechanisms involved in different aspects of physiological regulation of glucose have distinct genetic components, of which not all are related to the pathogenesis of type 2 diabetes. However, a minority of type 2 diabetes loci also overlap with several type 2 diabetes related traits such as *GCK* and *MTNR1B* (HbA1c, fasting glycaemia and type 2 diabetes), and *GCKR* (2 hour glucose, fasting insulinemia, fasting glycaemia and type 2 diabetes).

Some variants that confer type 2 diabetes susceptibility vary across different ethnic groups in terms of risk allele frequency or effects size. The type 2 diabetes associated variant at the *TCF7L2* locus which is present in 20–30% of European populations, is only present in 3–5% of Asian populations. The strongest type 2 diabetes associated variant in Asian populations occurs in KCNQ1,[29,30] but is rare in European populations. In Mexican populations sequence variants in *SLC16A11* and *SLC16A13* increased risk of type 2 diabetes,[31] while in those from Greenland, a new sequence variant in the gene *TBC1D4* has recently been identified which is associated with muscle insulin resistance, and markedly increased odds of type 2 diabetes (10.3 fold).[26] Most of these single nucleotide variants uncovered thus far may not be the actual causal variant, which needs to be pinpointed through fine-mapping, sequencing and functional studies.

Despite having similar symptoms, type 1 diabetes and type 2 diabetes do not share any genetic risk loci and represent two distinct diseases.[32]

13.2.3 Genetic Risk Prediction of Type 1 or Type 2 Diabetes

Currently, screening for immune or genetic markers of risk of type 1 diabetes is generally only undertaken in the context of defined research studies. The 5 year risk of a first degree relative with type 1 diabetes based on family history alone is 4–6% with an affected mother, 10–12% with an affected father and 4–6% with an affected sibling, compared to <0.4% background risk without an affected first degree relative. The 5 year risk of type 1 diabetes increases among a first degree family member of someone with type 1 diabetes if they have one persistent antibody (25%) or two persistent antibodies (45%), and with three persistent antibodies this risk is 70%. The corresponding 5 year risk of type 1 diabetes with knowledge of genetic DR3/3 status is 10%, DR4/4 is 17% and DR3/4 is 22%.[33] Hence, genetic testing is of limited value in family members in whom autoantibody levels are known. However, HLA class 2 typing may be useful in identifying infants for prospective follow up for the appearance of islet autoantibodies or for recruitment into clinical trials of interventions aimed at preventing the initiation of autoimmunity. Currently there is no means for preventing the disease, making routine screening of type 1 diabetes of limited clinical use.

The ability of the combined genetic variant information from up to 40 type 2 diabetes-related polymorphisms to predict type 2 diabetes status or incidence, provides an area under the receiver operating characteristic (ROC) curve of 0.54 to 0.63 from both cross-sectional and longitudinal studies (reviewed in ref. 34).[34] This compares poorly to clinical models of type 2 diabetes using different combinations of simple clinical and laboratory parameters (age, sex, BMI, fasting glucose and in some models family history of diabetes), with ROC curve of 0.61 to 0.95. Adding the genetic risk score in such models showed that in addition to clinical and laboratory parameters, genetic risk variants increase the predictive value only minimally. Hence genetic screening for the prediction of type 2 diabetes is currently of little value in clinical practice.

13.3 Environmental Contribution to Diabetes

13.3.1 Intrauterine Environment

Several intrauterine risk factors for type 1 diabetes have been reported, such as advanced maternal age,[35–38] excessive weight gain in pregnancy, amniocentesis,[37] pre-eclampsia[39] and gestational infections such as rubella,[40,41] enterovirus[42] and coxsackie B3.[43] Other gestational risk factors for type 1 diabetes includes first born birth order[44,45] and a large risk for gestational age birth weight.[46,47] The immunological mechanisms underlying these risk factors to the pathophysiology of type 1 diabetes are unclear and likely complex.

For type 2 diabetes, there are consistent associations between both low birth weight[48] and high birth weight.[49] The observations of low birth weight linked with later type 2 diabetes may be explained by underlying genetic influences that determine both early growth and subsequent type 2 diabetes risk, given that alleles that result in reduced insulin secretion or action contribute to type 2 diabetes risk in adulthood, as well as potentially compromising fetal growth due to the important role of insulin as a growth factor in early life.[50] Both ADCY5 and CDKAL1 gene variants have been linked with birth weight and type 2 diabetes.[51] In addition, the link with birth weight and type 2 diabetes may also be mediated by regulation of gene expression through chromatin modification or DNA methylation caused by altered intrauterine nutrition. Altered patterns of DNA methylation have been reported in individuals exposed to restricted maternal nutrition *in utero*, which have also been associated with increased risk for subsequent metabolic diseases. Altered methylation status of specific cytosine guanine dinucleotides (CpG) loci in blood in imprinted[52] and non-imprinted[53] genes have been associated with both exposure to famine *in utero* and metabolic disease risk in elderly individuals. Methylation levels at promoter regions for *RXRA* in umbilical cord tissue predicted >25% of the variation in age and sex adjusted fat mass in children at the age of 6 and 9 years and also inversely correlated with maternal carbohydrate intake during early pregnancy.[54] There may be a role of *RXRA* in lipid and carbohydrate metabolism, which provides a possible mechanism whereby maternal diet could have an impact on their child's adiposity long after birth.

13.3.2 Post-natal Environment

The increase of type 1 diabetes with time (approximately 3% increase in incidence annually), suggest that environmental factors may play an important role.[55,56] Various types of viral infections such as enterovirus, coxsackie, rotavirus, cytomegalovirus, mumps, rubella and others have long been implicated in the progression to type 1 diabetes.[57] They are thought to cause type 1 diabetes either through a direct cytolytic effect on the pancreatic beta cells or by triggering an autoimmune process leading gradually to beta cell destruction.

Increased linear growth and weight gain in infancy has been reported to be a risk factor for type 1 diabetes in childhood.[58–60] According to the accelerator hypothesis,[61] increasing prevalence of insulin resistance through increasing prevalence of obesity is an important factor affecting the rising incidence of both type 1 and type 2 diabetes, because the compensatory hyperinsulinemia is thought to promote beta cell fatigue and apoptosis.

The association of dietary factors such as breastfeeding, vitamin D, early introduction of cow milk proteins or fruits and berries with type 1 diabetes is controversial. The DAISY study found that a higher dietary glycemic index was associated with a more rapid progression from seroconversion to overt type 1 diabetes.[62] Only a few intervention trials modifying infant nutrition have been performed so far. The German BABYDIET study showed no difference in diabetes associated autoantibodies or clinical type 1 diabetes between late exposure to gluten beyond 12 months or routine exposure to gluten.[63] The pilot Trial to Reduce IDDM in the Genetically at Risk (TRIGR) recruited 230 Finnish infants with high risk HLA susceptibility and found that the group randomized to weaning on to extensively hydrolysed formula before the age of 6–8 months showed a reduction in most signs of beta cell autoimmunity compared to those weaned on to conventional cow's milk formula. The full scale TRIGR study is currently underway and will be completed in 2017.[64] Another pilot Nutritional Intervention for the Prevention of Type 1 Diabetes (NIP) clinical trial is currently underway to test the feasibility of a full scale clinical trial to test whether dietary supplementation with an omega 3 fatty acid during fetal life and in infancy will prevent beta cell autoimmunity in at risk infants.

Lifestyle and behavioral risk factors such as diet, obesity and physical inactivity are recognized as major determinants explaining the rapid increase in prevalence of type 2 diabetes in recent years. Studies of first and second generation migrant populations show that their level of acculturation (process of adaptation of behavioural patterns to the principal culture in the new country) is associated with higher prevalence of diabetes. This is thought to be due to the adoption of lifestyle changes which promote type 2 diabetes, such as increase in fast food consumption, smoking and lack of exercise, in addition to stress factors such as crowded living conditions, air pollution, depression and anxiety. These risk factors are discussed in Chapter 12.

13.3.3 Gut Microbiota

Microbial cells colonizing the human body vastly outnumber the cells that actually make up the human body, and the majority of these microbial cells reside in the gut. The gut microbiota is very complex and includes thousands of species that have a collective genome of approximately 5 million genes, termed the metagenome.[65] It has long been recognized that some intestinal microorganisms are associated with health and others with disease through characterization of individual microbes.[66] Now the availability of high-throughput sequencing has enabled the metagenome derived from stool samples to be sequenced, and when combined with clinical data, it is known as a metagenome-wide association study (MGWAS).

Animal studies support a causal role for gut microbiota in type 1 diabetes, since its development in non-obese diabetic mice can be prevented through modulation of the gut microbiota,[67] and exacerbated by losing the adaptor for multiple innate immune receptors responsible for recognizing microbial stimuli.[68] In contrast to healthy subjects, people with type 1 diabetes have a progressive increase in the proportion of *Bacteroidetes* phylum within their gut microbiota over time. In addition, people with type 1 diabetes tend to have less diversity of gut microbiota species which become less similar to each other compared to gut microbiota communities from controls.[69]

Aside from compositional changes in gut microbiota ascertained by 16S sequencing of the marker DNA, shotgun sequencing of the bacterial DNA enables the analysis of the entire functional gene capacity of the bacteria present in stool. Such metagenomic analysis showed functional differences among gut microbiota between cases with type 1 diabetes and controls, at the levels of community processes (less functionally diverse than healthy children), whole pathway (less butyrate producers), as well as for individual genes (more reads mapped to sulfur metabolism in people with type 1 diabetes, suggesting higher anaerobicity).[70] Butyrate is known as an anti-inflammatory short chain fatty acid that contributes to colon health,[71,72] mucin synthesis,[73] decreases bacterial transport across intestinal epithelia[74] and improves the intestinal barrier.[75] One hypothesis for a bacterial role in gut integrity leading to either a healthy state or autoimmunity and type 1 diabetes is the fate of lactate produced by lactic acid bacteria (*Lactobacillus, Bifidobacterium*), which if converted to butyrate (*e.g.* by *Faecalibacterium, Subdoligranulum*), results in more mucin synthesis and tighter junctions leading to improved gut integrity and health, while conversion to other short chain fatty acids (*e.g.* by *Alistipes, Bacteroides, Veillonella*) reduces mucin synthesis and tight junctions leading to a permeable gut and autoimmunity.[69,70]

For type 2 diabetes, the predictive power of MGWAS is much greater than GWAS of human genes. Two MGWAS show highly significant correlations of specific intestinal microbial genes with type 2 diabetes in a European cohort of 145 women[76] and in a Chinese cohort of men and women.[77] Both studies showed lower Clostridiales bacteria that produce the short-chain fatty acid butyrate (*Roseburia* species and *Faecalibacterium prausnitzii*) were highly discriminating of type 2 diabetes, supporting the known role of butyrate-producing bacteria as regulators of human glucose and lipid metabolism and in maintaining intestinal integrity.[78] Both studies found that genes involved in assembly of bacterial flagella or those involved in cofactors and vitamins were lower in type 2 diabetes. Only one study identified higher *Lactobacillus gasseri* and *Streptococous mutans,* usually found in the mouth and upper intestinal tract, in the European type 2 diabetes cohort,[76] while the Chinese cohort was found to have greater *Proteobacteria,* which may produce inflammatory lipopolysacharides that lead to endotoxaemia.[77]

Further support for a causal role of gut microbiota in type 2 diabetes comes from animal and human studies in which fecal transplantation from lean donors was sufficient to improve insulin resistance in recipients with metabolic syndrome.[79] However, much further work is needed before altered gut microbiota can be used in a diagnostic or therapeutic setting. Designing therapeutic bacteria that can be used to shift the gut microbiota composition towards that of healthy individuals is the goal.[80]

13.4 Gene–Environment Interactions in Diabetes

It is likely that the genetic factors that underlie individual susceptibility for diabetes are amplified in the presence of certain environmental triggers. A gene–environment interaction is defined, in this context, as when the combined risk attributable to genetic and environmental exposures is significantly greater or less than expected if their effects were simply additive. The presence of gene–environment interactions are likely during both disease processes that result in either type 1 diabetes or type 2 diabetes.

A link between polymorphism in the *PTPN22* gene and infant feeding has been reported, with the risk *PTPN22* variant associated with a higher rate of seroconversion to positive autoantibodies and progression to type 1 diabetes only among those who had been exposed to conventional cow's milk infant formulas before the age of 6 months.[81] Further studies are required to explore the interplay between environmental factors and genetic factors contributing to the pathogenesis of type 1 diabetes.

The detrimental effects of a Westernized diet on increased risk of type 2 diabetes may be enhanced by greater genetic susceptibility. A significant interaction between a Western dietary pattern derived from principal component analysis of 40 food groups and a genetic risk score of type 2 diabetes susceptibility based on 10 established loci was reported in men from the Health Professionals Follow-up Study ($p = 0.02$).[82] The multivariable odds ratios of diabetes across increasing quartiles of the Western dietary pattern were 1.00, 1.23, 1.49, and 2.06 among men with a higher genetic risk score of 12 or more risk alleles (p for trend $= 0.01$). Intake of processed meat, red meat, and heme iron characteristic of the Western dietary pattern each showed significant interactions with genetic risk score in relation to type 2 diabetes risk. Among those with a lower genetic risk score, the Western dietary pattern was not associated with diabetes risk.

The interaction of diet with increased risk of type 2 diabetes associated with *TCF7L2* variants have been examined according to carbohydrate quality and quantity among women in the Nurses Health Study.[83] The increased risk was more pronounced among women consuming a high glycemic load or high glycemic index diet: compared with the low risk genotype, multivariate adjusted odds ratios (95% CI) of diabetes associated with the high risk genotype were 2.71 (1.64, 4.46) and 2.69 (1.64, 4.43) among individuals in the highest tertile of glycemic load and glycemic index, respectively. However, the increased risk of type 2 diabetes associated with the *TCF7L2* risk variant was not completely abolished by consumption of a low glycemic load or low glycemic index diet. The risk associated with *TCF7L2* did not significantly differ with cereal fiber or carbohydrate intake.[83]

Total zinc intake has been associated with lower fasting glucose levels and an interaction between total zinc intake and the *SLC30A8* risk variant has also been reported, such that there is a stronger inverse association between total zinc intake and fasting glucose in individuals carrying the glucose raising risk allele compared with individuals who do not.[84] Higher zinc intake attenuating the glucose raising effect of the *SLC30A8* variant has recently been confirmed in another study where the benefit of higher zinc levels on the odds of diabetes decreased with increasing copies of the risk allele.[85]

13.5 Gene–Treatment Interactions in Diabetes

There are clear gene–treatment interactions for monogenic forms of diabetes, whereby those with an autosomal dominant mutation in the *HNF1A* or *HNF4A* genes causing MODY to respond extremely well to low dose sulfonylurea therapy, and those with a mutation in *KCNJ11* or *SUR1* causing neonatal diabetes to respond very well to high dose sulfonylurea therapy, rather than insulin therapy as is frequently given in absence of this genetic diagnosis.[86] Several pharmacogenetic interactions have been reported for metformin, sulphonylureas, repaglinide, thiazoledinediones and acarbose relating to type 2 diabetes. However, the potential individualization of diabetes prevention and treatment of type 2 diabetes based on genetic variation has not yet been realized.

If diet and physical activity changes are not effective in achieving glycemic control, metformin is recommended as the first glucose lowering drug in patients with type 2 diabetes, provided there are no contraindications such as significant renal impairment. However, the glucose lowering response to metformin can vary greatly between patients, with some patients responding poorly, which are not easily distinguished by clinical criteria such as body weight or age. Intensification of therapy generally consists of a second and third and in some cases more glucose lowering agents being added to metformin. However, if poor responders to metformin could be clearly identified, then this may be better stopped or never commenced. Genes encoding metformin transporters, *SLC22A1*, *SLC22A2*, and *SLC47A1* have been evaluated for predicting response to metformin therapy. Reduced diabetes risk over 3 years in response to metformin was found with rs683369 in *SLC22A1*, rs662301 in *SLC22A2*, and rs8065082 in *SLC47A1* in the diabetes prevention program study.[87] Other genetic variants at these three loci were not associated with glycemic response to metformin in people with type 2 diabetes.[88,89] The rs11212617 gene variant in *ATM*, implicated in the effect of metformin upstream of AMP-activated protein kinase, was linked with metformin treatment success.[90] Other minor interactions have been reported between metformin and genetic variants in genes encoding proteins associated with AMP-activated protein kinase-dependent inhibition of gluconeogenesis (*PRKAB2, PRKAA2, PRKAA1, STK11,*[91] *PCK1, PPARA,*[92] and *PPARGC1A*[87]) as well as genes associated with insulin secretion (*KCNJ11,*[93] *ABCC8, CDKN2A/B,*[94] *GCK, HNF4A* and *HNF1B*[87]) and insulin sensitivity (*ADIPOR2, ENPP1,*[95] *CAPN10*).

Pharmacogenetic interaction studies between sulfonylureas and genetic variants include two studies[96,97] which evaluated genetic variants in *CYP2C9*, the gene encoding the enzyme that metabolises sulfonylurea, of which one small study found significant associations with glycemic response.[96] Two genetic variants in *TCF7L2* linked with insulin secretion were associated with a glycemic response to sulfonylurea medication.[98] Sulfonylureas and repaglinide bind to the sufonylurea receptor (encoded by *ABCC8*), which then inhibits the function of the potassium channel encoded by *KCNJ11* and causes beta cell depolarization and insulin secretion. However, genetic variants within KCNJ11, which is critical for insulin, was not associated with sulfonylurea effect,[99] although it has been associated with repaglinide effect.[100]

Of the studies reporting genetic interactions with repaglinide, one of two genetic variants within the *SLC30A8* gene was associated with glycemic response.[101] Some glycemic outcomes after repaglinide were also predicted by genetic variants in IGF2BP2,[102] NEUROD1, PAX4,[103] and UCP2.[104]

The thiazolidinediones activate peroxisome proliferator-activated receptor gamma (PPARG) receptors which regulate expression of genes important for sensitivity to insulin, thus variation in PPARG would likely affect response to this class of medications. Only one[105] of two studies evaluating the effect of pioglitazone found this to be the case.[106]

Acarbose primarily reduces intestinal glucose absorption by inhibiting brush border alpha glucosidase enzymes that hydrolyse carbohydrates and does not seem to have obvious pharmacokinetic or pharmacodynamics targets for genetic interaction. Nonetheless, two of eleven genetic variants from the *PPARA* locus[107] and two of six genetic variants from the *HNF4A* locus were associated with response to acarbose.[108] A single genetic variant at the *PPARGC1A* loci[109] was also associated with acarbose response.

A novel strategy was recently used to identify a gene variant that confers relative unresponsiveness of type 2 diabetes to dipeptidyl peptidase-4 (DPP-4) inhibitor therapy. DPP4 inhibitors block the enzyme that degrades glucagon like peptide-1 (GLP-1) incretin hormone that is responsible for amplifying the insulin response. First, a metabochip (custom-designed illumina array containing top genetic variants from meta-analysed GWAS of a number of cardiometabolic phenotypes) was used to identify the specific genetic variants associated with GLP-1 induced insulin secretion among 232 non-diabetic participants studied with a hyperglycemic clamp.[110] The three genetic risk variants which exceeded the pre-specified statistical significance for association (set at $p<8.8\times10^{-7}$ based on 53 000 gene variants tested), were then genotyped in two cohorts of Dutch and Scottish patients with type 2 diabetes treated with either GLP-1 receptor agonists (22 in Netherlands and 151 in Scotland) or DPP-4 inhibitors (49 in Netherlands and 305 in Scotland). Only 1 genetic risk variant (presence of G allele at rs7202877) was found to confer relative unresponsiveness to DPP-4 inhibitor treatment. To elucidate the potential molecular mechanism of action of rs7202877, the expression of neighbouring genes (*CTRB1, CTRB2* and *BCAR1*) were assayed in 35 human pancreata and in islets isolated from 24 normoglycemic and 21 hyperglycemic subjects, which suggested that this SNP (rs7202877) acted as a marker of expression for CTRB1 and CTRB2 but not BCAR1. Since CTRB1 and 2 encode chymotrypsinogen, chymotrypsinogen activity was measured in stool samples of 80 participants, which supported the pancreas expression results in that the 40 G-allele carriers had increased chymotrypsin activity when compared with non-carriers. Stool chymotrypsin activity was reduced after DPP-4 treatment in TT homozygotes, but not in G allele carriers. Therefore, these data were used to derive a new pathophysiological model for type 2 diabetes in which the G-allele at rs7202877 represents a high expression locus for *CTRB1* and *CTRB2* genes, raising chymotrypsin activity, which might affect the delivery of nutrients into the gut, triggering previously unexplored changes in the incretin system that causes improved sensitivity of pancreatic beta cells to GLP-1, but reduced sensitivity to DPP-4. As a result, G-allele carriers experience higher insulin secretion after an oral glucose load, thereby lowering their risk of type 2 diabetes, but do not benefit from DPP-4 inhibition therapy (of which chymotrypsin is a known target). Interestingly, the G allele has previously been associated with increased risk of type 1 diabetes[19] but reduced risk of type 2 diabetes.[111] This pharmacogenetic finding, if confirmed, may be useful to stratify patients likely to benefit from DPP-4 inhibitors and potentially novel chymotrypsin enhancing therapies for type 2 diabetes, if this does not increase the risk of autoimmune type 1 diabetes.

Most of these biologically plausible gene–medication interactions reported for type 2 diabetes require follow up in larger, well-designed studies, before any of these genetic results can be applied to personalize therapy for type 2 diabetes prevention and management.

13.6 Future Perspectives

Current knowledge of genetic associations with type 1 and type 2 diabetes have several limitations in their application in risk prediction or clinical management. Future work is required to increase the precision of diabetes diagnosis beyond type 1 and type 2, identification of risk variants for type 2 diabetes in non-European populations, identification of low-frequency and rare genetic variants, increasing knowledge on structural variation, metagenomics and epigenetics, and by developing statistical techniques to evaluate gene–gene and gene–environment interactions.

13.6.1 Improved Precision of Diabetes Diagnosis

Imprecise classification of individuals with diabetes is likely to have weakened the power of previous genetic studies to detect an association. This is likely to also explain the lack of correlation between genetic variants linked with type 2 diabetes and component glucose homeostasis traits such as fasting insulin, glucose, HbA1c. The diagnosis of type 2 diabetes is a chronic hyperglycemic diagnosis of exclusion, that is not type 1 diabetes, or other subtypes of diabetes such as known monogenic defects of insulin secretion or insulin action, diseases of exocrine pancreas, due to endocrinopathies, drug or chemically induced diabetes, diabetes related to infections, rare forms of immune-mediated diabetes, other genetic syndromes associated with diabetes or gestational diabetes.[1] Differentiating between different subtypes of diabetes is challenging, especially between type 2 diabetes and late-onset and slowly developing type 1 diabetes, which are often erroneously diagnosed as type 2 diabetes.[112] The increasing number of monogenic forms of diabetes which are now able to be identified suggests that every cohort of type 2 diabetes patients may have a varying proportion, up to approximately 10%, who actually have either monogenic diabetes[113] or latent type 1 diabetes.[112] Further, several large scale studies include younger patients in the non-diabetic control group who may be yet to develop type 2 diabetes, and may already have several clinical risk factors. More careful phenotyping and classification of participants into accurate subtypes of diabetes are needed in future genetic studies.

13.6.2 Identification of Diabetes Risk Genes in Non-Europeans

The majority of GWAS of diabetes susceptibility have been undertaken in populations of European ancestry due to relatively poor coverage of common genetic variation in non-European ethnic groups in many genome-wide genotyping arrays, and lower number of samples available. However, the latest genome wide genotyping arrays are less biased towards Europeans, and more recent type 2 diabetes GWAS has been performed successfully in East Asian, South Asian, Mexican and African American populations, which provide evidence of overlap in type 2 diabetes susceptibility loci between ancestry groups. Transethnic meta-analyses have shown directional consistency of type 2 diabetes risk alleles across ancestry groups, and seven new type 2 diabetes loci were recently identified using this approach.[114] Differences in genetic architecture among the populations could help to identify variants that are relatively rare in Europeans but are more common in other ethnic groups.

13.6.3 Improved Sequencing Techniques to Identify Low-frequency Variants, Structural Variation, Epigenetics and Metagenomics

The common disease–common variant hypothesis states that common diseases are attributed to allelic variants present in more than 5% of the population, hence the genome wide association studies. However, in new technologies such as exome sequencing, custom made exome chips are needed to identify low frequency (<5%) or rare (<0.5%) variants having larger effect sizes. Furthermore, the genetic risk signals emerging from GWAS may not in fact be the causal variant or even in the causal gene, so fine mapping and sequencing efforts are required to identify these true genes and variants.

Structural variation in chromosomes such as copy number variants, including insertions and deletions, and inversions or translocations may also contribute to the risk of type 2 diabetes, however only affect up to 5% of the genome. This is under-investigated given that the current arrays and sequencing technologies are unable to capture the large genomic regions where these occur. Next generation sequencing may be better able to identify such structural variation.

The most investigated epigenetic variations are methylation of cytosine bases in DNA and histone modifications. Unfortunately, methods to estimate whole genome methylation are still under development and capture only a minor proportion of all methylation sites. Technical improvements are needed to make genome wide methylation scans more extensive and reliable.

High throughput sequencing assays have enabled the study of metagenomics which includes whole-genome shotgun analysis from the entire stool. Resulting sequences are able to be taxonomically assembled into open reading frames so that functional assignment of these sequences into enzymes and other gene products can occur and subsequent pathway analysis may allow inference of the overall metabolic potential of the gut bacterial community. Techniques to enable integration of these sequencing results, with information from other sources such as meta-transcriptomics (set of messenger RNA reflecting transcribed genes within the gut bacteria), meta-proteomics (set of peptides recovered from stool samples) and quantification of community metabolomics is still in its infancy. Further bioinformatics efforts are required to better understand the biology of these gut microbial communities and to see if insights from microbiota can be directly engineered using pharmaceuticals, prebiotics, probiotics or diet as a preventative or treatment for diabetes and other disorders.

13.6.4 Improved Statistical Techniques to Evaluate Gene–Gene and Gene–Environment Interactions

Currently, only very simple models of interaction have been examined, due to the complexities of such statistical modeling. Clinically relevant interaction effects for diabetes are likely to involve multiple genetic and non-genetic factors. It is believed that the larger the magnitude of a gene–gene effect or a gene–environment interaction, the greater the variance is associated with the main effects for the genetic component. This indicates that the gene variants that are most relevant for gene environment interactions may be those that rank poorly in most GWAS meta-analyses.

13.6.5 Human Systems Integration Approaches

Genome wide genetic variation is only one system level of the tools available to study diabetes. Other system levels include the transcriptome (messenger RNA), proteome (protein levels), and metabolome (metabolites), which apply to both the individual and their gut microbiota. Integrating such complex and large datasets is a challenge; however, both network analysis (knowledge driven) and computational modeling of all system-level data points (data-driven) approaches are being conducted. The ideal system-wide study would have multi-omic data for an individual collected at multiple time points along with exposure details and longitudinal follow up. A similar approach is currently being used in the IMI DIRECT consortium study which aims to discover and validate biomarkers that predict the rate of glycemic deterioration before and after type 2 diabetes, predict the response to diabetes therapies and help stratify type 2 diabetes into clearly definable disease subclasses that can be treated more effectively than without stratification.[115]

References

1. American Diabetes Association, *Diabetes Care*, 2013, **36**(Suppl 1), S67.
2. R. Murphy, S. Ellard and A. T. Hattersley, *Nat. Clin. Pract. Endocrinol. Metab.*, 2008, **4**, 200.

3. V. Hyttinen, J. Kaprio, L. Kinnunen, M. Koskenvuo and J. Tuomilehto, *Diabetes*, 2003, **52**, 1052.
4. A. K. Steck, K. J. Barriga, L. M. Emery, R. V. Fiallo-Scharer, P. A. Gottlieb and M. J. Rewers, *Diabetes Care*, 2005, **28**, 296.
5. J. Kaprio, J. Tuomilehto, M. Koskenvuo, K. Romanov, A. Reunanen, J. Eriksson, J. Stengard and Y. A. Kesaniemi, *Diabetologia*, 1992, **35**, 1060.
6. M. J. Redondo, M. Rewers, L. Yu, S. Garg, C. C. Pilcher, R. B. Elliott and G. S. Eisenbarth, *BMJ*, 1999, **318**, 698.
7. M. J. Redondo, L. Yu, M. Hawa, T. Mackenzie, D. A. Pyke, G. S. Eisenbarth and R. D. Leslie, *Diabetologia*, 2001, **44**, 354.
8. DIAMOND Project Group, *Diabetic Med.*, 2006, 23, 857.
9. J. Ji, K. Hemminki, J. Sundquist and K. Sundquist, *J. Clin. Endocrinol. Metab.*, 2010, **95**, 847.
10. J. L. Davies, Y. Kawaguchi, S. T. Bennett, J. B. Copeman, H. J. Cordell, L. E. Pritchard, P. W. Reed, S. C. Gough, S. C. Jenkins, S. M. Palmer, *et al.*, *Nature*, 1994, **371**, 130.
11. H. Erlich, A. M. Valdes, J. Noble, J. A. Carlson, M. Varney, P. Concannon, J. C. Mychaleckyj, J. A. Todd, P. Bonella, A. L. Fear, *et al.*, *Diabetes*, 2008, **57**, 1084.
12. M. J. Mason, C. Speake, V. H. Gersuk, Q. A. Nguyen, K. K. O'Brien, J. M. Odegard, J. H. Buckner, C. J. Greenbaum, D. Chaussabel and G. T. Nepom, *Diabetes*, 2014, **63**, 1789.
13. G. I. Bell, S. Horita and J. H. Karam, *Diabetes*, 1984, **33**, 176.
14. L. Nistico, R. Buzzetti, L. E. Pritchard, B. Van der Auwera, C. Giovannini, E. Bosi, M. T. Larrad, M. S. Rios, C. C. Chow, C. S. Cockram, *et al.*, *Hum. Mol. Genet.*, 1996, **5**, 1075.
15. N. Bottini, L. Musumeci, A. Alonso, S. Rahmouni, K. Nika, M. Rostamkhani, J. MacMurray, G. F. Meloni, P. Lucarelli, M. Pellecchia, *et al.*, *Nat. Genet.*, 2004, **36**, 337.
16. A. Vella, J. D. Cooper, C. E. Lowe, N. Walker, S. Nutland, B. Widmer, R. Jones, S. M. Ring, W. McArdle, M. E. Pembrey, *et al.*, *Am. J. Hum. Genet.*, 2005, **76**, 773.
17. J. P. Bradfield, H. Q. Qu, K. Wang, H. Zhang, P. M. Sleiman, C. E. Kim, F. D. Mentch, H. Qiu, J. T. Glessner, K. A. Thomas, *et al.*, *PLoS Genet.*, 2011, 7, e1002293.
18. J. D. Cooper, D. J. Smyth, A. M. Smiles, V. Plagnol, N. M. Walker, J. E. Allen, K. Downes, J. C. Barrett, B. C. Healy, J. C. Mychaleckyj, *et al.*, *Nat. Genet.*, 2008, **40**, 1399.
19. J. C. Barrett, D. G. Clayton, P. Concannon, B. Akolkar, J. D. Cooper, H. A. Erlich, C. Julier, G. Morahan, J. Nerup, C. Nierras, *et al.*, *Nat. Genet.*, 2009, **41**, 703.
20. M. Bakay, R. Pandey and H. Hakonarson, *Genes* 2013, **4**, 499.
21. E. Ferrannini, A. Natali, S. Camastra, M. Nannipieri, A. Mari, K. P. Adam, M. V. Milburn, G. Kastenmuller, J. Adamski, T. Tuomi, *et al.*, *Diabetes*, 2013, **62**, 1730.

22. Y. Horikawa, N. Oda, N. J. Cox, X. Li, M. Orho-Melander, M. Hara, Y. Hinokio, T. H. Lindner, H. Mashima, P. E. Schwarz, *et al.*, *Nat. Genet.*, 2000, **26**, 163.
23. D. Altshuler, J. N. Hirschhorn, M. Klannemark, C. M. Lindgren, M. C. Vohl, J. Nemesh, C. R. Lane, S. F. Schaffner, S. Bolk, C. Brewer, *et al.*, *Nat. Genet.*, 2000, **26**, 76.
24. A. L. Gloyn, M. N. Weedon, K. R. Owen, M. J. Turner, B. A. Knight, G. Hitman, M. Walker, J. C. Levy, M. Sampson, S. Halford, *et al.*, *Diabetes*, 2003, **52**, 568.
25. S. F. Grant, G. Thorleifsson, I. Reynisdottir, R. Benediktsson, A. Manolescu, J. Sainz, A. Helgason, H. Stefansson, V. Emilsson, A. Helgadottir, *et al.*, *Nat. Genet.*, 2006, **38**, 320.
26. I. Moltke, N. Grarup, M. E. Jorgensen, P. Bjerregaard, J. T. Treebak, M. Fumagalli, T. S. Korneliussen, M. A. Andersen, T. S. Nielsen, N. T. Krarup, *et al.*, *Nature*, 2014, 190–193.
27. K. J. Basile, M. E. Johnson, Q. Xia and S. F. Grant, *Int. J. Endocrinol.*, 2014, **2014**, 769671.
28. M. Vaxillaire, A. Bonnefond and P. Froguel, *Best Pract. Res. Clin. Endocrinol. Metab.*, 2012, **26**, 171.
29. J. S. Kooner, D. Saleheen, X. Sim, J. Sehmi, W. Zhang, P. Frossard, L. F. Been, K. S. Chia, A. S. Dimas, N. Hassanali, *et al.*, *Nat. Genet.*, 2011, **43**, 984.
30. Y. S. Cho, C. H. Chen, C. Hu, J. Long, R. T. Ong, X. Sim, F. Takeuchi, Y. Wu, M. J. Go, T. Yamauchi, *et al.*, *Nat. Genet.*, 2012, **44**, 67.
31. A. L. Williams, S. B. Jacobs, H. Moreno-Macias, A. Huerta-Chagoya, C. Churchhouse, C. Marquez-Luna, H. Garcia-Ortiz, M. J. Gomez-Vazquez, N. P. Burtt, C. A. Aguilar-Salinas, *et al.*, *Nature*, 2014, **506**, 97.
32. H. Q. Qu, S. F. Grant, J. P. Bradfield, C. Kim, E. Frackelton, H. Hakonarson and C. Polychronakos, *Diabetes*, 2008, **57**, 1983.
33. C. E. Taplin and J. M. Barker, *Autoimmunity*, 2008, **41**, 11.
34. V. Lyssenko and M. Laakso, *Diabetes Care*, 2013, **36**(Suppl 2), S120.
35. G. Dahlquist and B. Kallen, *Diabetologia*, 1992, **35**, 671.
36. L. Blom, G. Dahlquist, L. Nystrom, A. Sandstrom and S. Wall, *Diabetologia*, 1989, **32**, 7.
37. P. A. McKinney, R. Parslow, K. Gurney, G. Law, H. J. Bodansky and D. R. Williams, *Diabetologia*, 1997, **40**, 933.
38. G. G. Dahlquist, C. Patterson and G. Soltesz, *Diabetes Care*, 1999, **22**, 1698.
39. M. E. Jones, A. J. Swerdlow, L. E. Gill and M. J. Goldacre, *Int. J. Epidemiol.*, 1998, **27**, 444.
40. B. J. Freij, M. A. South and J. L. Sever, *Clin. Perinatol.*, 1988, **15**, 247.
41. F. Ginsberg-Fellner, M. E. Witt, B. Fedun, F. Taub, M. J. Dobersen, R. C. McEvoy, L. Z. Cooper, A. L. Notkins and P. Rubinstein, *Rev. Infect. Dis.*, 1985, 7(Suppl 1), S170.
42. G. G. Dahlquist, S. Ivarsson, B. Lindberg and M. Forsgren, *Diabetes*, 1995, **44**, 408.

43. G. Dahlquist, G. Frisk, S. A. Ivarsson, L. Svanberg, M. Forsgren and H. Diderholm, *Diabetologia*, 1995, **38**, 1371.
44. P. J. Bingley, I. F. Douek, C. A. Rogers and E. A. Gale, *BMJ*, 2000, **321**, 420.
45. L. C. Stene, P. Magnus, R. T. Lie, O. Sovik and G. Joner, *BMJ*, 2001, **323**, 369.
46. L. C. Stene, P. Magnus, R. T. Lie, O. Sovik and G. Joner, *BMJ*, 2001, **322**, 889.
47. G. Dahlquist, S. S. Bennich and B. Kallen, *BMJ*, 1996, **313**, 1174.
48. C. N. Hales and D. J. Barker, *Diabetologia*, 1992, **35**, 595.
49. D. Dabelea, R. L. Hanson, R. S. Lindsay, D. J. Pettit, G. Imperatore, M. M. Gabir, J. Roumain, P. H. Bennett and W. C. Knowler, *Diabetes*, 2000, **49**, 2208.
50. A. T. Hattersley and J. E. Tooke, *Lancet*, 1999, **353**, 1789.
51. M. Horikoshi, H. Yaghootkar, D. O. Mook-Kanamori, U. Sovio, H. R. Taal, B. J. Hennig, J. P. Bradfield, B. St Pourcain, D. M. Evans, P. Charoen, *et al.*, *Nat. Genet.*, 2013, **45**, 76.
52. B. T. Heijmans, E. W. Tobi, A. D. Stein, H. Putter, G. J. Blauw, E. S. Susser, P. E. Slagboom and L. H. Lumey, *Proc. Natl. Acad. Sci. U. S. A.*, 2008, **105**, 17046.
53. E. W. Tobi, L. H. Lumey, R. P. Talens, D. Kremer, H. Putter, A. D. Stein, P. E. Slagboom and B. T. Heijmans, *Hum. Mol. Genet.*, 2009, **18**, 4046.
54. K. M. Godfrey, A. Sheppard, P. D. Gluckman, K. A. Lillycrop, G. C. Burdge, C. McLean, J. Rodford, J. L. Slater-Jefferies, E. Garratt, S. R. Crozier, *et al.*, *Diabetes*, 2011, **60**, 1528.
55. C. C. Patterson, G. G. Dahlquist, E. Gyurus, A. Green and G. Soltesz, *Lancet*, 2009, **373**, 2027.
56. C. C. Patterson, E. Gyurus, J. Rosenbauer, O. Cinek, A. Neu, E. Schober, R. C. Parslow, G. Joner, J. Svensson, C. Castell, *et al.*, *Diabetologia*, 2012, **55**, 2142.
57. A. Kondrashova and H. Hyoty, *Int. Rev. Immunol.*, 2014, 284–295.
58. E. Hypponen, S. M. Virtanen, M. G. Kenward, M. Knip and H. K. Akerblom, *Diabetes Care*, 2000, **23**, 1755.
59. M. Knip, R. Veijola, S. M. Virtanen, H. Hyoty, O. Vaarala and H. K. Akerblom, *Diabetes*, 2005, **54**(Suppl 2), S125.
60. M. M. Lamb, X. Yin, G. O. Zerbe, G. J. Klingensmith, D. Dabelea, T. E. Fingerlin, M. Rewers and J. M. Norris, *Diabetologia*, 2009, **52**, 2064.
61. T. J. Wilkin, *Diabetologia*, 2001, **44**, 914.
62. M. M. Lamb, X. Yin, K. Barriga, M. R. Hoffman, A. E. Baron, G. S. Eisenbarth, M. Rewers and J. M. Norris, *J. Clin. Endocrinol. Metab.*, 2008, **93**, 3936.
63. S. Hummel, M. Pfluger, M. Hummel, E. Bonifacio and A. G. Ziegler, *Diabetes Care*, 2011, **34**, 1301.
64. H. K. Akerblom, J. Krischer, S. M. Virtanen, C. Berseth, D. Becker, J. Dupre, J. Ilonen, M. Trucco, E. Savilahti, K. Koski, *et al.*, *Diabetologia*, 2011, **54**, 627.

65. J. Qin, R. Li, J. Raes, M. Arumugam, K. S. Burgdorf, C. Manichanh, T. Nielsen, N. Pons, F. Levenez, T. Yamada, *et al.*, *Nature*, 2010, **464**, 59.
66. J. A. Mendelsohn, *Hist. Philos. Life Sci.*, 2002, **24**, 3.
67. C. King and N. Sarvetnick, *PLoS One*, 2011, **6**, e17049.
68. L. Wen, R. E. Ley, P. Y. Volchkov, P. B. Stranges, L. Avanesyan, A. C. Stonebraker, C. Hu, F. S. Wong, G. L. Szot, J. A. Bluestone, *et al.*, *Nature*, 2008, **455**, 1109.
69. A. Giongo, K. A. Gano, D. B. Crabb, N. Mukherjee, L. L. Novelo, G. Casella, J. C. Drew, J. Ilonen, M. Knip, H. Hyoty, *et al.*, *ISME J.*, 2011, **5**, 82.
70. C. T. Brown, A. G. Davis-Richardson, A. Giongo, K. A. Gano, D. B. Crabb, N. Mukherjee, G. Casella, J. C. Drew, J. Ilonen, M. Knip, *et al.*, *PLoS One*, 2011, **6**, e25792.
71. H. M. Hamer, D. Jonkers, K. Venema, S. Vanhoutvin, F. J. Troost and R. J. Brummer, *Aliment. Pharmacol. Ther.*, 2008, **27**, 104.
72. P. Louis and H. J. Flint, *FEMS Microbiol. Lett.*, 2009, **294**, 1.
73. N. Burger-van Paassen, A. Vincent, P. J. Puiman, M. van der Sluis, J. Bouma, G. Boehm, J. B. van Goudoever, I. van Seuningen and I. B. Renes, *Biochem. J.*, 2009, **420**, 211.
74. K. Lewis, F. Lutgendorff, V. Phan, J. D. Soderholm, P. M. Sherman and D. M. McKay, *Inflamm. Bowel Dis.*, 2010, **16**, 1138.
75. L. Peng, Z. R. Li, R. S. Green, I. R. Holzman and J. Lin, *J. Nutr.*, 2009, **139**, 1619.
76. F. H. Karlsson, V. Tremaroli, I. Nookaew, G. Bergstrom, C. J. Behre, B. Fagerberg, J. Nielsen and F. Backhed, *Nature*, 2013, **498**, 99.
77. J. Qin, Y. Li, Z. Cai, S. Li, J. Zhu, F. Zhang, S. Liang, W. Zhang, Y. Guan, D. Shen, *et al.*, *Nature*, 2012, **490**, 55.
78. A. Vrieze, F. Holleman, E. G. Zoetendal, W. M. de Vos, J. B. Hoekstra and M. Nieuwdorp, *Diabetologia*, 2010, **53**, 606.
79. A. Vrieze, E. Van Nood, F. Holleman, J. Salojarvi, R. S. Kootte, J. F. Bartelsman, G. M. Dallinga-Thie, M. T. Ackermans, M. J. Serlie, R. Oozeer, *et al.*, *Gastroenterology*, 2012, **143**, 913.
80. R. S. Kootte, A. Vrieze, F. Holleman, G. M. Dallinga-Thie, E. G. Zoetendal, W. M. de Vos, A. K. Groen, J. B. Hoekstra, E. S. Stroes and M. Nieuwdorp, *Diabetes, Obes. Metab.*, 2012, **14**, 112.
81. J. Lempainen, O. Vaarala, M. Makela, R. Veijola, O. Simell, M. Knip, R. Hermann and J. Ilonen, *J. Autoimmun.*, 2009, **33**, 155.
82. L. Qi, M. C. Cornelis, C. Zhang, R. M. van Dam and F. B. Hu, *Am. J. Clin. Nutr.*, 2009, **89**, 1453.
83. M. C. Cornelis, L. Qi, P. Kraft and F. B. Hu, *Am. J. Clin. Nutr.*, 2009, **89**, 1256.
84. S. Kanoni, J. A. Nettleton, M. F. Hivert, Z. Ye, F. J. van Rooij, D. Shungin, E. Sonestedt, J. S. Ngwa, M. K. Wojczynski, R. N. Lemaitre, *et al.*, *Diabetes*, 2011, **60**, 2407.
85. Z. Shan, W. Bao, Y. Zhang, Y. Rong, X. Wang, Y. Jin, Y. Song, P. Yao, C. Sun, F. B. Hu, *et al.*, *Diabetes*, 2014, **63**, 1796.

86. S. A. Greeley, R. N. Naylor, L. H. Philipson and G. I. Bell, *Curr. Diabetes Rep.*, 2011, **11**, 519.
87. K. A. Jablonski, J. B. McAteer, P. I. de Bakker, P. W. Franks, T. I. Pollin, R. L. Hanson, R. Saxena, S. Fowler, A. R. Shuldiner, W. C. Knowler, *et al.*, *Diabetes*, 2010, **59**, 2672.
88. J. H. Choi, S. W. Yee, A. H. Ramirez, K. M. Morrissey, G. H. Jang, P. J. Joski, J. A. Mefford, S. E. Hesselson, A. Schlessinger, G. Jenkins, *et al.*, *Clin. Pharmacol. Ther.*, 2011, **90**, 674.
89. I. Tkac, L. Klimcakova, M. Javorsky, M. Fabianova, Z. Schroner, H. Hermanova, E. Babjakova and R. Tkacova, *Diabetes, Obes. Metab.*, 2013, **15**, 189.
90. K. Zhou, C. Bellenguez, C. C. Spencer, A. J. Bennett, R. L. Coleman, R. Tavendale, S. A. Hawley, L. A. Donnelly, C. Schofield, C. J. Groves, *et al.*, *Nat. Genet.*, 2011, **43**, 117.
91. L. Gong, S. Goswami, K. M. Giacomini, R. B. Altman and T. E. Klein, *Pharmacogenet. Genomics*, 2012, **22**, 820.
92. M. Rakhshandehroo, B. Knoch, M. Muller and S. Kersten, *PPAR Res.*, 2010, **2010**, 1–20.
93. J. C. Florez, K. A. Jablonski, S. E. Kahn, P. W. Franks, D. Dabelea, R. F. Hamman, W. C. Knowler, D. M. Nathan and D. Altshuler, *Diabetes*, 2007, **56**, 531.
94. A. F. Moore, K. A. Jablonski, J. B. McAteer, R. Saxena, T. I. Pollin, P. W. Franks, R. L. Hanson, A. R. Shuldiner, W. C. Knowler, D. Altshuler, *et al.*, *Diabetes*, 2008, **57**, 2503.
95. A. F. Moore, K. A. Jablonski, C. C. Mason, J. B. McAteer, R. F. Arakaki, B. J. Goldstein, S. E. Kahn, A. E. Kitabchi, R. L. Hanson, W. C. Knowler, *et al.*, *J. Clin. Endocrinol. Metab.*, 2009, **94**, 449.
96. K. Suzuki, T. Yanagawa, T. Shibasaki, N. Kaniwa, R. Hasegawa and M. Tohkin, *Diabetes Res. Clin. Pract.*, 2006, **72**, 148.
97. M. L. Becker, L. E. Visser, P. H. Trienekens, A. Hofman, R. H. van Schaik and B. H. Stricker, *Clin. Pharmacol. Ther.*, 2008, **83**, 288.
98. E. R. Pearson, L. A. Donnelly, C. Kimber, A. Whitley, A. S. Doney, M. I. McCarthy, A. T. Hattersley, A. D. Morris and C. N. Palmer, *Diabetes*, 2007, **56**, 2178.
99. A. L. Gloyn, Y. Hashim, S. J. Ashcroft, R. Ashfield, S. Wiltshire and R. C. Turner, *Diabetic Med.*, 2001, **18**, 206.
100. Y. Y. He, R. Zhang, X. Y. Shao, C. Hu, C. R. Wang, J. X. Lu, Y. Q. Bao, W. P. Jia and K. S. Xiang, *Acta Pharmacol. Sin.*, 2008, **29**, 983.
101. Q. Huang, J. Y. Yin, X. P. Dai, J. Wu, X. Chen, C. S. Deng, M. Yu, Z. C. Gong, H. H. Zhou and Z. Q. Liu, *Eur. J. Clin. Pharmacol.*, 2010, **66**, 1207.
102. Q. Huang, J. Y. Yin, X. P. Dai, Q. Pei, M. Dong, Z. G. Zhou, X. Huang, M. Yu, H. H. Zhou and Z. Q. Liu, *Acta Pharmacol. Sin.*, 2010, **31**, 709.
103. Z. C. Gong, Q. Huang, X. P. Dai, G. H. Lei, H. B. Lu, J. Y. Yin, X. J. Xu, J. Qu, Q. Pei, M. Dong, *et al.*, *Br. J. Clin. Pharmacol.*, 2012, **74**, 501.
104. S. Wang, Y. M. Se, Z. Q. Liu, M. X. Lei, B. Hao, Z. X. Sun, S. D. Nie, X. M. Zeng and J. Wu, *Pharmazie*, 2012, **67**, 74.

105. Q. Pei, Q. Huang, G. P. Yang, Y. C. Zhao, J. Y. Yin, M. Song, Y. Zheng, Z. H. Mo, H. H. Zhou and Z. Q. Liu, *Acta Pharmacol. Sin.*, 2013, **34**, 255.
106. M. Bluher, G. Lubben and R. Paschke, *Diabetes Care*, 2003, **26**, 825.
107. L. Andrulionyte, T. Kuulasmaa, J. L. Chiasson and M. Laakso, *Diabetes*, 2007, **56**, 1181.
108. L. Andrulionyte, O. Laukkanen, J. L. Chiasson and M. Laakso, *J. Mol. Med.*, 2006, **84**, 701.
109. L. Andrulionyte, J. Zacharova, J. L. Chiasson and M. Laakso, *Diabetologia*, 2004, **47**, 2176.
110. L. M. t Hart, A. Fritsche, G. Nijpels, N. van Leeuwen, L. A. Donnelly, J. M. Dekker, M. Alssema, J. Fadista, F. Carlotti, A. P. Gjesing, *et al.*, *Diabetes*, 2013, **62**, 3275.
111. A. P. Morris, B. F. Voight, T. M. Teslovich, T. Ferreira, A. V. Segre, V. Steinthorsdottir, R. J. Strawbridge, H. Khan, H. Grallert, A. Mahajan, *et al.*, *Nat. Genet.*, 2012, **44**, 981.
112. M. K. Andersen, V. Lundgren, J. A. Turunen, C. Forsblom, B. Isomaa, P. H. Groop, L. Groop and T. Tuomi, *Diabetes Care*, 2010, **33**, 2062.
113. S. Johansson, H. Irgens, K. K. Chudasama, J. Molnes, J. Aerts, F. S. Roque, I. Jonassen, S. Levy, K. Lima, P. M. Knappskog, *et al.*, *PLoS One*, 2012, 7, e38050.
114. A. Mahajan, M. J. Go, W. Zhang, J. E. Below, K. J. Gaulton, T. Ferreira, M. Horikoshi, A. D. Johnson, M. C. Ng, I. Prokopenko, *et al.*, *Nat. Genet.*, 2014, **46**, 234.
115. R. W. Koivula, A. Heggie, A. Barnett, H. Cederberg, T. H. Hansen, A. D. Koopman, M. Ridderstrale, F. Rutters, H. Vestergaard, R. Gupta, *et al.*, *Diabetologia*, 2014, **57**, 1132.

CHAPTER 14

Drug Design and Therapeutic Development for Diabetes Mellitus

SHAHRAD TAHERI,*[a,b,c,d,e] HADEEL ZAGHLOOL[c] AND MARIA PALLAYOVA[f]

[a] Department of Medicine, Weill Cornell Medical College, New York, USA; [b] Department of Medicine, Weill Cornell Medical College, Doha, Qatar; [c] Clinical Research Core, Weill Cornell Medical College, Doha, Qatar; [d] Diabetes Centre and Specialist Weight Management Service, Hamad Medical Corporation, Doha, Qatar; [e] Department of Medicine, King's College London, London, UK; [f] Department of Human Physiology, Faculty of Medicine, Pavol Jozef Safarik University, Kosice, Slovak Republic
*Email: staheri@me.com

14.1 Introduction

Diabetes mellitus has emerged as one of the greatest health challenges worldwide. The World Health Organization (WHO) has estimated that about 347 million individuals are afflicted with diabetes worldwide, with the majority (90%) of cases having type 2 diabetes.[1] WHO has further predicted that by 2030, diabetes will emerge as the 7th leading cause of death in the world.[1] The International Diabetes Federation (IDF) recently estimated that globally over 382 million individuals had diabetes in 2013 (about 50% undiagnosed) and that this number will rise to 592 million by 2035.[2] The healthcare costs

RSC Drug Discovery Series No. 45
Cardiovascular and Metabolic Disease: Scientific Discoveries and New Therapies
Edited by Philip Peplow, James Adams and Tim Young

Published by the Royal Society of Chemistry, www.rsc.org

of diabetes worldwide were estimated to be about 548 billion USD in 2013, and are predicted to rise to 627 billion USD in 2035.[2]

The main forms of diabetes are type 1 and type 2 diabetes. The prevalence of both types of diabetes is increasing, but more so for type 2 diabetes, which is closely linked to greater adiposity (overweight and obesity) and lifestyle factors such as poor diet and lack of physical activity.[3] Type 1 diabetes is characterized by long-term absolute insulin deficiency with total dependence on insulin for survival. Type 1 diabetes has an autoimmune pathophysiology and usually presents in childhood and adolescence with rapid development of symptoms (weight loss, polydipsia, polyuria) and tendency for diabetic keto-acidosis, a major cause of mortality in type 1 diabetes. Type 2 diabetes has a complex pathophysiology based on disease stage; early type 2 diabetes is characterized by insulin resistance and insulin hypersecretion, while with loss of pancreatic beta cell reserve as the disease progresses, there will be a relative to total insulin insufficiency. Type 2 diabetes has a strong, although complex, genetic basis with strong familial aggregation. It may have a more varied clinical presentation and is increasingly diagnosed through simple blood tests (glucose levels or glycated haemoglobin, HbA1c) in asymptomatic patients. With the rise in obesity, there are increasing cases of type 1 diabetes, which show insulin resistance requiring higher doses of insulin ("double diabetes").[4]

The burden of diabetes on the affected individual includes the complications of the disease itself and the complications of the treatment of the disease. Diabetes is characterized by microvascular and macrovascular complications. Microvascular complications include retinopathy, neuropathy and nephropathy. Diabetic nephropathy is the major cause of end-stage renal disease requiring renal dialysis and transplantation. Diabetes is the major cause of blindness, significantly affecting quality of life. Diabetic neuropathy is associated with increased risk of amputation, and specifically, cardiac autonomic neuropathy increases the risk of cardiovascular death. Diabetic macrovascular complications include ischaemic heart disease (angina and myocardial infarction), stroke, and peripheral arterial disease. The macrovascular complications are responsible for increased mortality in type 2 diabetes. Other diabetes complications include depression, cognitive decline, and several major cancers, *e.g.* breast cancer. While the majority of diabetes complications are related to dysglycaemia, other pathophysiological mechanisms and pathways are involved. Thus, diabetes is not just a "sugar" imbalance disease and requires a holistic multifactorial treatment approach including treatment of dyslipidaemia and hypertension, and encouragement of lifestyle change.

The main treatment approaches employed for hyperglycaemia include insulin replacement, enhancement of insulin secretion and release, improvement of insulin sensitivity, and reduced absorption of glucose from the intestinal tract. Recently, manipulation of renal glucose excretion has been introduced to diabetes treatment. The aim of this chapter is to review the advances in the treatment of type 1 and type 2 diabetes, with a particular emphasis on treatments for glycaemic control.

14.2 Importance of Glycaemic Control and Multifactorial Treatment in Diabetes

The benefits of intensive glycaemic control for diabetes microvascular complications has been clearly demonstrated by several large studies. For type 1 diabetes, the Diabetes Control and Complications Trial (DCCT), included 1441 patients who had no retinopathy (to gauge incidence) and mild retinopathy (to gauge progression). Patients were randomized to intensive insulin treatment (3 or more insulin injections or insulin pump) or conventional treatment (once or twice daily insulin) with a mean follow-up of 6.5 years.[5] DCCT observed that there was a 76% (95% CI: 62–85%) reduction in retinopathy adjusted mean risk in the intensive treatment group.[5] There was a 54% (95% CI: 39–66%) reduction in retinopathy progression in the intensive treatment group.[5] Also for the intensive treatment group, DCCT observed a 60% (95% CI: 38–74%) reduction in neuropathy risk, and 54% (95% CI: 19–74%) reduction in severe albuminuria (≥300 mg per 24 hours).[5] The benefits of intensive glycaemic control were accompanied by a 2–3 fold increase in hypoglycaemia, a major complication of diabetes treatment.[5] In the DCCT follow-up study, EDIC (Epidemiology of Diabetes Interventions and Complications), with a mean follow-up of 17 years, there was a reduction in any cardiovascular event by 42% (95% CI: 9–63%) in the intensive treatment group.[6] EDIC has also reported that despite equivalent HbA1c levels after the DCCT trial ended, the intensive treatment group continued to have significant benefits.[6] The underlying mechanisms for this continued benefit (called "metabolic memory") are unknown, but may involve epigenetic changes.

For type 2 diabetes, the UKPDS (United Kingdom Prospective Diabetes Study) examined the impact of intensive glycaemic control on diabetes complications in 4209 newly diagnosed diabetic patients over a median follow-up of 10 years. Patients underwent intensive glycaemic control with sulphonylurea and insulin treatment or conventional therapy.[7] The intensive group achieved a median HbA1c of 7% *vs.* 7.9% in the conventional group.[8] The lower HbA1c was achieved at the cost of greater treatment burden, greater hypoglycaemic episodes, and greater weight gain.[8] In the UKPDS study, intensive glycaemic control was associated with a 25% risk reduction in microvascular end-points.[8] A recent systematic review and meta-analysis, including over 28 000 patients from 7 trials, reported that intensive glycaemic control was associated with reduction in microalbuminuria and macroalbuminuria (risk ratios 0.86 95% CI: 0.76–0.96 and 0.74 95% CI: 0.65–0.85, respectively).[9]

The Danish Steno-2 trial[10–12] adopted a multifactorial intervention to tackle cardiovascular mortality and microalbuminuria associated with type 2 diabetes, addressing behaviour modification and pharmacotherapy for hyperglycaemia, hypertension, dyslipidaemia, and microalbuminuria. The control arm received conventional treatment. After 7.8 years follow-up, those in the intensive treatment group had a significantly lower risk of nephropathy (hazard ratio, HR 0.39 95% CI: 0.17–0.87), retinopathy (HR 0.42 95% CI: 0.21–0.86), autonomic neuropathy (HR 0.37 95% CI: 0.18–0.79), and cardiovascular disease (HR 0.47 95% CI: 0.24–0.73). The ADDITION (Anglo-Danish-Dutch study of Intensive Treatment In PeOple with screeN) trial of 2861 screening-detected patients with type 2 diabetes in primary care implemented a multifactorial intervention with set targets and compared this with the routine care group.[13] After 5 years, 22.7% and 24.4% of the intensive treatment and routine care groups had some degree of albuminuria. Retinopathy developed in 10.2% of the intensive treatment group and 12.1% of the routine care group. Neuropathy was observed in 4.9% of the intensive group and 5.9% of the routine care group. There was no significant difference between groups for any of the microvascular complications.[13] This may be due to the very small differences in glycaemia, blood pressure, and lipids between the groups and the relatively short follow-up duration.

Cardiovascular disease (CVD) is common in diabetes, and diabetes is a major risk factor for CVD, increasing the risk twofold.[14,15] CVD accounts for about 75% of deaths in patients with type 2 diabetes.[14] Type 2 diabetes belongs to the cluster of the metabolic syndrome, which includes central adiposity, dysglycaemia, insulin resistance, hypertension, dyslipidaemia, and pro-thrombotic and pro-inflammatory states. An additional factor that has been proposed as part of the metabolic syndrome is Obstructive Sleep Apnea (OSA). In the look AHEAD (Action for Health in Diabetes) study, intensive lifestyle intervention and reduction in adiposity has recently been shown to improve CVD risk factors in diabetes, but was not associated with a reduction in mortality.[16] Blood pressure treatment is key to reducing CVD risk. The Anglo-Scandinavian Cardiac Outcomes Trial (ASCOT) study showed that controlling blood pressure in patients with diabetes reduced cardiovascular events by 16% and stroke by 23%.[17] Reduction in blood pressure below 130/80 mmHg is usually recommended but reaching this blood pressure is more likely through combination therapy than maximizing dosage of single agents. However, requirement for multiple medications is associated with reduced adherence and a negative impact on quality of life.[18] The ACCORD (Action to Control Cardiovascular Risk in Diabetes) trial examined the role of lowering blood pressure further. The intensively treated patients required more than 3 medications to achieve an average blood pressure of 119/64 mmHg. This had no significant effect on the primary outcome, which was a composite of non-fatal myocardial infarction (MI), nonfatal stroke, and CVD mortality.[19] The secondary end-point of annual rates of stroke was, however, reduced with a hazard ratio of 0.59 (95% CI: 0.39–0.89).[19] The benefits of lowering of blood pressure were suggested to be greater in those who were in the usual care rather than the intensive glycaemic control group, suggesting an interaction between glycaemia and hypertension. Cholesterol reduction is an essential measure in the prevention of CVD in patients with type 2 diabetes. A large meta-analysis has shown that for every mmol L^{-1} reduction in LDL-cholesterol, there is a 12% (95% CI: 9–16%) reduction in all-cause mortality, 19% (95% CI: 15–24%)

reduction in coronary mortality, 23% (95% CI: 20–26%) reduction in myocardial infarction or coronary death, and 17% (95% CI: 12–22%) reduction in stroke.[20] In the CARDS (Collaborative Atorvastatin Diabetes Study) study, a randomized controlled trial of 2838 patients with diabetes and no history of CVD, examining the effect of atorvastatin, LDL-cholesterol reduced from 3 mmol L^{-1} to 2.1 mmol L^{-1} resulted in a 37% relative risk reduction in cardiovascular events.[21] A positive linear relationship appears to exist between cardiovascular event rates and LDL-cholesterol levels. The evidence that statin combination therapy with ezetimibe, fibrates or niacin in CVD risk reduction over statin therapy is either not available for hard end-points or inconclusive.[22,23] The ACCORD study showed no benefit for CVD in lowering triglycerides in the intensive therapy (targeting an HbA1c level below 6.0%) group,[24] while the FIELD (Fenofibrate Intervention and Event Lowering in Diabetes) study showed benefit for secondary outcomes of total CVD events and coronary revascularization.[25] Both ACCORD and FIELD showed a benefit for diabetic retinopathy and nephropathy.[24–26] Fenofibrate may thus be useful in that it is effective in prevention of retinopathy progression in type 2 diabetes patients with pre-proliferative diabetic retinopathy and/or diabetic maculopathy, and those requiring laser therapy with macular oedema.

UKPDS did show a 16% reduction in myocardial infarction in the intervention group as opposed to the conventional treatment group, but this was not statistically significant ($p = 0.052$).[27] In UKPDS, a subsample of overweight patients ($n = 753$) were randomized to diet or metformin treatment. The metformin-treated group benefitted from the following reductions: 32% in diabetes-related end-points, 42% diabetes-related death ($p = 0.017$), 36% all cause mortality ($p = 0.011$), 39% myocardial infarction ($p = 0.010$), and 30% combined macrovascular events ($p = 0.020$).[28]

The findings from the main UKPDS study group suggested that glycaemic control might not be as important as other factors in macrovascular disease associated with type 2 diabetes. However, further follow-up of participants revealed that although glycaemic control between the 2 groups converged, over the next 10 years, there was still a 15% reduction in myocardial infarction and 13% reduction in all-cause mortality in the intensively-treated group. For the metformin group, there was a 33% reduction in myocardial infarction and 27% reduction in all-cause mortality.[27] The findings of the UKPDS follow-up study suggest that early glycaemic control in newly diagnosed patients has a beneficial long-term effect on CVD and mortality, termed the "legacy" or metabolic memory effect.[7] The mechanisms for this effect remain to be determined but may involve epigenetic changes.

The role of intensive glycaemic control in type 2 diabetes for macrovascular complications remains controversial given findings from recent studies. The PROactive study (Prospective Pioglitazone Clinical Trial in Macrovascular Events) included 5238 patients and observed that treatment with pioglitazone and concomitant glucose lowering was associated with a significant reduction in myocardial infarction and stroke.[29] Several recent large clinical trials have examined the impact of intensive diabetes control on macrovascular outcomes. These include ACCORD,[30] VADT (Veteran Affairs Diabetes Trial)[31] and ADVANCE (Action in Diabetes and Vascular Disease: Preterax and Diamicron Modified Release Controlled Evaluation)[32] trials, with over 25 000 diabetes patients enrolled. None of these studies showed a cardiovascular benefit for intensive glycaemic control. In fact, the intensive groups tended to have greater risk of hypoglycaemia and also weight gain. However, unlike UKPDS, these studies were conducted in patients with more established diabetes. A recent report using UK primary care data showed that in type 2 diabetes patients over 50 years old, both low and high HbA1c levels were associated with higher mortality.[33] Thus, the role of glycaemic control in macrovascular disease is unclear; however, based on UKPDS, intensive control in early diabetes is likely to be beneficial. The findings from major trials have resulted in guidelines recommending a more tailored approach to diabetes treatment taking into account multiple factors[34] (Table 14.1).

14.3 Type 1 Diabetes

Type 1 diabetes accounts for 5%–15% of diabetes cases in high-income countries.[2] Of the estimated 497 100 children living with type 1 diabetes, 26% live in the Europe Region and 22% in the North America and Caribbean Region.[2] Some 79 100 children under 15 years are estimated to develop type 1 diabetes annually worldwide.[2] The number of new cases of diagnosed type 1 diabetes every year is increasing rapidly worldwide, although not all data from around the world are reliable. This rise in incidence may be due to a combination of factors, including genetic predisposition, prenatal influences, diet early in life, viral infections, transforming environments, and others.[35]

The main goal of treatment for type 1 diabetes is to prevent acute and chronic diabetic complications by maintaining safe asymptomatic glucose control with normal or nearly normal blood glucose levels. Timely and long-term intensive glycaemic control may effectively delay the onset and slow the progression of chronic diabetic complications as shown by DCCT and EDIC trials.[5,36] However, at the same time, intensive glycaemic control places people with diabetes at increased risk of hypoglycaemia that can be disabling or even life threatening.[5,36] Therefore, an urgent need exists to discover new improved therapies for treating type 1 diabetes.

Purified animal-sourced insulin was the initial insulin that was available[37–39] until the first synthetic insulin was produced *via* recombinant DNA technology using *Escherichia coli* bacterial cell as a vector.[40,41] Later, genetically engineered human insulin was produced using yeast, usually *Saccharomyces cerevisiae.*[42] Biosynthetic human insulin is chemically identical to natural insulin, which reduces potential allergic reactions, as observed with animal-derived insulin, and other complications resulting from anti-insulin antibody production.[40,41]

Table 14.1 The modern approach to personalization of diabetes treatments. Several factors need to be considered in tailoring patient treatment.

Factor	Notes – some examples
Age	• Younger patients will experience longer exposure to diabetes and greater complications – more intensive approach is required • Older patients are at high risk of hypoglycaemia • Older patients are more likely to have accompanying co-morbidities *e.g.* renal dysfunction, visual impairment, stroke *etc.*
Gender	• Potential pregnancy needs to be considered
Ethnicity	Diet and cultural practices need to be considered
Body weight	Medications may exacerbate or improve overweight/ obesity
Diabetes duration	• Longer duration is associated with greater co-morbidities • Early intensive intervention has beneficial effects on diabetes complications • Early disease has a potential for reversibility with lifestyle change, weight loss, and bariatric surgery
Diabetes complications	• Intensive glycaemic control may exacerbate diabetes complications • Early intensive intervention has beneficial effects on diabetes complications • Retinopathy and visual impairment make use of injectable medications difficult
Co-morbidities	Renal disease, liver disease, cardiac dysfunction, previous stroke, cancer *etc.* will affect drug selection and treatment intensity
Patient preference	Patient preference will determine uptake of and adherence to treatment
Psychological	• Medication may exacerbate psychological problems associated with diabetes (*e.g.* depression) • Psychological factors may affect adherence to medication
Social factors	• Occupation • Stigma • Social support
Service provision	More complex treatment regimens require greater support, education, and follow-up

Insulin analogues based on human insulin have been developed through altering insulin's amino acid sequence to modify (and optimise) pharmacokinetic and pharmacodynamic properties.[43] By improving rapidity or duration of action, insulin analogues aim to aid patients to achieve better diabetes control with less hypo- and hyper-glycaemic excursions. Indeed, major progress in reducing the risks of hypoglycaemia has occurred with the development of insulin analogues.[44] There are several types of insulin analogues, including rapid-, intermediate-, long-, and ultralong-acting analogues and premixed insulin analogues (Figure 14.1). The rapid-acting insulin analogues (lispro, aspart, glulisine) are absorbed rapidly, and have

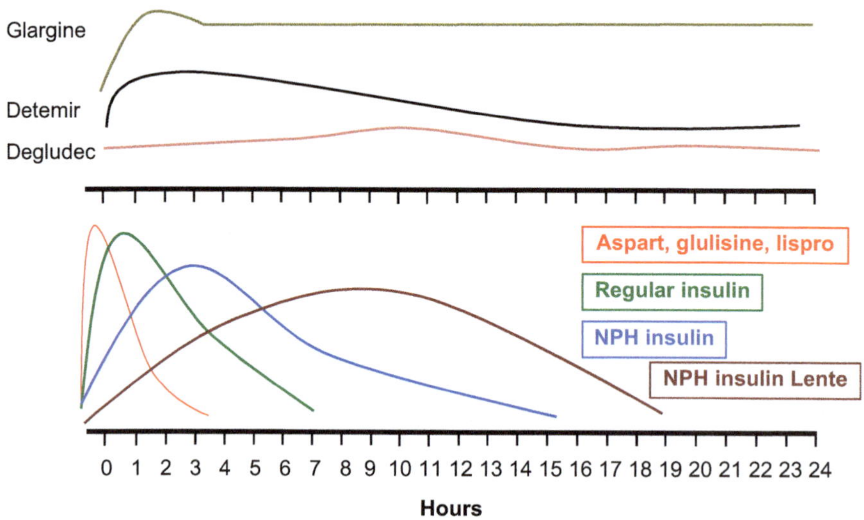

Figure 14.1 The profiles of action of current human insulins. Longer acting insulins (glargine, detemir and degludec) are not associated with a significant insulin rise. NPH (neutral protamine hagedorn) insulins are associated with a peak that can result in hypoglycaemia. Modern short acting insulins (aspart, glulisine, lispro) act rapidly and have a more predictable shorter profile compared to regular human insulin.

faster onset and shorter duration of action than regular human insulin. This provides increased flexibility for patients, promotes compliance, and reduces side effects. These insulins provide the bolus insulin for mealtimes (prandial insulin), for high blood glucose correction (correction boluses), or for use in the insulin pump. The intermediate-(detemir), long-(glargine), and ultralong-acting (degludec) insulin analogues aim to replace the basal level of physiological insulin, and may have a duration of action of between 8 and 72 hours (Figure 14.1). By providing relatively peak-less and more physiologic insulin replacement than neutral protaminated Hagedorn insulin, they avoid the risk of hypoglycaemia.[45] The two most recent basal insulin analogues (insulin degludec and a pegylated insulin lispro) have been tested for the treatment of both type 1 and type 2 diabetes. Clinical trials assessing the efficacy and safety have reported good tolerability with reduced hypoglycaemia risk. Glucose control was comparable to the currently available basal insulin analogues, and rates of nocturnal hypoglycaemia have been reported to be somewhat lower.[45] The advances in basal-bolus insulin replacement with use of insulin analogues may offer flexibility to patients, and important advantages over the existing options for insulin treatment.

There are currently many different types of insulin delivery devices available, including disposable plastic insulin syringes, disposable insulin pens pre-filled with insulin, reusable insulin pens using insulin cartridges, external insulin pumps, implantable insulin pumps, insulin injection aids (spring-loaded syringe holders, insulin injection ports, infusion sets, *etc.*), and insulin jet injectors[46] (that send a fine spray of insulin through the skin using a high-pressure air current as opposed to a needle). Also, digital insulin pens with a multiple-dose memory feature,[47] as well as built-in insulin dose calculators in some of the newer insulin pumps and glucose meters make existing diabetes devices smarter. Insulin can also be inhaled, although the first inhalable insulin, Exubera (Pfizer, New York, NY), a powdered form of recombinant human insulin, approved by the U.S. Food and Drug Administration (FDA)[48] was available only from September 2006 to October 2007. At the time of Exubera's withdrawal for commercial reasons (October 2007), other companies were pursuing inhaled insulin. Only recently, the U.S. FDA approved Afrezza Inhalation Powder (manufactured by MannKind Corporation, Danbury, Connecticut), a rapid-acting inhaled mealtime insulin to improve glycaemic control in adults with diabetes mellitus[49] with a label restriction for patients having asthma, active lung cancer or COPD.[50] Other alternative modes of insulin delivery are currently at different stages of availability and development throughout the world (*e.g.* ocular, nasal, buccal, gastrointestinal, rectal, other transmucosal, transdermal and intradermal insulin delivery).[51,52]

14.3.1 Artificial Pancreas

The aim of the artificial pancreas[53] is to mimic normal physiology by responding automatically to glucose levels to avoid both hyper- and hypoglycaemia that commonly accompany the usual treatment approaches. Currently, the continuous subcutaneous insulin infusion (CSII; insulin pumps and pods)[54] best imitates the physiological basal-bolus release of insulin. Coupling CSII with continuous glucose monitoring (CGM)[55] and a set of algorithms led to the development of the first commercial closed-loop bedside device, the Biostator (Miles Laboratories, Elkhart, IN, USA). The Biostator was a computerized, feedback glucose-controlled infusion system for dynamic control of blood glucose concentrations in diabetes.[56] The Biostator adopted a coupled intravascular sensing with the computer-controlled infusion pump programmed using a set of algorithms to deliver insulin and (or) dextrose to the patient depending on the dynamic and (or) static blood-glucose concentrations.[56] Early Biostator systems were introduced in 1974.[57,58]

Advances in diabetes technology and control algorithms, ongoing computer revolution and electronic miniaturization, as well as broadening knowledge about type 1 diabetes pathophysiology over the past decades have helped develop more advanced artificial pancreas devices. Besides the intravascular route, other means of insulin delivery and glucose sensing have been studied, with the focus on the subcutaneous route for measurement of interstitial glucose and subcutaneous drug delivery.[59]

Two major approaches to achieving closed-loop control of blood glucose in type 1 diabetes are currently available. Both rely on complex control algorithms to provide for safety of the user. Uni-hormonal (insulin alone) to reduce blood glucose, and bi-hormonal approaches (both insulin and

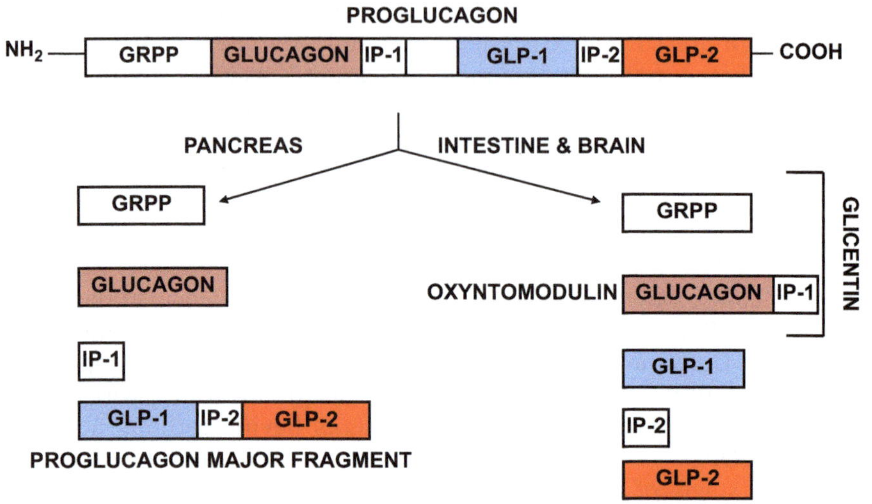

Figure 14.2 The proglucagon gene derived peptides. Differential processing occurs in the pancreas and the brain and gut. Glucagon-like peptide-1 (GLP-1) is the main incretin hormone in humans. Glucagon is a major defence against hypoglycaemia. Oxyntomodulin has been shown to have satiety effects. GRPP = glicentin related pancreatic peptide, IP-1 = intervening peptide 1, IP-2 = intervening peptide 2, GLP-2 = glucagon-like peptide-2.

glucagon to lower and raise blood glucose) have been employed.[60] Pancreatic glucagon, a proglucagon derived peptide (Figure 14.2), released from the pancreatic islet alpha cells is an important line of defence against hypoglycaemia as it mobilises glycogen stores from the liver. The external closed-loop systems are based on subcutaneous drug delivery and glucose monitoring, which is not directly reflective of on-going physiology. Although the concept of a fully implanted closed-loop system is also feasible, the first clinical experience with a prototype of implanted artificial beta-cell showed that improvements in the sensor structure to increase its longevity and decrease sensor delay at mealtimes are needed.[61]

Substantial progress has been made in the field of artificial pancreas research in recent years. In the case of the uni-hormonal approach, overnight closed-loop control[62–66] may be achievable.[60] Only recently, Russell and colleagues reported results of their study on the safety and effectiveness of the artificial pancreas in the outpatient setting.[66] They found that in adolescents and adults with type 1 diabetes, a wearable bi-hormonal closed-loop system improved glucose control with less hypoglycaemia episodes as compared with an insulin pump under unrestricted outpatient conditions.[66]

Although the artificial pancreas seems to be within reach, successful development of effective commercial artificial pancreas devices faces a number of challenges. These include considerations for the effect of exercise, concurrent illness, large carbohydrate meals, and the pharmacokinetics of current subcutaneous insulin.[60]

14.3.2 Novel Non-insulin Drugs and Adjunct Treatment of Type 1 Diabetes

Despite advances in diabetes treatments, including the availability of a variety of insulin analogues, the majority of patients with type 1 diabetes are not reaching glycaemic targets (an HbA1c value $<$7% based on DCCT trial).[67,68] Increased risks of hypoglycaemia, weight gain, and marked post-prandial glucose excursions are common drawbacks of intensive insulin treatment. Available anti-hyperglycaemic therapy in type 1 diabetes is currently restricted to insulin, pramlintide, and pancreas or islet cell transplantation.

Pramlintide is a novel synthetic analogue of the pancreatic beta cell hormone amylin. In type 1 diabetes, pramlintide inhibits postprandial glucagon secretion,[69–71] slows gastric emptying,[70,72] effectively decreases postprandial hyperglycaemia,[69–73] as well as 24 h glucose concentrations,[74] and also improves long-term glycaemic control,[73,75] while reducing mealtime insulin doses and body weight.[73,76] In different studies, pramlintide had been administered either *via* pre-prandial subcutaneous injections or by a dual pump system, which infuses basal and prandial insulin and pramlintide *via* independent subcutaneous pumps (continuous simultaneous subcutaneous basal-bolus insulin and pramlintide infusion).[71] Even under closed-loop conditions [closed-loop insulin delivery], pramlintide improves postprandial meal-related hyperglycaemic excursion by delaying gastric emptying, allowing a better match between carbohydrate and insulin absorptions.[72] The role of glucagon in controlling glycaemic excursions has been highlighted in a study of children with type 1 diabetes, which not only demonstrated that postprandial glucagon modulation with pramlintide as an adjunct to insulin therapy reduces post-meal glycaemic swings, but also included the use of glucagon to treat late postprandial hypoglycaemia.[69]

In the past decade, two new classes of drugs for the non-insulin-based adjunct treatment of type 1 diabetes were introduced, including incretin mimetics and sodium–glucose co-transporter 2 (SGLT2) inhibitors (gliflozins). The goal of the adjunct treatment is to address some of the pathophysiological mechanisms of type 1 diabetes, including beta-cell destruction and increased post-prandial alpha-cell glucagon secretion.[77,78]

The observation that orally administered glucose results in a robust secretion of insulin suggested the existence of a hormone released from the gastrointestinal tract that stimulated insulin secretion from pancreatic beta cells.[79] The incretin hormones include the proglucagon derived peptide glucagon-like peptide-1 (GLP-1; Figure 14.2) and gastric inhibitory peptide (GIP).[80] GLP-1 is released from intestinal L cells in the ileum and colon, which co-secrete the satiety hormone peptide tyrosine tyrosine (PYY). GIP is released from small intestinal K cells.[80] The main human incretin hormone is GLP-1, which acts on pancreatic beta cells *via* the glucagon-like peptide-1 receptor (GLP-1R).[81] GLP-1 also delays gastric emptying and, *via* the central nervous system, promotes satiety[81] (Figure 14.3). The half-life of GLP-1 is about 2 minutes as it is broken down by the enzyme dipeptidyl peptidase 4[82]

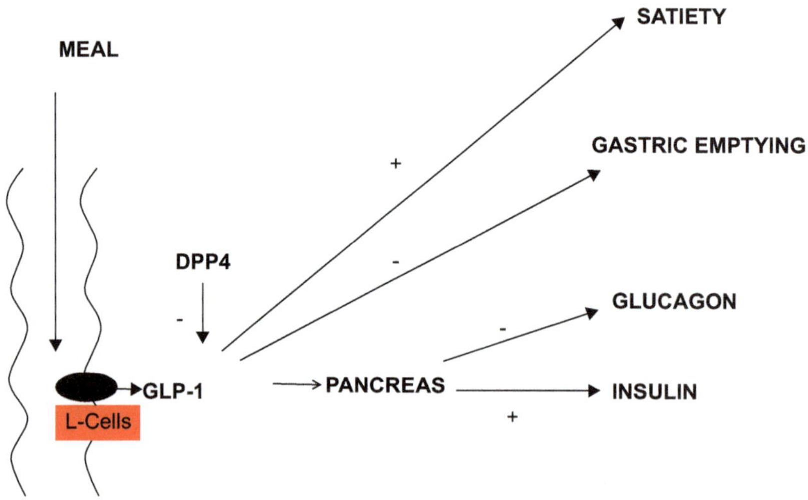

Figure 14.3 The physiological actions of glucagon-like peptide-1 (GLP-1). GLP-1 has a short half-life in the circulation as it is inactivated by dipeptidyl peptidase 4 (DPP4).

(DPP4; see below). Several GLP-1 analogues have been developed that are resistant to DPP4 degradation, resulting in longer half-life. Exenatide (exendin-4; see below) is a peptide incretin mimetic. In adolescents with type 1 diabetes, adjunctive exenatide therapy has been reported to reduce both post-prandial hyperglycaemia and insulin dosage.[83] Traina and colleagues also reported that in type 1 diabetes, the addition of once-weekly exenatide (see below) to basal-bolus insulin therapy led to significant improvements in HbA1c, body weight, and insulin requirements.[84]

The kidney regulates glucose levels through gluconeogenesis and through glucose reabsorption *via* sodium–glucose co-transporters SGLT1 and SGLT2 (see below). Gliflozins are inhibitors of SGLT2[85,86] (*e.g.* dapagliflozin, canagliflozin, and empagliflozin) and represent a new class of oral anti-diabetic agent for the management of type 2 diabetes mellitus. Gliflozins are currently under investigation as a novel non-insulin-mediated add-on therapy of type 1 diabetes. Although short-term SGLT inhibition in patients with type 1 diabetes has been associated with reduced insulin requirements, improved glycaemic control, weight loss, and reduced cardiovascular risk,[87–89] it might also lead to an under-insulinized state with an increased risk of potentially dangerous euglycaemic diabetic ketoacidosis.[88,89]

Metformin, a well-established biguanide anti-diabetic drug for type 2 diabetes (see below), also belongs to drugs that have been investigated as a possible adjunct therapy in type 1 diabetes. Recent findings from the randomized, double-blinded placebo-controlled study of 74 pubertal adolescents showed that low-dose metformin improves insulin sensitivity and fat distribution in type 1 diabetes adolescents, as indicated by significant decreases in total daily insulin dose, BMI *z*-score, and waist circumference.[90]

14.4 Type 2 Diabetes

Type 2 diabetes has a complex pathophysiology, involving the disturbance of several mechanisms (Figure 14.4), which, *via* gluco-lipotoxicity, conspire to result in micro- and macro-vascular complications. These include: (i) insulin resistance, (ii) early hyperinsulinaemia and subsequent relative insulin insufficiency, (iii) alpha cell over-function, (iv) reduced incretin response, (v) insulin resistance in liver and muscle, (vi) increased hepatic gluconeogenesis, (vii) increased lipolysis, (viii) increased renal glucose reabsorption, and (ix) hypothalamic dysfunction. To address type 2 diabetes, there is a need to address all these mechanisms (Table 14.2). Critically, gluco- and lipotoxicity accelerate several pathophysiological pathways. Thus, reducing glucotoxicity has been a key therapeutic goal. Previously, the approach to the treatment of diabetes took a "one size fits all" approach aiming to lower HbA1c towards the normal range. Unfortunately, this approach, in combination with the use of a limited array of drugs, resulted in patient weight gain, increased side effects (particularly hypoglycaemia), and increased patient psychological burden. With the advent of precision and personalised medicine, this approach has been replaced by individualized plans (Table 14.1). Thus, holistic diabetes treatment has to choose the right patient at the right time, at the right place with the right clinical team, select the right treatment at the right time, and provide the right support and follow-up. A key cornerstone of diabetes care is lifestyle change with metformin treatment. Thereafter, other drugs are added to this based on consideration of multiple factors

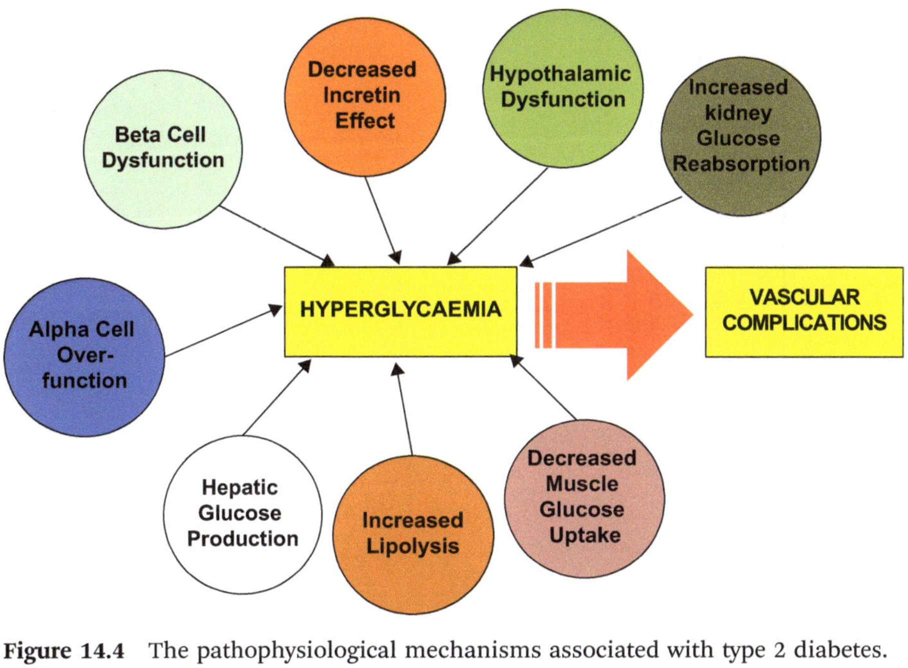

Figure 14.4 The pathophysiological mechanisms associated with type 2 diabetes.

Table 14.2 The current diabetes drugs, their mechanisms, efficacy, and side effects.

Class	Example compounds	Cellular mechanisms	Physiological mechanisms	Hba1c (%) reduction monotherapy	Benefits	Risks/side effects	Notes
Biguanides	Metformin	• cAMP-Kinase activation • Several other mechanisms	Reduction in gluconeogenesis	1.0–2.0	• Wide experience with drug • No hypoglycaemia • Weight neutral or reducing • Potential beneficial effects: CVD, Cancer • Inexpensive	• Gastrointestinal side effects • Vitamin B12 deficiency • Care with reduced renal function • Care with dehydration (especially in elderly) • Lactic acidosis (very rare)	• Cornerstone of type 2 diabetes treatment after lifestyle change • Phenformin (discontinued because of lactic acidosis) • Metformin can be used in gestational diabetes • Long acting formulations of metformin available to help adherence • Combination with other medications available to help adherence • Used in polycystic ovarian syndrome (PCOS) • Avoid in conditions that predispose to lactic acidosis *e.g.* alcoholism, liver cirrhosis • Can exacerbate irritable bowel syndrome

Alpha-glucosidase inhibitors	• Acarbose • Miglitol	Inhibit intestinal alpha-glucosidase	Delay intestinal carbohydrate uptake	0.5–0.8	• Wide experience with drug • Intra-intestinal effects, not systemic • Reduces post-prandial glucose • Inexpensive	• Gastrointestinal side effects • No hypoglycaemia • Modest HbA1c reduction • Multiple dosing required	Not commonly used because of gastrointestinal side effects (flatulence, diarrhoea)
Sulphonylureas	• Glyburide (glibenclamide) • Glipizide • Glimepride • Gliclazide • Others	Pancreatic beta cell membrane K_{ATP} closure	Increase in insulin secretion from pancreas	1.0–2.0	• Wide experience with drug class • Extensive clinical trial data • Relatively inexpensive	• Hypoglycaemia • Weight gain	• Modified release gliclazide is weight neutral • HbA1c reducing effects tend to fade over time (low durability) • Can cause prolonged hypoglycaemia in elderly • Potential issue of some compounds with reduced renal function
Meglitinides (Glinides)	• Repaglinide • Nateglinide	Pancreatic beta cell membrane K_{ATP} closure	Increase in insulin secretion from pancreas	0.5–1.5	• Post-prandial glycaemic control • Flexible dosing	• Hypoglycaemia • Weight gain • Frequent dosing • Expensive	• Useful with irregular eating habits • Useful with high post-prandial glucose

Table 14.2 (*Continued*)

Class	Example compounds	Cellular mechanisms	Physiological mechanisms	Hba1c (%) reduction monotherapy	Benefits	Risks/side effects	Notes
Thiazolidene-diones (TZDs)	• Rosiglitazone • Pioglitazone	Activate nuclear transcription factor PPAR-gamma	Increase insulin sensitivity	0.5–1.4	• Wide experience with drug class • Low hypoglycaemia risk • Reduced triglycerides (pioglitazone) • Reduced cardiovascular events (pioglitazone)	• Weight gain • Peripheral oedema • Exacerbation of heart failure • Bone fractures • Potential risks with cardiovascular disease (rosiglitazone) and bladder cancer (pioglitazone) • Osteoporosis • Bladder cancer risk (pioglitazone) • Macular oedema • Eczema (rosiglitazone)	• Use has fluctuated based on negative findings • Can be useful for insulin resistance, metabolic syndrome, non-alcoholic fatty liver
Dipeptidyl peptidese 4 (DPP-4) inhibitors	• Vildagliptin • Sitagliptin • Saxagliptin • Linagliptin • Alogliptin	Inhibit DPP-4 enzyme that inactivated glucagon-like peptide-1 (GLP-1) – increase endogenous GLP-1	Increase insulin secretion (glucose-dependent) Reduce glucagon secretion (glucose-dependent)	0.5–0.8	• No hypoglycaemia risk • Weight neutral • Once daily dosing (unless combined with metformin)	• Modest HbA1c lowering effect • Allergic reactions • Skin reactions • Nasopharyngitis • Upper respiratory infections • Urinary tract infections • More expensive than older drugs • DPP4 inhibitors have immune effects	• Rare hepatotoxicity (Vildagliptin) • Combinations with metformin available to improve adherence • HbA1c lowering appears to be durable, but longer term data required • Concerns regarding pancreatitis and pancreatic cancer risk

Sodium–Glucose co-transporter 2 (SGLT2) inhibitors	• Dapagliflozin • Canagliflozin • Empagliflozin	Inhibit renal proximal tubule sodium–glucose transporter	Increase renal glucose excretion	0.5–0.7	• No hypoglycaemia • Independent of diabetes aetiology • Modest HbA1c lowering • Modest weight loss • Blood pressure reduction	• Urinary tract infections • Genital fungal infections	• Urinary problems are common in type 2 diabetes and may be exacerbated • Early concerns regarding bladder and breast cancer • Not effective at lower renal function
Insulin	Multiple available: Short acting, long acting, ultra long acting, pre-mixed Inhaled short acting insulin	Activation of insulin receptors	Increase glucose disposal Inhibit hepatic gluconeogenesis	1.5–3.5	• Predictable and durable action • Consistent HbA1c lowering • Reduce diabetes microvascular risk	• Mainly injectable (multiple may be needed) • Hypoglycaemia • Weight gain • Psychological burden	• Some evidence that early insulin initiation may be beneficial in type 2 diabetes, however, there are risks *e.g.* hypoglycaemia and weight gain • Elderly at high risk of hypoglycaemia • Insulin has a potential mitogenic effect • Allergic reactions may occur • Implications for driving and occupation • Patient adherence is problematic • U500 (concentrated) Insulin is available for highly insulin resistant patients

Table 14.2 (*Continued*)

Class	Example compounds	Cellular mechanisms	Physiological mechanisms	Hba1c (%) reduction monotherapy	Benefits	Risks/side effects	Notes
GLP-1 receptor agonists	• Exenatide • Liraglutide • Lixisenatide	Activation of GLP-1 receptors	Increase insulin secretion (glucose-dependent) Reduce glucagon secretion (glucose-dependent) Slow gastric emptying Increase satiety	0.5–1.0	• No/low hypoglycaemia risk • Weight loss • Blood pressure reduction	• Injectable • Gastrointestinal side effects (nausea, vomiting)	• Gastrointestinal side effects usually improve after first few weeks • Longer duration formulations available and under development • Potential beta cell preservation • Potential cardiovascular benefits • Concerns regarding pancreatitis and pancreatic neoplasia • Concerns regarding thyroid C-cell hyperplasia and medullary thyroid cancer (animals) • Caution with reduced renal function • Allergic reactions may occur
Amylin mimetics	Pramlintide	Activate amylin receptors	Reduce glucagon secretion Increase satiety Delay gastric emptying	0.5–1.0	• Weight reduction • Reduce post-prandial glucose levels	• Injectable • Modest HbA1c reduction • Gastrointestinal side effects • Hypoglycaemia • Frequent dosing • Expensive	• Adherence problems since injectable

as outlined in Table 14.1. It should be emphasized that treatments change as disease either improves or progresses. The various commonly used drugs available are discussed in more detail below. Table 14.2 lists all the current major drugs, and their efficacy, benefits, and side effects.

14.4.1 Metformin

The biguanides consist of phenformin (now withdrawn because of risk of lactic acidosis) and metformin (dimethylbiguanide). Metformin has been shown to improve vascular outcomes in diabetes, it is weight friendly with no hypoglycaemia, and may have other benefits such as potential reduction in cancer incidence.[91–93] The root of guanidine-derived drugs goes back to *Galega officinalis* (goat's-rue or French lilac),[94,95] whose active ingredient was used as the base for phenformin and metformin (composed of two guanidine groups).[95,96] Unlike phenformin,[95,96] lactic acidosis is rare with metformin and may occur in the context of dehydration, sepsis, and alcohol abuse.[95,97] Metformin is orally administered and 30–60% bioavailable. It has a half-life in plasma of around 2 hours and a half-life of elimination of 4–8.7 hours.[98,99] The gastrointestinal tract is a major site for metformin accumulation, explaining gastrointestinal side effects including diarrhoea, flatulence and abdominal discomfort.[100] An extended-release formulation of metformin developed to reduce the gastrointestinal side effects is now available, permitting once-daily dosing.[101]

Metformin is water-soluble and renally cleared un-metabolised.[98,99] Thus, it may accumulate in conditions where glomerular filtration is reduced.[102] Metformin enters cells through organic cation transporters (OCT 1–3) which, with various transporters, has tissue specific expression, as well as the plasma membrane monoamine transporter (PMAT).[103] The precise mechanisms of action of metformin are unknown. The frequently cited cellular target for metformin is AMP-activated protein kinase (AMPK). The major effect of metformin is the inhibition of hepatic gluconeogenesis.[104–107] Metformin inhibits gene expression of phosphophenolpyruvate carboxylase (PEPCK) and glucose-6-phosphatase, the rate limiting enzymes in gluconeogenesis.[108,109] Other effects include increased peripheral insulin sensitivity by enhancing insulin-regulated GLUT-4-mediated glucose transport into striated muscles and adipocytes,[104,105] and alterations of the gut microbiome.[110]

Metformin is the recommended first-line treatment in type 2 diabetes.[96,111,112] It is weight neutral or aids in weight loss with lifestyle change.[100,113,114] It has low risk for hypoglycaemia in monotherapy.[115] In metformin intolerance, the choice of first line drug should be individualized taking into account multiple factors (Table 14.1), degree of hyperglycaemia as measured by HbA1c, and drug efficacy. Thus, a symptomatic patient with HbA1c of ≥10% cannot be treated to acceptable targets without initial insulin treatment. Once the HbA1c is reduced towards goal with insulin, however, it may be possible to introduce other treatments and withdraw insulin. Metformin in not used in patients with low glomerular filtration rates (estimated GFR <30 mL min^{-1}).[102] The risk of lactic acidosis with metformin is higher in elderly patients, who are at risk of dehydration, and those who have other co-morbidities predisposing to hypoxaemia.[95,97]

14.4.2 Sulphonylureas

Sulphonylureas are the oldest oral drugs for type 2 diabetes treatment.[116] Some of the early sulphonylureas such as chlorpropamide had multiple side effects that limited their use. Chlorpropamide was also long acting with the danger of prolonged hypoglycaemia. Newer sulphonylureas are much better tolerated.[117,118] Sulphonylureas are glucose-independent insulin secretagogues that act at the SUR1 regulatory subunit of the ATP-sensitive potassium channels (K_{ATP}) channel on pancreatic beta cells resulting in cell depolarization, opening of voltage-gated calcium channels, calcium influx, exocytosis of insulin secretory granules, and secretion of insulin.[119,120] Thus, they are most effective early in disease when there is a pool of functioning beta cells.[118] Cardiac and vascular K_{ATP} channels are associated with different SUR subunits.[119] Action of some sulphonylureas at cardiac SUR subunits can explain the reported negative cardiovascular effects (*e.g.* abolishing the protective effects of myocardial ischaemic preconditioning) of some sulphonylurea drugs.[121] While glimepride, gliclazide and tolbutamide are highly tissue selective and are thus associated with a lower cardiovascular risk, glibenclamide has been associated with a higher all-cause and cardiovascular mortality.[120,122] A retrospective analysis of nearly 24 000 patients found an increased cardiovascular risk in patients treated with sulphonylureas (glipizide, glyburide and glimepiride) compared to those treated of metformin.[123] The addition of a sulphonylurea to monotherapy with metformin is a common practice. A meta-analysis found that combination therapy of metformin with a sulphonylurea significantly increased the relative risk of the composite of cardiovascular hospitalization and mortality independent of the control group (diet therapy, metformin monotherapy or sulphonylurea monotherapy). However, there was no significant difference of the combination for cardiovascular or all-cause mortality independently.[124] Despite these findings, along with the mechanistic hypotheses, an increase in cardiovascular risk was not observed in UKPDS or ADVANCE trials.[32,118,125] Although the addition of sulphonylurea to metformin has long been a second step in management of type 2 diabetes, the recent emergence of drugs such as incretins, gliptins, and gliflozins (see below) is changing prescribing practice.

The two most common adverse effects associated with sulphonylureas are hypoglycaemia and weight gain.[116] Hypoglycaemia is of particular concern in the elderly, those with irregular meal schedules, and those with worsening renal function.[8,116,126,127] Diamicron MR is a formulation of gliclazide with modified-release characteristics. It has a hydrophilic hypromellose-based polymer matrix, which allows slow release through once-daily dosing at breakfast.[128] The GUIDE study (GlUcose control In type 2 diabetes: Diamicron MR *versus* glimEpiride) was a double-blind, 27 week, parallel-group study of 845 type 2 diabetic patients.[129] GUIDE showed that HbA1c decreased by 1.2% in newly diagnosed patients treated with gliclazide MR monotherapy and 1% with glimepiride, showing non-inferiority. There was, however, a reduced risk of hypoglycaemia (about one third) with gliclazide MR.[129] In the ADVANCE study, the intensive arm patients receiving gliclazide MR had a low risk of severe hypoglycaemia (only 2.7%).[32] The ADVANCE study also demonstrated that there was no weight gain in the group with intensive glucose control based on gliclazide MR, even in obese patients.[130] Thus, unlike other sulphonylureas, gliclazide MR is weight neutral.

14.4.3 The Meglitinides (Glinides)

The meglitinides are non-sulphonylurea insulin secretagogues with shorter half-lives than sulphonylureas resulting in brief episodic insulin secretion.[131] Repaglinide is a benzoic acid derivative (non-sulphonylurea moiety of gliburide and salicylate derivative), whereas nateglinide is a D-phenylalanine derivative, and both have a mechanism of action similar to sulphonylureas: interaction with K_{ATP} channels on beta cells, but at a site distinct from sulphonylureas binding sites.[116,132] The side effects of glinides are similar to sulphonylureas (hypoglycaemia and weight gain), but are probably less pronounced.[116] The need for multiple dosing along with the risk of hypoglycaemia, however, limits the use of glinides.[131] Although not absolutely selective, the higher potency and efficacy of glinides in inhibiting the pancreatic than the cardiovascular K_{ATP} channels supports their clinical use.[133]

14.4.4 The Thiazolidinediones (TZDs) or Glitazones

Thiazolidinediones are peroxisome proliferator-activated receptor γ (PPARγ) agonists. PPARγ receptors are found throughout the body, including in muscle and fat cells, pancreatic beta cells, liver cells, and endothelial cells.[121,134] Activation of these receptors increases insulin sensitivity in fat and muscle. PPARγ receptor activation reduces hepatic gluconeogenesis and inflammation, and may reduce or delay atherosclerosis, as well as reducing lipotoxicity and enhancing the redistribution of fat from muscle, liver and pancreas to subcutaneous stores.[121,135] Potentially due to their effects on improved insulin sensitivity and glucose homeostasis, the thiazolidenediones also reduce blood pressure and microalbuminuria.[136] However, they do not increase insulin secretion.[121] The action of these drugs takes several weeks to take effect clinically. Currently, this class of drugs is represented by two drugs: rosiglitazone and pioglitazone. Rosiglitazone is a pure PPARγ receptor agonist while pioglitazone acts at both PPARγ and PPARα receptors, which may explain the favourable lipid effects of pioglitazone (LDL-cholesterol is increased with rosiglitazone). PPARα receptors are found mainly in the liver, heart, skeletal muscle, and vascular walls.[135] These drugs improve insulin sensitivity in muscle and liver and, when taken as monotherapy, do not cause significant hypoglycaemia.[131,137] Rosiglitazone and pioglitazone agents are orally effective and have a half-life of up to 3 and 24 hours, respectively.[121] These drugs are associated with weight gain through stimulation of central PPARγ receptors (increasing appetite) and fluid retention.[138] Because of fluid retention, these drugs are contraindicated in congestive cardiac failure. There is evidence suggesting that thiazolidinediones may promote a reduction in bone density and predispose to fractures, especially in women.[139,140] There are data to suggest a link between pioglitazone and bladder cancer.[141] This class of drugs can also result in macular oedema.[142]

After approval of rosiglitazone, a meta-analysis raised concerns regarding increased risk of myocardial infarction and death from cardiovascular disease.[143] This led to restrictions for using rosiglitazone in patients with cardiovascular disease and cardiovascular recommendations for the use of thiazolidinediones. Importantly, the U.S. FDA proposed guidance for approval of all diabetes drugs included ruling out excess cardiovascular risk, defined as an upper bound of the two-sided 95% confidence interval for major adverse cardiovascular events (MACE) of less than 1.80 and less than 1.30, respectively.[144] Thus, all new diabetes drugs are assessed for cardiovascular outcomes.

14.4.5 Novel Therapies for the Treatment of Type 2 Diabetes

14.4.5.1 Incretin-system Based Therapies

Incretin hormones (Figure 14.3) are released from the gastrointestinal tract and act on pancreatic beta cells to stimulate insulin secretion. The major incretin hormones are gastric inhibitory peptide (GIP) released from intestinal K cells and glucagon-like peptide-1 (GLP-1), a glucagon-derived peptide (Figure 14.2), which is released from lower intestinal L cells.[80] GLP-1 levels are diminished in type 2 diabetes. Thus, enhancing GLP-1 action was seen as an important mechanism for the treatment of type 2 diabetes.[145]

14.4.5.1.1 Glucagon-like Peptide-1 Receptor Agonists (Incretin Mimetics). GLP-1 analogues were developed to overcome the rapid degradation of endogenous GLP-1 by DDP4.[145] Like insulin, these peptide drugs have to be delivered by injection to avoid gastrointestinal degradation. The incretin mimetics acting through the glucagon-like peptide-1 receptors (GLP-1R) have several advantages: (i) their glucose-dependent effects reduce the risk of hypoglycaemia;[80,146] (ii) they not only increase insulin secretion, but reduce glucagon secretion which drives endogenous glucose production and secretion;[81] (iii) they aid weight loss through delaying gastric emptying (feeling of fullness) and reducing appetite through central nervous system receptors;[81,147] (iv) increasingly available longer acting preparations allow better patient tolerance and adherence;[148] (v) based on animal studies,

GLP-1 analogues may help prevent the beta cell decline observed with type 2 diabetes;[149] (vi) GLP-1 analogues may have favourable blood pressure and cardiovascular effects.[29,136,150–153]

Exenatide is the synthetic form of Exendin-4, a GLP-1R agonist isolated from the saliva of the Gila monster, a poisonous lizard found in the Arizona desert, USA.[81] Exendin-4 has 50% amino-acid sequence homology to mammalian GLP-1. Exenatide has a half-life of 60–90 minutes and is administered as a twice-daily injection because plasma concentrations last between 4–6 hours.[81] Exenatide (Byetta) is injected twice daily, 40–60 min before breakfast and dinner. About 40–50% of patients receiving exenatide develop antibodies, but these rarely affect the response to the drug.[154,155] Liraglutide (Victoza) is a modified human GLP-1 with 97% homology, administered as a once-daily injection.[148] The modifications in GLP-1 structure for liraglutide are shown in Figure 14.5. Liraglutide is slowly absorbed and has a half-life of about 11–13 h after subcutaneous administration[81] allowing once daily administration. Lixisenatide is another exendin-4 based agent, with a deletion of a proline and addition of six lysine amino acids at the C-terminus. Lixisenatide has a half-life of 2–3 hours after subcutaneous injection.[156] In a head to head comparison of lixisenatide and exenatide twice daily, lixisenatide demonstrated non-inferiority with regard to Hb1Ac reduction. Lxisenatide was slightly better tolerated than exenatide. However, lixisenatide treatment was slightly inferior to exenatide for weight loss.[157] Longer acting GLP-1 analogues have been developed. A once-weekly formulation of exenatide (Exenatide LAR, Bydureon) has also become available with similar effects. Exenatide LAR consists of encapsulated molecules of exenatide in injectable microspheres.[149] Other long acting analogues include albiglutide and dulaglutide, which are both administered once weekly.[158]

In general, GLP-1 analogues achieve an HbA1c reduction of about 1%.[81] They are well tolerated. The main side effects are gastrointestinal (nausea, diarrhoea, and rarely vomiting). Starting gradually at a small dose may prevent these side effects, which tend to settle down in 2–3 weeks after

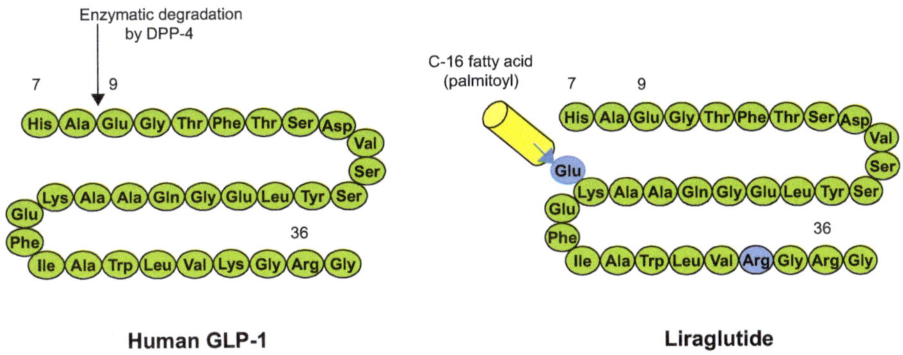

Figure 14.5 The structures of human glucagon-like peptide-1 (GLP-1) and liraglutide.

therapy initiation.[154,155] Although there has been concern about these drugs increasing the risk of pancreatitis,[159,160] there is currently insufficient evidence to support this.[161] Similarly, the concern regarding the development of thyroid C-cell carcinoma with liraglutide treatment observed from animal studies, has not been replicated in humans.[162]

Due to the presence of GLP-1 receptors in multiple tissues, GLP-1 receptor agonists can exert many important non-glycaemic effects. GLP-1 receptor agonists have been reported to have beneficial effects on cardiovascular function including improving left ventricular ejection fraction and cardiac wall motion score indexes in patients with acute myocardial infarction and reduced left ventricular ejection fraction after successful primary angioplasty. These benefits were independent of acute myocardial infarction location or diabetes history.[151] GLP-1 receptors are also present on vascular cells, monocytes, and macrophages; thus, the protective cardiovascular effects may be due to oxidative stress and vascular inflammation reduction, and improvement in endothelial function.[151–153] There are on-going studies to examine the role of GLP-1 analogues in several disorders including non-alcoholic fatty liver disease[163] and neurodegenerative disorders.[164]

14.4.5.1.2 Dipeptidyl Peptidase-4 Inhibitors (Incretin Enhancers). With GLP-1 levels being diminished in type 2 diabetes, enhancing endogenous GLP-1 through DPP4 inhibition is a potentially useful approach.

DDP-4 (also known as CD26) is a membrane-spanning 766 amino-acid cell surface amino-peptidase expressed in several tissues including liver, lungs, kidney, intestinal brush border, lymphocytes and endothelial cells.[81,165–168] The extracellular domain of DPP4 may be cleaved to circulate in the plasma, while retaining activity.[81] DPP4 preferentially cleaves peptides with a proline or alanine in the second amino terminal position and thus has several substrates including several gut hormones (including GLP-1 and GIP),[166,169] neuropeptides, cytokines, and chemokines.[167,169] DPP4 inhibitors (gliptins) have several advantages: (i) they are taken orally, usually as a single dose;[170] (ii) because of glucose-dependent effects, they are not associated with hypoglycaemia; (iii) they inhibit glucagon secretion; (iv) they are weight neutral;[81,171] (v) because of lack of effects on gastric emptying,[81,171] they have less gastrointestinal side effects that GLP-1 analogues.

DPP4 inhibitors include a diverse group of compounds that include those that mimic the DPP4 dipeptide substrate structure (sitagliptin, vildagliptin and saxagliptin), and those that are non-peptidomimetic (alogliptin and linagliptin).[170] All current compounds are reversible competitive inhibitors.[170,172,173] However, there are differences in how these compounds interact with DPP4. Whereas sitagliptin, alogliptin and linagliptin form non-covalent interactions with residues in the catalytic site, vildagliptin and saxagliptin act through a two-step process that involves formation of a reversible covalent bond with slow dissociation, which may explain the longer action of these drugs.[170,174,175] Gliptins have generally high bioavailability (about 87% for sitagliptin,[176] 85% for vildagliptin,[177] 67% for saxagliptin[178] and 30% for linagliptin[179]). Preclinical studies have shown that these agents will have high concentrations in the liver[170,180] and do not cross the blood brain barrier.[178,180] There are only minor differences in the pharmacodynamic and pharmacokinetic properties of these molecules. Significant inhibition of plasma DDP-4 activity is seen as quickly as within 5 minutes of administration.[170] Sitagliptin, alogliptin and linagliptin are largely eliminated unchanged, whereas vildagliptin and saxagliptin undergo extensive liver and other tissue metabolism *via* a CYP_{450}–independent mechanism.[170]

When compared with GLP-1R agonists, DDP-4 antagonists were found to be inferior with regards to HbA1c reduction and weight loss.[181] A 26 week, randomized, double-dummy superiority trial found that treatment with exenatide reduced HbA1c (least square mean −1.5%, 95% CI: −1.7 to −1.4) significantly more than sitagliptin (−0.9%, 95% CI: −1.1 to −0.7). Weight loss with exenatide (−2.3 kg, 95% CI: −2.9 to −1.7) was significantly greater than with sitagliptin (difference −1.5 kg, 95% CI: −2.4 to −0.7).[182] When compared to metformin as monotherapy in drug-naïve patients who did not achieve glycaemic control with diet and exercise, metformin showed greater reductions in both Hb1Ac and body weight, but with less gastrointestinal tolerability.[183–187]

There are several unwanted effects with the gliptins that need to be considered. There is a risk of hepatic adverse events with vildagliptin.[170,188] There is an increased risk of upper respiratory tract infections and nasopharyngitis (sitagliptin and vildagliptin) as well as headache (more with vildagliptin).[189] Updated safety analyses, however, have been published showing no increased risk for infections or headache with these agents.[190,191] An analysis of 12 randomised trials with study durations up to 108 weeks suggests that the effect of DPP4 inhibitors may decrease during the second year of treatment. Thus, future research should focus on the characteristics of patients who will most benefit from these agents in terms of glycaemic durability.[192] A recent large study of over 16 000 patients, did not find any adverse consequences with the use of saxagliptin.[193] Numerous clinical trials are currently underway to determine the cardiovascular effects of gliptins.[194,195]

14.4.5.2 Sodium–Glucose Co-transporter 2 Inhibitors (SGLT2 Inhibitors)

The kidney has emerged as an important organ in glucose homeostasis. The kidney is involved in gluconeogenesis, and glucose filtration, reabsorption and excretion. About 180 g of glucose is filtered daily by the kidney[196] (Figure 14.6). The majority (90%) of filtered glucose is absorbed through the sodium–glucose co-transporter 2 in the proximal tubule, with the rest reabsorbed by the sodium–glucose co-transporter 1 distally in the proximal tubule.[197,198] There are 6 sodium–glucose co-transporters, but SGLT-1 and 2 have an important glucose transport function. SGLT-2 is mainly found in the kidney while SGLT-1 is found in the kidney and the intestine. SGLT-2 is a

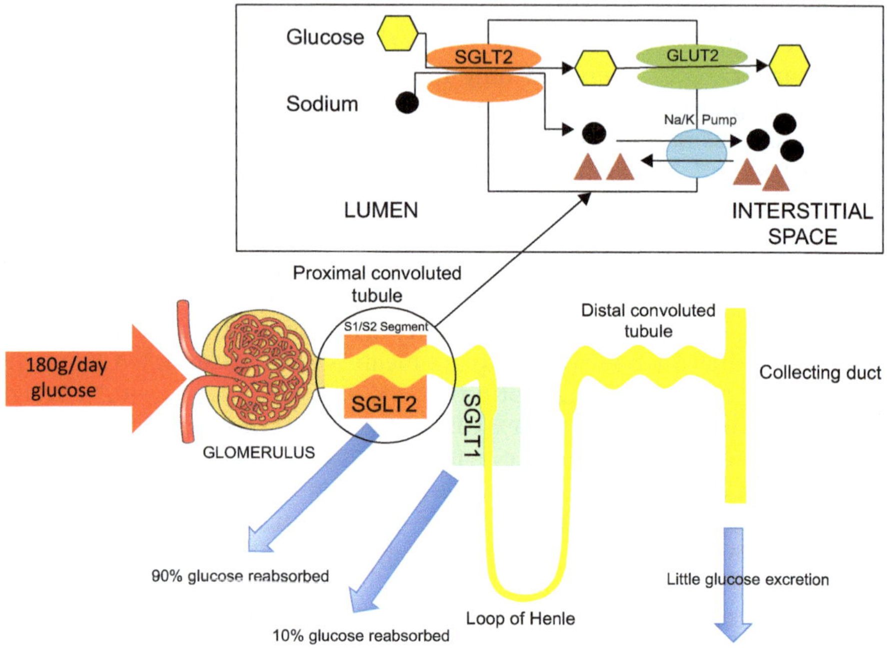

Figure 14.6 The physiological function of sodium–glucose co-transporters in the kidney.

Table 14.3 The benefits of sodium–glucose co-transporter 2 inhibitors in diabetes.

Benefits of treatment	Description
Glucose lowering	1. Reduce glucotoxicity 2. Reduced risk of hypoglycaemia (as glucose lowers, renal excretion is reduced) 3. Prevention of diabetes complications
Insulin-independent action	1. Can be used in both type 1 and type 2 diabetes 2. Independent of stage of disease 3. Can augment other anti-diabetic medications
Osmotic diuresis	1. Weight loss 2. Reduction in blood pressure
Excess urinary calorie loss	1. Weight loss and weight loss maintenance 2. Reduce weight gain with other medications

high capacity low affinity transporter while SGLT-1 is a low capacity high affinity transporter.[197,198] Both SGLT-2 and GLUT-2 required for renal reabsorption of glucose (Figure 14.6) are upregulated in type 2 diabetes. SGLT-2 inhibition is attractive for the treatment of type 2 diabetes for several reasons (Table 14.3). The first SGLT inhibitor was phlorizin, isolated from the apple tree bark.[199] Because of its effects, on both SGLT-2 and SGLT-1, it had hypoglycaemic actions, but was associated with diarrhoea because of action on SGLT-1.[199,200] Newer agents (dapagliflozin, canagliflozin, empagliflozin) are more selective to SGLT-2. Phlorizin consists of a glucose ring and two aromatic rings joined by a glycosidic oxygen.[85] More orally stable SGLT-2 inhibitors are *C*-glucosides with current agents being *C*-aryl.[201]

The use of SGLT-2 inhibitors, besides increasing renal glucose excretion, has been found to improve insulin sensitivity as well as reduce blood pressure and weight. However, SGLT-2 inhibitors may potentially result in dehydration. Furthermore, there have been reports of enhanced endogenous glucose production, presumably a compensatory mechanism to the increased glucose excretion in urine.[202]

Symptoms of urgency and frequency of urination were found to be common with these drugs. Additionally, there was an increased risk of urinary tract and genital infections.[203] Because of their potential risk for dehydration (especially in elderly patients) and a reduced glucose lowering effect in patients with renal impairment, appropriate patient selection and close monitoring are important.[204] The role of these drugs in the treatment of diabetes remains to be determined. The burgeoning type 2 diabetes treatment options have allowed greater personalization of diabetes.[205]

14.4.6 The Role of Insulin in the Treatment of Type 2 Diabetes

Insulin is the most effective treatment for reduction of blood glucose levels and lowering of HbA1c (Table 14.2). Insulin has several roles in the treatment of type 2 diabetes. In early diabetes, with poor glycaemic control, insulin can be used to control symptoms and improve glycaemia. This insulin can later be withdrawn once oral treatments have taken effect and glycaemia is better controlled. Insulin is also important in later stages of type 2 diabetes, where oral treatments have not shown durability and there is a need to address the potential beta cell decline as type 2 diabetes progresses. There are several controversies in insulin use in type 2 diabetes. The use in later stages of the disease can be problematic. This is because it is not possible to clinically evaluate patients who have reached sufficient beta cell decline to become insulin-dependent. The observed poor glycaemic control in these patients may be a reflection of poor adherence to oral medication. Insulin introduction at this stage can be deleterious resulting in hypoglycaemic symptoms and weight gain. Furthermore, it has been shown that it is possible to improve diabetes or put it into remission with bariatric surgery, even in those with long duration of diabetes and those on insulin.[206–208] Generally, the success of bariatric surgery in reversing diabetes depends upon the patient age (younger), diabetes duration (shorter), and diabetes medications (less medication and no insulin).

Another controversy is the use of early insulin and treatment intensification in type 2 diabetes. The UKPDS study showed that early intensification results in better outcomes and that beneficial outcomes persisted beyond the trial (the "legacy effect").[7,28] The ORIGIN (Outcome Reduction with

Initial Glargine INtervention) study examined once daily insulin glargine (Lantus) in type 2 diabetes. The study did not show any harm from this approach, but there was no great benefit.[209] Not supporting these findings are data from of recent large trials, such as ACCORD, that showed weight gain, increased hypoglycaemia, and no cardiovascular benefit.[30] These studies were conducted in different patient populations *i.e.* early *vs.* later diabetes. Their findings have instigated the current guidelines to treatment individualization.[34] All available insulin and its analogues have been used in the treatment of type 2 diabetes. When necessary, insulin is usually started using a long acting insulin or an insulin mixture. Care is taken with adjustment of other medication to avoid hypoglycaemia.

The latest long-acting insulin is insulin degludec. Degludec is an ultra-long-acting insulin identical to human insulin but modified such that the last amino acid from the B-chain is removed and glutamyl is added to link LysB29 to a hexadecanedioic fatty acid. These additions result in the development of multihexamers at the injection site with slow release of insulin monomers[210] (Figure 14.7). Because of the modification, degludec can provide a predictable and steady insulin concentration over a prolonged period (up to 72 hours). Unlike other long acting insulins (glargine and detemir), another property of degludec is that it can be mixed with shorter prandial insulins without alterations in pharmacokinetics.

In a study to compare insulin degludec with glargine for efficacy and safety in insulin-naive type 2 diabetes, inadequately controlled with oral drugs, HbA1c reduction between degludec and glargine were similar (1.06% and

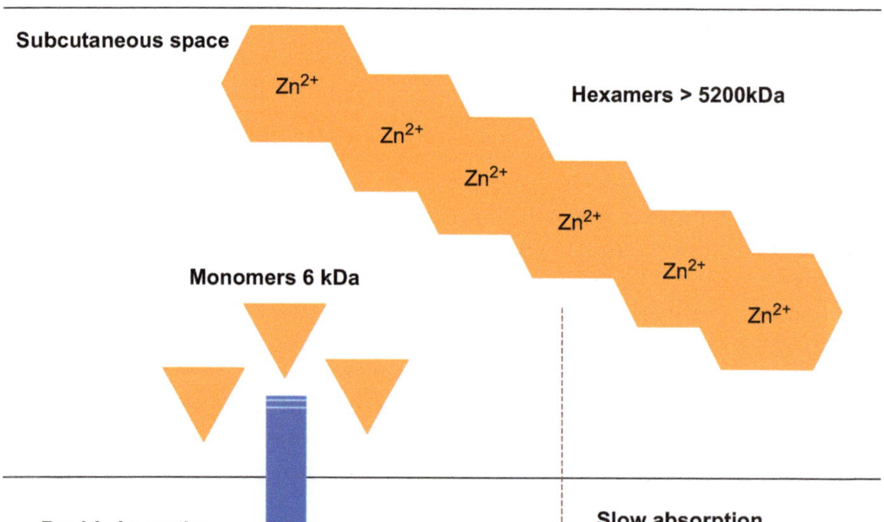

Figure 14.7 Insulin degludec forms hexamers resulting in slow release into the circulation.

1.19%, respectively), but hypoglycaemia was lower in the degludec group.[211] There are currently uncertainties regarding potential negative cardiovascular effects of degludec.[121]

14.4.7 Beta Cell Preservation

Pancreatic beta cell loss is a key problem of the pathophysiology of type 2 diabetes. Interestingly, mouse models of obesity and insulin resistance, such as the ob/ob leptin deficient mouse, demonstrate beta cell proliferation suggesting that insulin resistance triggers beta cell proliferation. Liver specific insulin receptor gene knockout mice also display beta cell proliferation highlighting the importance of the liver in diabetes and suggesting that the liver releases a factor that induces beta cell proliferation.[212–215] Parabiosis experiments have confirmed the transfer of such a factor from liver specific insulin receptor gene knockout mice to wild type mice. Recently, using an insulin receptor antagonist, pancreatic beta cell proliferation was observed.[216] A direct action of the antagonist on the pancreas was excluded. This proliferation was associated with increased expression of betatrophin in the liver and white adipose tissue. Betatrophin, also called lipasin, is a 22 kDa 198 amino acid peptide secreted by the liver that has been proposed as a key factor in beta cell proliferation. In humans betatrophin is mainly expressed in the liver. The role of betatrophin in humans is unclear and several cross-sectional studies have reported conflicting results regarding the association of circulating betatrophin in type 2 diabetes and obesity.[217] Understanding factors that affect beta (and also alpha) cell proliferation will provide novel therapeutic approaches for tackling type 2 diabetes. These factors are also important in preserving beta cells in early type 1 diabetes and potentially in the survival of islet cell transplants and stem cell treatments for diabetes.

14.4.8 The Gut Microbiome

The role of the gut microbiota in obesity, insulin resistance, diabetes, and metabolic disorders is gaining considerable interest.[218–220] Changes in the gut microbiome could influence intestinal hormone secretion including secretion of the incretin hormone glucagon-like peptide 1 from intestinal L-cells (GLP-1).[221] Since inflammation, mediated by inflammatory markers, is a key mechanism in diabetic complications, the gut microbial environment may contribute to the development and progression of diabetes complications through local and systemic (endotoxemia) mechanisms. From a metagenome-wide association study in Chinese patients,[222] it has been reported that in patients with type 2 diabetes there was a bacterial dysbiosis with an increase in opportunistic pathogens and a decrease in butyrate-producing bacteria. Altering the gut microbial environment could have beneficial effects for treatment of diabetes and its complications.

14.5 Future Developments

Diabetes is a chronic disorder associated with long-term micro- and macrovascular complications. Most serious complications of diabetes develop over a long time and there is a need to study the effect of drugs beyond early observations. The safety of drugs also requires long-term observation. Several alarms regarding diabetes drugs have been raised through post-marketing surveillance. Concerns have been raised regarding cardiovascular disease (rosiglitazone) and insulin glargine (cancer), and these have not been fully resolved. There is still a need to develop approaches to address the various pathophysiological pathways. Some insights will come from understanding how bariatric surgery procedures improve or resolve diabetes, resulting in novel therapeutic approaches. Recent interest in gut microbiome could also open new avenues for treatment of diabetes and its complications. The move towards personalized and precision medicine is highly important in diabetes, and with technological advances this goal is increasingly achievable.

References

1. http://www.who.int/mediacentre/factsheets/fs312/en/.
2. International Diabetes Federation. *IDF Diabetes Atlas*, 6th edn, Brussels, Belgium, 2012, vol. 8, p. 228.
3. L. Chen, D. J. Magliano and P. Z. Zimmet, *Nat. Rev. Endocrinol.*, 2012, **8**, 228.
4. P. Pozzilli and R. Buzzetti, *Trends Endocrinol. Metab.*, 2007, **18**, 52.
5. The Diabetes Control and Complications Trial Research Group, *N. Engl. J. Med.*, 1993, **329**(14), 977–986.
6. D. M. Nathan, P. A. Cleary, J. Y. Backlund, S. M. Genuth, J. M. Lachin, T. J. Orchard, P. Raskin, B. Zinman and Diabetes Control and Complications Trial/Epidemiology of Diabetes Interventions and Complications Study Research Group, *N. Engl. J. Med.*, 2005, **353**, 2643.
7. J. Chalmers and M. E. Cooper, *N. Engl. J. Med.*, 2008, **359**, 1618.
8. *Lancet*, 1998, 352, 837.
9. S. G. Coca, F. Ismail-Beigi, N. Haq, H. M. Krumholz and C. R. Parikh, *Arch. Intern. Med.*, 2012, **172**, 761.
10. P. Gaede, H. Lund-Andersen, H. H. Parving and O. Pedersen, *N. Engl. J. Med.*, 2008, **358**, 580.
11. P. Gaede, L. Tarnow, P. Vedel, H. H. Parving and O. Pedersen, *Nephrol., Dial., Transplant.*, 2004, **19**, 2784.
12. P. Gaede, P. Vedel, H. H. Parving and O. Pedersen, *Lancet*, 1999, **353**, 617.
13. A. Sandbaek, S. J. Griffin, S. J. Sharp, R. K. Simmons, K. Borch-Johnsen, G. E. Rutten, M. van den Donk, N. J. Wareham, T. Lauritzen, M. J. Davies and K. Khunti, *Diabetes Care*, 2014, **37**, 2015.
14. S. M. Grundy, B. Howard, S. Smith, Jr., R. Eckel, R. Redberg and R. O. Bonow, *Circulation*, 2002, **105**, 2231.

15. W. B. Kannel and D. L. McGee, *J. Am. Med. Assoc.*, 1979, **241**, 2035.
16. A. R. G. Look, R. R. Wing, P. Bolin, F. L. Brancati, G. A. Bray, J. M. Clark, M. Coday, R. S. Crow, J. M. Curtis, C. M. Egan, M. A. Espeland, M. Evans, J. P. Foreyt, S. Ghazarian, E. W. Gregg, B. Harrison, H. P. Hazuda, J. O. Hill, E. S. Horton, V. S. Hubbard, J. M. Jakicic, R. W. Jeffery, K. C. Johnson, S. E. Kahn, A. E. Kitabchi, W. C. Knowler, C. E. Lewis, B. J. Maschak-Carey, M. G. Montez, A. Murillo, D. M. Nathan, J. Patricio, A. Peters, X. Pi-Sunyer, H. Pownall, D. Reboussin, J. G. Regensteiner, A. D. Rickman, D. H. Ryan, M. Safford, T. A. Wadden, L. E. Wagenknecht, D. S. West, D. F. Williamson and S. Z. Yanovski, *N. Engl. J. Med.*, 2013, **369**, 145.
17. B. Dahlof, P. S. Sever, N. R. Poulter, H. Wedel, D. G. Beevers, M. Caulfield, R. Collins, S. E. Kjeldsen, A. Kristinsson, G. T. McInnes, J. Mehlsen, M. Nieminen, E. O'Brien and J. Ostergren, ASCOT Investigators, *Lancet*, 2005, **366**, 895.
18. R. R. Rubin, *Am. J. Med.*, 2005, **118**(Suppl 5A), 27S.
19. A. S. Group, W. C. Cushman, G. W. Evans, R. P. Byington, D. C. Goff, Jr., R. H. Grimm, Jr., J. A. Cutler, D. G. Simons-Morton, J. N. Basile, M. A. Corson, J. L. Probstfield, L. Katz, K. A. Peterson, W. T. Friedewald, J. B. Buse, J. T. Bigger, H. C. Gerstein and F. Ismail-Beigi, *N. Engl. J. Med.*, 2010, 362 , 1575.
20. C. Baigent, A. Keech, P. M. Kearney, L. Blackwell, G. Buck, C. Pollicino, A. Kirby, T. Sourjina, R. Peto, R. Collins and R. Simes, Cholesterol Treatment Trialists' (CTT) Collaborators, *Lancet*, 2005, **366**, 1267.
21. H. M. Colhoun, D. J. Betteridge, P. N. Durrington, G. A. Hitman, H. A. Neil, S. J. Livingstone, M. J. Thomason, M. I. Mackness, V. Charlton-Menys and J. H. Fuller, CARDS Investigators, *Lancet*, 2004, **364**, 685.
22. J. Lounsbery, S. Prasad and K. Rowland, *J. Fam. Pract.*, 2010, **59**, 582.
23. J. A. Farmer, *Curr. Atheroscler. Rep.*, 2011, **13**, 104.
24. A. S. Group, H. N. Ginsberg, M. B. Elam, L. C. Lovato, J. R. Crouse, 3rd, L. A. Leiter, P. Linz, W. T. Friedewald, J. B. Buse, H. C. Gerstein, J. Probstfield, R. H. Grimm, F. Ismail-Beigi, J. T. Bigger, D. C. Goff, Jr., W. C. Cushman, D. G. Simons-Morton and R. P. Byington, *N. Engl. J. Med.*, 2010, **362**, 1563.
25. A. Keech, R. J. Simes, P. Barter, J. Best, R. Scott, M. R. Taskinen, P. Forder, A. Pillai, T. Davis, P. Glasziou, P. Drury, Y. A. Kesaniemi, D. Sullivan, D. Hunt, P. Colman, M. d'Emden, M. Whiting, C. Ehnholm and M. Laakso; FIELD Study Investigators, *Lancet*, 2005, **366**, 1849.
26. A. S. Group, A. E. S. Group, E. Y. Chew, W. T. Ambrosius, M. D. Davis, R. P. Danis, S. Gangaputra, C. M. Greven, L. Hubbard, B. A. Esser, J. F. Lovato, L. H. Perdue, D. C. Goff, Jr., W. C. Cushman, H. N. Ginsberg, M. B. Elam, S. Genuth, H. C. Gerstein, U. Schubart and L. J. Fine, *N. Engl. J. Med.*, 2010, **363**, 233.
27. R. R. Holman, S. K. Paul, M. A. Bethel, D. R. Matthews and H. A. Neil, *N. Engl. J. Med.*, 2008, **359**, 1577.

28. *Lancet*, 1998, **352**, 854.
29. J. A. Dormandy, B. Charbonnel, D. J. Eckland, E. Erdmann, M. Massi-Benedetti, I. K. Moules, A. M. Skene, M. H. Tan, P. J. Lefebvre, G. D. Murray, E. Standl, R. G. Wilcox, L. Wilhelmsen, J. Betteridge, K. Birkeland, A. Golay, R. J. Heine, L. Koranyi, M. Laakso, M. Mokan, A. Norkus, V. Pirags, T. Podar, A. Scheen, W. Scherbaum, G. Schernthaner, O. Schmitz, J. Skrha, U. Smith and J. Taton, PROactive Investigators, *Lancet*, 2005, **366**, 1279.
30. Action to Control Cardiovascular Risk in Diabetes Study Group, H. C. Gerstein, M. E. Miller, R. P. Byington, D. C. Goff, Jr., J. T. Bigger, J. B. Buse, W. C. Cushman, S. Genuth, F. Ismail-Beigi, R. H. Grimm, Jr., J. L. Probstfield, D. G. Simons-Morton and W. T. Friedewald, *N. Engl. J. Med.*, 2008, **358**, 2545.
31. W. Duckworth, C. Abraira, T. Moritz, D. Reda, N. Emanuele, P. D. Reaven, F. J. Zieve, J. Marks, S. N. Davis, R. Hayward, S. R. Warren, S. Goldman, M. McCarren, M. E. Vitek, W. G. Henderson and G. D. Huang, VADT Investigators, *N. Engl. J. Med.*, 2009, **360**, 129.
32. A. C. Group, A. Patel, S. MacMahon, J. Chalmers, B. Neal, L. Billot, M. Woodward, M. Marre, M. Cooper, P. Glasziou, D. Grobbee, P. Hamet, S. Harrap, S. Heller, L. Liu, G. Mancia, C. E. Mogensen, C. Pan, N. Poulter, A. Rodgers, B. Williams, S. Bompoint, B. E. de Galan, R. Joshi and F. Travert, *N. Engl. J. Med.*, 2008, **358**, 2560.
33. C. J. Currie, J. R. Peters, A. Tynan, M. Evans, R. J. Heine, O. L. Bracco, T. Zagar and C. D. Poole, *Lancet*, 2010, **375**, 481.
34. American Diabetes Association, *Diabetes Care*, 2014, **37**(Suppl 1), S14.
35. American Diabetes Association, *Diabetes Care*, 2008, **31**(Suppl 1), S55.
36. D. M. Nathan and D. E. R. Group, *Diabetes Care*, 2014, **37**, 9.
37. F. Sanger, *Biochem. J.*, 1945, **39**, 507.
38. F. Sanger, *Biochem. J.*, 1949, **45**, 563.
39. F. Sanger, *Adv. Protein Chem.*, 1952, 7, 1.
40. Genentech, 06 September 1978.
41. http://www.littletree.com.au/dna.htm, Accessed 13 July 2014.
42. T. Kjeldsen, *Appl. Microbiol. Biotechnol.*, 2000, **54**, 277.
43. T. Heise and L. Heinemann, *Curr. Pharm. Des.*, 2001, 7, 1303.
44. G. Grunberger, *Diabetes Obes. Metab.*, 2013, **15**(Suppl 1), 1.
45. B. Zinman, *Diabetes Obes. Metab.*, 2013, **15**(Suppl 1), 6.
46. M. P. Golden, M. Haymond, D. A. Hinnen, D. F. Kruger and O. P. Schumacher, *Diabetes Care*, 1988, **11**, 600.
47. T. Danne, T. Forst, J. Deinhard, L. Rose, E. Moennig and A. Haupt, *J. Diabetes Sci. Technol.*, 2012, **6**, 1392.
48. http://www.fda.gov/NewsEvents/Newsroom/PressAnnouncements/2006/ucm108585.htm.
49. http://www.fda.gov/NewsEvents/Newsroom/PressAnnouncements/ucm403122.htm, Accessed 13 July 2014. [press release].

50. http://www.mannkindcorp.com/Collateral/Documents/English-US/, Accessed 13 July 2014.
51. G. Goomber and R. Kumria, *J. Diabetol.*, 2011, **2**, 1.
52. M. Hultstrom, N. Roxhed and L. Nordquist, *J. Diabetes Sci. Technol.*, 2014, **8**, 453.
53. http://www.fda.gov/downloads/MedicalDevices/DeviceRegulationandGuidance/GuidanceDocuments/UCM259305.pdf, Accessed 13 July 2014.
54. J. Pickup and H. Keen, *Diabetes Care*, 2002, **25**, 593.
55. D. C. Klonoff, *Diabetes Care*, 2005, **28**, 1231.
56. R. W. Samuels, *Clin. Toxicol.*, 1978, **12**, 543.
57. E. F. Pfeiffer, C. Thum and A. H. Clemens, *Horm. Metab. Res.*, 1974, **6**, 339.
58. A. H. Clemens, P. H. Chang and R. W. Myers, *Horm. Metab. Res.*, 1977, (Suppl 7), 23.
59. R. Hovorka, *Diabet. Med.*, 2006, **23**, 1.
60. T. Peyser, E. Dassau, M. Breton and J. S. Skyler, *Ann. N. Y. Acad. Sci.*, 2014, **1311**, 102.
61. E. Renard, *Arq. Bras. Endocrinol. Metabol.*, 2008, **52**, 349.
62. R. Hovorka, K. Kumareswaran, J. Harris, J. M. Allen, D. Elleri, D. Xing, C. Kollman, M. Nodale, H. R. Murphy, D. B. Dunger, S. A. Amiel, S. R. Heller, M. E. Wilinska and M. L. Evans, *Br. Med. J.*, 2011, **342**, d1855.
63. K. Kumareswaran, D. Elleri, J. M. Allen, J. Harris, D. Xing, C. Kollman, M. Nodale, H. R. Murphy, S. A. Amiel, S. R. Heller, M. E. Wilinska, C. L. Acerini, M. L. Evans, D. B. Dunger and R. Hovorka, *J. Diabetes Sci. Technol.*, 2011, **5**, 1352.
64. D. Elleri, J. M. Allen, M. Biagioni, K. Kumareswaran, L. Leelarathna, K. Caldwell, M. Nodale, M. E. Wilinska, C. L. Acerini, D. B. Dunger and R. Hovorka, *Pediatr. Diabetes*, 2012, **13**, 449.
65. D. Elleri, J. M. Allen, K. Kumareswaran, L. Leelarathna, M. Nodale, K. Caldwell, P. Cheng, C. Kollman, A. Haidar, H. R. Murphy, M. E. Wilinska, C. L. Acerini, D. B. Dunger and R. Hovorka, *Diabetes Care*, 2013, **36**, 838.
66. S. J. Russell, F. H. El-Khatib, M. Sinha, K. L. Magyar, K. McKeon, L. G. Goergen, C. Balliro, M. A. Hillard, D. M. Nathan and E. R. Damiano, *N. Engl. J. Med.*, 2014, **371**, 313–325.
67. Diabetes Control and Complications Trial/Epidemiology of Diabetes Interventions and Complications Research Group, D. M. Nathan, B. Zinman, P. A. Cleary, J. Y. Backlund, S. Genuth, R. Miller and T. J. Orchard, *Arch. Intern. Med.*, 2009, **169**, 1307.
68. M. Calvert, A. Shankar, R. J. McManus, H. Lester and N. Freemantle, *Br. Med. J.*, 2009, **338**, b1870.
69. R. A. Heptulla, L. M. Rodriguez, L. Bomgaars and M. W. Haymond, *Diabetes*, 2005, **54**, 1100.

70. H. P. Chase, K. Lutz, R. Pencek, B. Zhang and L. Porter, *J. Pediatr.*, 2009, **155**, 369.
71. R. A. Heptulla, L. M. Rodriguez, K. J. Mason and M. W. Haymond, *J. Clin. Endocrinol. Metab.*, 2009, **94**, 1608.
72. S. A. Weinzimer, J. L. Sherr, E. Cengiz, G. Kim, J. L. Ruiz, L. Carria, G. Voskanyan, A. Roy and W. V. Tamborlane, *Diabetes Care*, 2012, **35**, 1994.
73. C. M. Kishiyama, P. L. Burdick, E. C. Cobry, V. L. Gage, L. H. Messer, K. McFann and H. P. Chase, *Pediatrics*, 2009, **124**, 1344.
74. R. G. Thompson, J. Peterson, A. Gottlieb and J. Mullane, *Diabetes*, 1997, **46**, 632.
75. F. Whitehouse, D. F. Kruger, M. Fineman, L. Shen, J. A. Ruggles, D. G. Maggs, C. Weyer and O. G. Kolterman, *Diabetes Care*, 2002, **25**, 724.
76. K. Herrmann, J. P. Frias, S. V. Edelman, K. Lutz, K. Shan, S. Chen, D. Maggs and O. G. Kolterman, *Postgrad. Med.*, 2013, **125**, 136.
77. R. S. Sherwin, M. Fisher, R. Hendler and P. Felig, *N. Engl. J. Med.*, 1976, **294**, 455.
78. P. Raskin and R. H. Unger, *N. Engl. J. Med.*, 1978, **299**, 433.
79. K. Kazafeos, *Diabetes Res. Clin. Pract.*, 2011, **93**(Suppl 1), S32.
80. S. Taheri, M. A. Ghatei and S. R. Bloom, in *Endocrinology, Adult and pediatric*, ed. J. L. Jameson and L. J. De Groot, Elsevier Saunders, Philadelphia, 6th edn, 2010, p. 2759.
81. D. J. Drucker and M. A. Nauck, *Lancet*, 2006, **368**, 1696.
82. J. J. Holst, *Physiol. Rev.*, 2007, **87**, 1409.
83. V. S. Raman, K. J. Mason, L. M. Rodriguez, K. Hassan, X. Yu, L. Bomgaars and R. A. Heptulla, *Diabetes Care*, 2010, **33**, 1294.
84. A. N. Traina, M. E. Lull, A. C. Hui, T. M. Zahorian and J. Lyons-Patterson, *Can. J. Diabetes*, 2014, **38**, 269.
85. E. M. Wright, D. D. Loo and B. A. Hirayama, *Physiol. Rev.*, 2011, **91**, 733.
86. C. S. Hummel, C. Lu, D. D. Loo, B. A. Hirayama, A. A. Voss and E. M. Wright, *Am. J. Physiol. Cell. Physiol.*, 2011, **300**, C14.
87. http://www.lexgen.com/news/press-releases/2321-lx4211-achieves-positive-results-in-type-1-diabetes-clinical-trial.html, Accessed June 27, 2014. [press release].
88. B. A. Perkins, D. Z. Cherney, H. Partridge, N. Soleymanlou, H. Tschirhart, B. Zinman, N. M. Fagan, S. Kaspers, H. J. Woerle, U. C. Broedl and O. E. Johansen, *Diabetes Care*, 2014, **37**, 1480.
89. E. M. Lamos, L. M. Younk and S. N. Davis, *Expert Opin. Invest. Drugs*, 2014, **23**, 875.
90. K. J. Nadeau, K. Chow, S. Alam, K. Lindquist, S. Campbell, K. McFann, G. Klingensmith and P. Walravens, *Pediatr. Diabetes*, 2014, 1.
91. G. Libby, L. A. Donnelly, P. T. Donnan, D. R. Alessi, A. D. Morris and J. M. Evans, *Diabetes Care*, 2009, **32**, 1620.
92. S. Ali and V. Fonseca, *Expert Opin. Pharmacother.*, 2012, **13**, 1797.
93. I. Pernicova and M. Korbonits, *Nat. Rev. Endocrinol.*, 2014, **10**, 143.
94. C. J. Bailey and C. Day, *Diabetes Care*, 1989, **12**, 553.

95. C. J. Bailey and R. C. Turner, *N. Engl. J. Med.*, 1996, **334**, 574.
96. G. Rena, E. R. Pearson and K. Sakamoto, *Diabetologia*, 2013, **56**, 1898.
97. C. J. Dunn and D. H. Peters, *Drugs*, 1995, **49**, 721.
98. P. J. Pentikainen, P. J. Neuvonen and A. Penttila, *Eur. J. Clin. Pharmacol.*, 1979, **16**, 195.
99. G. T. Tucker, C. Casey, P. J. Phillips, H. Connor, J. D. Ward and H. F. Woods, *Br. J. Clin. Pharmacol.*, 1981, **12**, 235.
100. D. Kirpichnikov, S. I. McFarlane and J. R. Sowers, *Ann. Intern. Med.*, 2002, **137**, 25.
101. L. Blonde, G. E. Dailey, S. A. Jabbour, C. A. Reasner and D. J. Mills, *Curr. Med. Res. Opin.*, 2004, **20**, 565.
102. A. Frid, G. N. Sterner, M. Londahl, C. Wiklander, A. Cato, E. Vinge and A. Andersson, *Diabetes Care*, 2010, **33**, 1291.
103. A. T. Nies, H. Koepsell, K. Damme and M. Schwab, *Handb. Exp. Pharmacol.*, 2011, 105.
104. M. D. Fullerton, S. Galic, K. Marcinko, S. Sikkema, T. Pulinilkunnil, Z. P. Chen, H. M. O'Neill, R. J. Ford, R. Palanivel, M. O'Brien, D. G. Hardie, S. L. Macaulay, J. D. Schertzer, J. R. Dyck, B. J. van Denderen, B. E. Kemp and G. R. Steinberg, *Nat. Med.*, 2013, **19**, 1649.
105. M. R. Owen, E. Doran and A. P. Halestrap, *Biochem. J.*, 2000, **348**(Pt 3), 607.
106. J. Yang and G. D. Holman, *Endocrinology*, 2006, **147**, 2728.
107. B. Viollet, B. Guigas, N. Sanz Garcia, J. Leclerc, M. Foretz and F. Andreelli, *Clin. Sci.*, 2012, **122**, 253.
108. C. Minassian, S. Tarpin and G. Mithieux, *Biochem. Pharmacol.*, 1998, **55**, 1213.
109. L. Yuan, R. Ziegler and A. Hamann, *Chin. Med. J.*, 2002, **115**, 1843.
110. F. Cabreiro, C. Au, K. Y. Leung, N. Vergara-Irigaray, H. M. Cocheme, T. Noori, D. Weinkove, E. Schuster, N. D. Greene and D. Gems, *Cell*, 2013, **153**, 228.
111. A. Giaccari, C. B. Giorda, G. Riccardi, A. De Micheli, G. Bruno, L. Monge and S. Frontoni, *Diabetes Care*, 2012, **35**, e71, author reply e72.
112. D. M. Nathan, J. B. Buse, M. B. Davidson, E. Ferrannini, R. R. Holman, R. Sherwin, B. Zinman, American Diabetes Association and European Association for the Study of Diabetes, *Diabetes Care*, 2009, **32**, 193.
113. K. Johansen, *Diabetes Care*, 1999, **22**, 33.
114. E. Haupt, B. Knick, T. Koschinsky, H. Liebermeister, J. Schneider and H. Hirche, *Diabete Metab.*, 1991, **17**, 224.
115. M. Stumvoll, N. Nurjhan, G. Perriello, G. Dailey and J. E. Gerich, *N. Engl. J. Med.*, 1995, **333**, 550.
116. S. E. Inzucchi, *J. Am. Med. Assoc.*, 2002, **287**, 360.
117. K. L. Cohen and S. Harris, *Diabetes Care*, 1987, **10**, 555.
118. C. L. Aquilante, *Expert Rev. Cardiovasc. Ther.*, 2010, **8**, 359.
119. P. Proks, F. Reimann, N. Green, F. Gribble and F. Ashcroft, *Diabetes*, 2002, **51**(Suppl 3), S368.

120. K. Nagashima, A. Takahashi, H. Ikeda, A. Hamasaki, N. Kuwamura, Y. Yamada and Y. Seino, *Diabetes Res. Clin. Pract.*, 2004, **66**(Suppl 1), S75.
121. C. R. Triggle and H. Ding, *Ther. Adv. Chronic Dis.*, 2014, **5**, 245.
122. S. Bo, A. Castiglione, E. Ghigo, L. Gentile, M. Durazzo, P. Cavallo-Perin and G. Ciccone, *Eur. J. Endocrinol.*, 2013, **169**, 117.
123. K. M. Pantalone, M. W. Kattan, C. Yu, B. J. Wells, S. Arrigain, A. Jain, A. Atreja and R. S. Zimmerman, *Diabetes Obes. Metab.*, 2012, **14**, 803.
124. A. D. Rao, N. Kuhadiya, K. Reynolds and V. A. Fonseca, *Diabetes Care*, 2008, **31**, 1672.
125. R. J. Stevens, R. L. Coleman, A. I. Adler, I. M. Stratton, D. R. Matthews and R. R. Holman, *Diabetes Care*, 2004, **27**, 201.
126. B. R. Zimmerman, *Endocrinol. Metab. Clin. North Am.*, 1997, **26**, 511.
127. C. Kilo, A. Meenan and Z. Bloomgarden, *Clin. Ther.*, 1992, **14**, 801.
128. P. J. Guillausseau and W. Greb, *Diabetes Metab.*, 2001, **27**, 133.
129. G. Schernthaner, A. Grimaldi, U. Di Mario, J. Drzewoski, P. Kempler, M. Kvapil, A. Novials, R. Rottiers, G. E. Rutten and K. M. Shaw, *Eur. J. Clin. Invest.*, 2004, **34**, 535.
130. S. Zoungas, J. Chalmers, A. P. Kengne, A. Pillai, L. Billot, B. de Galan, M. Marre, B. Neal, S. Harrap, N. Poulter and A. Patel, *Diabetes Res. Clin. Pract.*, 2010, **89**, 126.
131. C. C. Quianzon and I. E. Cheikh, *J Community Hosp. Intern. Med. Perspect.*, 2012, **2**, online publication.
132. R. Perfetti and A. Ahmad, *Trends Endocrinol. Metab.*, 2000, **11**, 218.
133. D. Stephan, M. Winkler, P. Kuhner, U. Russ and U. Quast, *Diabetologia*, 2006, **49**, 2039.
134. L. Michalik, J. Auwerx, J. P. Berger, V. K. Chatterjee, C. K. Glass, F. J. Gonzalez, P. A. Grimaldi, T. Kadowaki, M. A. Lazar, S. O'Rahilly, C. N. Palmer, J. Plutzky, J. K. Reddy, B. M. Spiegelman, B. Staels and W. Wahli, *Pharmacol. Rev.*, 2006, **58**, 726.
135. W. W. Cheatham, *Am. J. Clin. Nutr.*, 2010, **91**, 262S.
136. R. A. Defronzo, R. J. Mehta and J. J. Schnure, *Hosp. Pract. (1995)*, 2013, **41**, 132.
137. M. N. Ghazzi, J. E. Perez, T. K. Antonucci, J. H. Driscoll, S. M. Huang, B. W. Faja and R. W. Whitcomb, *Diabetes*, 1997, **46**, 433.
138. R. W. Nesto, D. Bell, R. O. Bonow, V. Fonseca, S. M. Grundy, E. S. Horton, M. Le Winter, D. Porte, C. F. Semenkovich, S. Smith, L. H. Young, R. Kahn, American Heart Association and American Diabetes Association, *Circulation*, 2003, **108**, 2941.
139. D. J. Betteridge, *Diabet. Med.*, 2011, **28**, 759.
140. C. Meier, M. E. Kraenzlin, M. Bodmer, S. S. Jick, H. Jick and C. R. Meier, *Arch. Intern. Med.*, 2008, **168**, 820.
141. R. Mamtani, K. Haynes, W. B. Bilker, D. J. Vaughn, B. L. Strom, K. Glanz and J. D. Lewis, *J. Natl. Cancer Inst.*, 2012, **104**, 1411.
142. E. H. Ryan, Jr., D. P. Han, R. C. Ramsay, H. L. Cantrill, S. R. Bennett, S. Dev and D. F. Williams, *Retina*, 2006, **26**, 562.

143. S. E. Nissen and K. Wolski, *N. Engl. J. Med.*, 2007, **356**, 2457.
144. http://www.fda.gov/downloads/Drugs/GuidanceComplianceRegulatoryInformation/Guidances/ucm071627.pdf.
145. W. Kim and J. M. Egan, *Pharmacol. Rev.*, 2008, **60**, 470.
146. P. J. Larsen and J. J. Holst, *Regul. Pept.*, 2005, **128**, 97.
147. A. Flint, C. Kapitza and M. Zdravkovic, *Diabetes Obes. Metab.*, 2013, **15**, 958.
148. A. Lund, F. K. Knop and T. Vilsboll, *Eur. J. Intern. Med.*, 2014, **25**, 407.
149. Y. Li, T. Hansotia, B. Yusta, F. Ris, P. A. Halban and D. J. Drucker, *J. Biol. Chem.*, 2003, **278**, 471.
150. J. Zhong, X. Rao and S. Rajagopalan, *Atherosclerosis*, 2013, **226**, 305.
151. L. A. Nikolaidis, S. Mankad, G. G. Sokos, G. Miske, A. Shah, D. Elahi and R. P. Shannon, *Circulation*, 2004, **109**, 962.
152. T. Mita and H. Watada, *Cardiovasc. Hematol. Agents Med. Chem.*, 2012, **10**, 309.
153. H. Oeseburg, R. A. de Boer, H. Buikema, P. van der Harst, W. H. van Gilst and H. H. Sillje, *Arterioscler., Thromb., Vasc. Biol.*, 2010, **30**, 1407.
154. R. A. DeFronzo, R. E. Ratner, J. Han, D. D. Kim, M. S. Fineman and A. D. Baron, *Diabetes Care*, 2005, **28**, 1092.
155. J. B. Buse, R. R. Henry, J. Han, D. D. Kim, M. S. Fineman, A. D. Baron and G. Exenatide-113 Clinical Study, *Diabetes Care*, 2004, **27**, 2628.
156. M. Christensen, F. K. Knop, T. Vilsboll and J. J. Holst, *Expert Opin. Invest. Drugs*, 2011, **20**, 549.
157. J. Rosenstock, D. Raccah, L. Koranyi, L. Maffei, G. Boka, P. Miossec and J. E. Gerich, *Diabetes Care*, 2013, **36**, 2945.
158. J. M. Trujillo and W. Nuffer, *Pharmacotherapy*, 2014, **34**, 1174.
159. A. E. Butler, M. Campbell-Thompson, T. Gurlo, D. W. Dawson, M. Atkinson and P. C. Butler, *Diabetes*, 2013, **62**, 2595.
160. S. Singh, H. Y. Chang, T. M. Richards, J. P. Weiner, J. M. Clark and J. B. Segal, *JAMA Intern. Med.*, 2013, **173**, 534.
161. A. G. Egan, E. Blind, K. Dunder, P. A. de Graeff, B. T. Hummer, T. Bourcier and C. Rosebraugh, *N. Engl. J. Med.*, 2014, **370**, 794.
162. L. Hegedus, A. C. Moses, M. Zdravkovic, T. Le Thi and G. H. Daniels, *J. Clin. Endocrinol. Metab.*, 2011, **96**, 853.
163. J. Liu, G. Wang, Y. Jia and Y. Xu, *Diabetes Metab. Res. Rev.*, 2014, Epub.
164. T. Foltynie and I. Aviles-Olmos, *Alzheimers Dement.*, 2014, **10**, S38.
165. A. Barnett, *Int. J. Clin. Pract.*, 2006, **60**, 1454.
166. R. Mentlein, *Regul. Pept.*, 1999, **85**, 9.
167. I. De Meester, C. Durinx, G. Bal, P. Proost, S. Struyf, F. Goossens, K. Augustyns and S. Scharpe, *Adv. Exp. Med. Biol.*, 2000, **477**, 67.
168. I. de Meester, A. M. Lambeir, P. Proost and S. Scharpe, *Adv. Exp. Med. Biol.*, 2003, **524**, 3.
169. T. J. Kieffer, C. H. McIntosh and R. A. Pederson, *Endocrinology*, 1995, **136**, 3585.
170. C. F. Deacon, *Diabetes Obes. Metab.*, 2011, **13**, 7.

171. S. Colagiuri, *Diabetes Obes. Metab.*, 2010, **12**, 463.
172. D. J. Augeri, J. A. Robl, D. A. Betebenner, D. R. Magnin, A. Khanna, J. G. Robertson, A. Wang, L. M. Simpkins, P. Taunk, Q. Huang, S. P. Han, B. Abboa-Offei, M. Cap, L. Xin, L. Tao, E. Tozzo, G. E. Welzel, D. M. Egan, J. Marcinkeviciene, S. Y. Chang, S. A. Biller, M. S. Kirby, R. A. Parker and L. G. Hamann, *J. Med. Chem.*, 2005, **48**, 5025.
173. B. F. Burkey, P. K. Hoffmann, U. Hassiepen, J. Trappe, M. Juedes and J. E. Foley, *Diabetes Obes. Metab.*, 2008, **10**, 1057.
174. I. Brandt, J. Joossens, X. Chen, M. B. Maes, S. Scharpe, I. De Meester and A. M. Lambeir, *Biochem. Pharmacol.*, 2005, **70**, 134.
175. Y. B. Kim, L. M. Kopcho, M. S. Kirby, L. G. Hamann, C. A. Weigelt, W. J. Metzler and J. Marcinkeviciene, *Arch. Biochem. Biophys.*, 2006, **445**, 9.
176. G. A. Herman, P. P. Stein, N. A. Thornberry and J. A. Wagner, *Clin. Pharmacol. Ther.*, 2007, **81**, 761.
177. Y. L. He, B. M. Sadler, R. Sabo, S. Balez, Y. Wang, J. Campestrini, A. Laurent, M. Ligueros-Saylan and D. Howard, *Clin. Pharmacokinet.*, 2007, **46**, 787.
178. A. Fura, A. Khanna, V. Vyas, B. Koplowitz, S. Y. Chang, C. Caporuscio, D. W. Boulton, L. J. Christopher, K. D. Chadwick, L. G. Hamann, W. G. Humphreys and M. Kirby, *Drug Metab. Dispos.*, 2009, **37**, 1164.
179. C. F. Deacon and J. J. Holst, *Expert Opin. Invest. Drugs*, 2010, **19**, 133.
180. H. Fuchs, R. Binder and A. Greischel, *Biopharm. Drug Dispos.*, 2009, **30**, 229.
181. A. J. Scheen, *Diabetes Metab.*, 2012, **38**, 89.
182. R. M. Bergenstal, C. Wysham, L. Macconell, J. Malloy, B. Walsh, P. Yan, K. Wilhelm, J. Malone, L. E. Porter and for the DURATION-2 Study Group, *Lancet*, 2010, **376**, 431.
183. A. Schweizer, A. Couturier, J. E. Foley and S. Dejager, *Diabet. Med.*, 2007, **24**, 955.
184. B. J. Goldstein, M. N. Feinglos, J. K. Lunceford, J. Johnson, D. E. Williams-Herman and Sitagliptin 036 Study Group, *Diabetes Care*, 2007, **30**, 1979.
185. D. Williams-Herman, J. Johnson, R. Teng, E. Luo, M. J. Davies, K. D. Kaufman, B. J. Goldstein and J. M. Amatruda, *Curr. Med. Res. Opin.*, 2009, **25**, 569.
186. D. Williams-Herman, J. Johnson, R. Teng, G. Golm, K. D. Kaufman, B. J. Goldstein and J. M. Amatruda, *Diabetes Obes. Metab.*, 2010, **12**, 442.
187. B. Goke, K. Hershon, D. Kerr, A. Calle Pascual, A. Schweizer, J. Foley, Q. Shao and S. Dejager, *Horm. Metab. Res.*, 2008, **40**, 892.
188. http://cws.huginonline.com/N/134323/PR/200802/1187381_5_2.html.
189. B. Richter, E. Bandeira-Echtler, K. Bergerhoff and C. L. Lerch, *Cochrane Database Syst. Rev.*, 2008, CD006739.
190. M. Ligueros-Saylan, J. E. Foley, A. Schweizer, A. Couturier and W. Kothny, *Diabetes Obes. Metab.*, 2010, **12**, 495.

191. D. Williams-Herman, S. S. Engel, E. Round, J. Johnson, G. T. Golm, H. Guo, B. J. Musser, M. J. Davies, K. D. Kaufman and B. J. Goldstein, *BMC Endocr. Disord.*, 2010, **10**, 7.
192. K. Esposito, P. Chiodini, M. I. Maiorino, G. Bellastella, A. Capuano and D. Giugliano, *BMJ Open*, 2014, **4**, e005442.
193. B. M. Scirica, D. L. Bhatt, E. Braunwald, P. G. Steg, J. Davidson, B. Hirshberg, P. Ohman, D. L. Price, R. Chen, J. Udell and I. Raz, *Am. Heart J.*, 2011, **162**, 818.
194. A. J. Scheen, *Nat. Rev. Cardiol.*, 2013, **10**, 73.
195. A. J. Scheen, *Postgrad. Med.*, 2013, **125**, 7.
196. *Guyton & Hall Textbook of Medical Physiology*, ed. A. C. Guyton and J. E. Hall, Elsevier Saunders, Philadelphia, 11th edn, 2006, p. 327.
197. C. Meyer, J. M. Dostou, S. L. Welle and J. E. Gerich, *Am. J. Physiol. Endocrinol. Metab.*, 2002, **282**, E419.
198. J. E. Gerich, *Diabetes Obes. Metab.*, 2000, **2**, 345.
199. J. R. Ehrenkranz, N. G. Lewis, C. R. Kahn and J. Roth, *Diabetes Metab. Res. Rev.*, 2005, **21**, 31.
200. D. Dimitrakoudis, M. Vranic and A. Klip, *J. Am. Soc. Nephrol.*, 1992, **3**, 1078.
201. W. N. Washburn, *J. Med. Chem.*, 2009, **52**, 1785.
202. A. Merovci, C. Solis-Herrera, G. Daniele, R. Eldor, T. V. Fiorentino, D. Tripathy, J. Xiong, Z. Perez, L. Norton, M. A. Abdul-Ghani and R. A. DeFronzo, *J. Clin. Invest.*, 2014, **124**, 509.
203. T. P. Angelopoulos and J. Doupis, *Adv. Ther.*, 2014, **31**, 579.
204. R. M. Guthrie, *Postgrad. Med.*, 2013, **125**, 21.
205. A. J. Scheen and L. F. Van Gaal, *Lancet Diabetes Endocrinol.*, 2014, **2**(11), 911–922.
206. S. E. Kahn, M. E. Cooper and S. Del Prato, *Lancet*, 2014, **383**, 1068.
207. N. Saeidi, L. Meoli, E. Nestoridi, N. K. Gupta, S. Kvas, J. Kucharczyk, A. A. Bonab, A. J. Fischman, M. L. Yarmush and N. Stylopoulos, *Science*, 2013, **341**, 406.
208. M. A. Stefater, H. E. Wilson-Perez, A. P. Chambers, D. A. Sandoval and R. J. Seeley, *Endocr. Rev.*, 2012, **33**, 595.
209. ORIGIN trial investigators, R. E. Gilbert, J. F. Mann, M. Hanefeld, G. Spinas, J. Bosch, S. Yusuf and H. C. Gerstein, *Diabetologia*, 2014, **57**, 1325.
210. T. Heise, L. Nosek, S. G. Bottcher, H. Hastrup and H. Haahr, *Diabetes Obes. Metab.*, 2012, **14**, 944.
211. B. Zinman, A. Philis-Tsimikas, B. Cariou, Y. Handelsman, H. W. Rodbard, T. Johansen, L. Endahl and C. Mathieu; on behalf of the NN1250-3579 (BEGIN Once Long) Trial Investigators, *Diabetes Care*, 2012, **35**, 2464.
212. A. El Ouaamari, D. Kawamori, E. Dirice, C. W. Liew, J. L. Shadrach, J. Hu, H. Katsuta, J. Hollister-Lock, W. J. Qian, A. J. Wagers and R. N. Kulkarni, *Cell Rep.*, 2013, **3**, 401.

213. S. N. Flier, R. N. Kulkarni and C. R. Kahn, *Proc. Natl. Acad. Sci. U. S. A.*, 2001, **98**, 7475.
214. M. D. Michael, R. N. Kulkarni, C. Postic, S. F. Previs, G. I. Shulman, M. A. Magnuson and C. R. Kahn, *Mol. Cell*, 2000, **6**, 87.
215. M. Pallayova, V. Wilson, R. John and S. Taheri, *Diabetes Care*, 2013, **36**, e97.
216. P. Yi, J. S. Park and D. A. Melton, *Cell*, 2013, **153**, 747.
217. H. Hu, W. Sun, S. Yu, X. Hong, W. Qian, B. Tang, D. Wang, L. Yang, J. Wang, C. Mao, L. Zhou and G. Yuan, *Diabetes Care*, 2014, **37**, 2718.
218. P. D. Cani, *Briefings Funct. Genomics*, 2013, **12**, 381.
219. N. M. Delzenne and P. D. Cani, *Curr. Diab. Rep.*, 2011, **11**, 154.
220. A. Everard and P. D. Cani, *Best Pract. Res. Clin. Gastroenterol.*, 2013, **27**, 73.
221. A. Everard and P. D. Cani, *Rev. Endocr. Metab. Disord.*, 2014, **15**, 189.
222. J. Qin, Y. Li, Z. Cai, S. Li, J. Zhu, F. Zhang, S. Liang, W. Zhang, Y. Guan, D. Shen, Y. Peng, D. Zhang, Z. Jie, W. Wu, Y. Qin, W. Xue, J. Li, L. Han, D. Lu, P. Wu, Y. Dai, X. Sun, Z. Li, A. Tang, S. Zhong, X. Li, W. Chen, R. Xu, M. Wang, Q. Feng, M. Gong, J. Yu, Y. Zhang, M. Zhang, T. Hansen, G. Sanchez, J. Raes, G. Falony, S. Okuda, M. Almeida, E. LeChatelier, P. Renault, N. Pons, J. M. Batto, Z. Zhang, H. Chen, R. Yang, W. Zheng, S. Li, H. Yang, J. Wang, S. D. Ehrlich, R. Nielsen, O. Pedersen, K. Kristiansen and J. Wang, *Nature*, 2012, **490**, 55.

Future Perspectives

CHAPTER 15

Insights and Future Directions

PHILIP V. PEPLOW*[a] AND JAMES D. ADAMS JR[b]

[a] Department of Anatomy, University of Otago, Dunedin, New Zealand; [b] Department of Pharmacology and Pharmaceutical Sciences, School of Pharmacy, University of Southern California, Los Angeles, CA 90089, USA
*Email: phil.peplow@otago.ac.nz

15.1 Genetic, Molecular and Physiological Mechanisms of Obesity

15.1.1 CNS Control of Food Intake

It was proposed in 1953 that inhibitory signals generated in proportion to body fat stores act in the brain to reduce food intake.[1] Thus, when weight loss induced by caloric restriction reduces the level of these inhibitory signals, food intake increases until the energy deficit is corrected. To explain how energy intake is controlled during individual meals, it was proposed that signals generated during a meal, termed "satiety factors", which included peptides secreted from the gastrointestinal tract, provide information to the brain that inhibits feeding and leads to meal termination.[2] Studies have demonstrated that insulin and leptin fulfill criteria as adiposity signals; administration of either peptide hormone directly into the brain reduces food intake whereas deficiency of either hormone increases food intake.[3,4] Different mechanisms underlie the association of insulin and leptin with body fat content. The effect of weight gain to reduce insulin sensitivity seems to explain how insulin but not leptin varies according to body fat stores. As weight increases, insulin secretion must increase in both the basal state and in response to meals to compensate for insulin resistance

RSC Drug Discovery Series No. 45
Cardiovascular and Metabolic Disease: Scientific Discoveries and New Therapies
Edited by Philip Peplow, James Adams and Tim Young

Published by the Royal Society of Chemistry, www.rsc.org

if normal glucose homeostasis is to be maintained.[5] Failure of pancreatic β-cells to achieve this increase in insulin secretion causes hyperglycemia, and probably contributes to the association of type 2 diabetes with obesity. The rate of insulin-stimulated glucose utilization in adipocytes is a key factor linking leptin secretion to body fat mass.[6] Several studies have suggested that leptin has a more important role than insulin in the CNS control of energy homeostasis. Leptin deficiency causes severe obesity, with hyperphagia that persists despite high insulin levels. In contrast, obesity is not caused by insulin deficiency. However, such comparisons are complicated by the critical role that insulin has in promoting both fat storage and leptin synthesis by fat cells. Some cases of human obesity may be due to reduced leptin action in the brain.[7] Resistance to leptin is shown in mice and rats bearing mutant leptin receptors.[8,9]

Several distinct hypothalamic neuropeptide-containing pathways have emerged as candidate mediators of leptin and insulin action in the CNS. Prominent among anabolic effector pathways is a circuit containing neuropeptide Y (NPY). Continuous or repeated central administration of NPY stimulates food intake and leads readily to obesity.[10] Catabolic effector signalling molecules have an opposite set of characteristics. Melanocortins such as α-melanocyte-stimulating hormone (α-MSH), corticotropin-releasing hormone (CRH), thyrotropin-releasing hormone (TRH) promote negative energy balance by suppressing food intake.

It has been proposed that hypothalamic pathways involved in energy homeostasis interact with a distinctly different set of pathways involved in the response to satiety signals.[11,12] While the timing of meal initiation can be influenced by many external and internal variables, meal termination tends to be a more biologically controlled process. Signals involved in energy homeostasis may control food intake primarily by adjusting the size of individual meals. One way that this could be accomplished is by modulating the response to satiety signals in brain areas that process this information. Satiety information generated during the course of a meal is largely conveyed to the hindbrain by means of afferent fibres of the vagus nerve and by afferents passing into the spinal cord from the upper gastrointestinal tract.[13] Such information converges in the nucleus tractus solitarius (NTS), an area of the caudal brainstem that integrates information from the gastrointestinal tract and abdominal viscera, as well as taste information from the oral cavity.[14] Satiety-inducing signals that reach the NTS are initiated by mechanical or chemical stimulation of the stomach and small intestine during food ingestion, neural input related to energy metabolism in the liver, and humoral signals such as cholecystokinin (CCK) that are released upon nutrient stimulation of neuroendocrine secretory cells lining the intestinal lumen.[15,16] The suggestion that forebrain responses to adiposity signals coupled to changes in the size of single meals ultimately involve an interaction with hindbrain areas that control satiety is supported by the ability of both leptin and insulin to enhance the satiating effect of CCK.[17,18]

A more detailed understanding of the pathogenesis of human obesity may ultimately guide treatment of individuals. An increased ability to diagnose the pathophysiological basis of human obesity will have direct applications to its treatment. A multi-drug regimen that targets multiple sites within the weight-regulatory system may be necessary to achieve and sustain weight loss in many individuals.[19]

15.1.2 Brain Reward Mechanisms in Obesity and Genetic Vulnerability

In addition to metabolic systems, brain reward systems also play an important role in feeding behavior.[20,21] Bland tasting foods are not eaten to excess, whereas palatable foods are often consumed even after energy requirements have been met. Ease of access to palatable energy-dense food is a major environmental risk factor for obesity, and overconsumption of palatable food a major factor contributing to the recent surge in obesity.[22] Macronutrients in palatable food can intensely stimulate brain reward systems independent of their caloric value and the motivation to consume palatable food can be extremely high even in the absence of homeostatic energy requirements.[23] The influence of intrinsic or diet-induced alterations on the responsiveness of brain reward systems, and how these contribute to overeating and obesity remains unclear.

Human brain imaging studies have shown that food and food-related visual and olfactory cues activate corticolimbic and mesoaccumbans brain circuits implicated in reward, most notably the orbitofrontal cortex (OFC), insula, amygdala, hypothalamus, striatum, and midbrain regions including the ventral tegmental area (VTA) and substantia nigra (SN).[24] The striatum, insula, anterior cingulate cortex, and midbrain structures encode the subjective value of rewards regardless of their type (*e.g.* food, sex, monetary rewards), consistent with a role for this neuronal network in general hedonic representation.[25] The OFC appears to play a prominent role in representations related to the value of specific types of rewards including palatable food.[25–27] Hunger can enhance palatable food-induced activation of corticolimbic and midbrain regions in humans.[28] The intensity of activation of the ventral striatum, amygdala, insula, and OFC in response to high-calorie palatable food was far greater when human subjects were hungry rather than well fed.[29] Conversely, overfeeding can reduce neuronal responses to palatable food, particularly in the insular cortex and hypothalamus.[30] Thus, the hedonic value of food is influenced by metabolic state, suggesting that regulators of metabolism such as leptin and ghrelin may influence the activity of hedonic systems in the brain. Leptin receptors are expressed on midbrain dopamine neurons in the VTA and SN,[31] suggesting that leptin may influence hedonic aspects of feeding behavior through modulation of mesostriatal dopamine transmission. Leptin infusions into the VTA inhibited the activity of dopamine neurons and decreased food intake in rats.[32] Conversely, knockdown of leptin receptors in the VTA increased food intake, enhanced locomotor activity, and increased preference for palatable food in rats.[32] The hunger-related hormone ghrelin was shown to potentiate the activation of hedonic systems in the brain in response to food cues.[33] Specifically, ghrelin enhanced the activation of OFC, amygdala, insula, striatum, VTA and SN in obese individuals in response to pictures of highly palatable food.[33] Ghrelin stimulates midbrain dopamine systems in rats and mice.[34,35] These findings show that palatable food activates brain reward systems and that hormonal regulators of appetite can influence food intake in part by modulating hedonic responses to food.

Functional MRI in response to imagined intake of palatable foods shows that future increases in body mass can be predicted by weaker brain activation of specific brain areas, particularly in individuals carrying low functioning variants of dopamine receptor genes, such as the DRD2 TaqlA A1 allele or the DRD4-7R allele.[36,37] Among women with seasonal affective disorder, who experienced both carbohydrate craving and weight gain during winter depressive episodes, the 7-repeat allele of DRD4 was associated with childhood dysphoria and both binge eating and obesity in adulthood.[38] The 7R allele of DRD4 influenced macronutrient intakes and specific food choices in children as early as 4 years of age.[39]

It is very striking that similar brain systems and common adaptive responses are triggered in response to consumption of both palatable food and addictive drugs. Overconsumption of palatable food or drugs of abuse induces similar deficits in the responsiveness of brain reward systems. Common genetic vulnerabilities in brain reward systems may predispose individuals to overeating and obesity and also drug use and addiction.[40,41]

15.1.3 Adipocytes as Regulators of Energy Balance and Glucose Homeostasis

Adipocytes integrate a wide array of homeostatic processes that are coordinated mainly through the synthesis and release of peptide hormones, termed adipokines, by these cells. Adipocytes also release fatty acids into the circulation which are used by most organs for fuel when glucose is limiting. These fatty acids are produced by breaking down triacylglycerols, which contain nine times more energy per unit mass than do carbohydrates.

Most mammals have stereotypical depots of adipose tissue located throughout the body. Some of these depots provide mechanical support but contribute relatively little to energy homeostasis *e.g.* fat pads of heels, fingers, toes, and periorbital fat. Other adipocytes exist in loose association with the skin, and are termed the subcutaneous tissue. There are also several distinct depots within the body cavities, surrounding the heart and other organs, associated with the intestinal mesenteries, and retroperitoneally in association with the posterior abdominal wall. Adipocytes and preadipocytes from different depots have different replicative potential, and different responses to hormonal signals. In addition, there is a further distinction between brown and white adipocytes. Brown adipocytes are found only in mammals, and contain less total lipid than white adipocytes and are particularly rich in mitochondria. They differ from the more typical white adipocytes in that they express uncoupling protein-1 (UCP-1), which dissipates the proton gradient across the inner mitochondrial membrane that is produced by the action of the electron transport chain. This generates heat at the expense of ATP. Rodents have a distinct brown fat pad which lies in the interscapular region. In humans, brown adipose tissue surrounds the heart and great vessels in infancy, but tends to disappear over time until only scattered cells are found within white fat pads.[42]

Energy balance in animals is often expressed as a simple equation: Energy intake = energy burned + energy stored.[43] Lipid storage in adipose tissue represents excess energy consumption relative to energy expenditure. Energy homeostasis depends upon the balance between caloric intake and energy expenditure. Although caloric intake is almost entirely due to the consumption of food, energy expenditure has more components including basal metabolism, physical activity, and adaptive thermogenesis. Adipose tissues are critical integrators of energy balance through the regulation of both food intake and energy expenditure.

Leptin was the first adipokine found to have a role in modulating adiposity. Leptin is secreted almost exclusively by fat cells and acts as a major "adipostat" by repressing food intake and promoting energy expenditure. The leptin receptor is expressed at low levels in many tissues, but is present at high levels in the mediobasal hypothalamus, particularly the arcuate nucleus, ventromedial nucleus and dorsomedial nucleus.[44,45] Leptin receptor activation at these sites leads to repression of orexigenic pathways (*e.g.* those involving neuropeptide Y and agouti-related peptide) and induction of anorexigenic pathways (*e.g.* those involving pro-opiomelanocortin, and cocaine and amphetamine-regulated transcript). Another way in which adipocytes can regulate whole body energy balance is through alterations in their own metabolism. Brown adipocytes are highly specialized to perform uncoupled respiration, which dissipates chemical energy in the form of heat. Mice lacking brown adipose tissue exhibit reduced energy expenditure and are prone to diet-induced obesity.[46,47]

Alterations in adiposity have profound implications for glucose homeostasis with too much fat (obesity) and too little fat (lipodystrophy) both associated with insulin resistance and hyperglycemia. In addition to its role in energy balance, leptin has effects on glucose homeostasis. Leptin improves glucose homeostasis in lipodystrophic mice, and in humans with lipodystrophy or congenital leptin deficiency.[48,49] The anti-hyperglycemic action of leptin is mediated through several different organs. Leptin improves insulin sensitivity in muscles and in the liver, functioning in part to reduce intracellular triacylglycerol levels.

Adiponectin given to obese diabetic mice stimulates AMP kinase activity in the liver and skeletal muscle, with profound effects on fatty acid oxidation and insulin sensitivity. Adiponectin had no effect on insulin secretion in pancreatic islets from healthy mice or humans, but increased glucose-stimulated insulin secretion in islets from mice with diet-induced obesity.[50] Interestingly, the anti-diabetic action of thiazolidinediones (TZDs) requires adiponectin, with plasma adiponectin levels being raised through TZD effects on synthesis and secretion.[51,52]

Tumor necrosis factor-α (TNF-α) is another adipokine shown to have effects on glucose homeostasis.[53] TNF-α levels are raised in obesity, and blockade of TNF-α action by biochemical or genetic means restores insulin sensitivity. TNF-α appears to be derived from cell types other than adipocytes themselves, with macrophages being implicated in TNF-α production from murine fat pads.[54]

15.2 Molecular Mechanisms of Obesity-associated Hypertension and Cardiovascular Disease

15.2.1 Mechanisms Linking Obesity to Hypertension

Obese adults have a 3.5-fold increased likelihood of having hypertension, and it had been estimated that 60–70% of hypertension in adults may be directly attributable to adiposity.[43,55] Similarly, the prevalence of hypertension is 3-fold higher in obese children than in non-obese children.[56] Most studies suggest that centrally located body fat is a more important determinant of blood pressure elevation than peripheral body fat in both men and women.[57,58] Visceral obesity increases the risks for insulin resistance and dyslipidemia. Furthermore, insulin resistance and obesity are associated with vascular endothelial dysfunction.[59,60]

Obesity-related hypertension may represent a genetically distinct hypertensive phenotype. For example, some genes associated with adiposity may also contribute to the development of hypertension in overweight and obese individuals.[61] In a relatively isolated French Canadian population, in 120 extended families with at least one sib pair affected with early onset hypertension and/or dyslipidemia, a total genome scan identified a cluster of overlapping quantitative trait loci with significant logarithm (base 10) of odds scores on chromosome 1 for the phenotypes BMI, fasting insulin, leptin, diastolic blood pressure.[62]

Environmental (*e.g.* diet, physical activity, level of "stress"), physiological, and genetic factors influence the impact of obesity on arterial pressure. Putative mechanisms of obesity-related hypertension include the following.

15.2.1.1 Increased Sympathetic Nervous System Activity

Most evidence indicates that sympathetic nervous system activity is increased in obesity, particularly sympathetic activity to the kidney and skeletal muscle.[63,64] Neural activity to skeletal muscle is more closely related to abdominal visceral fat than to total fat mass or abdominal subcutaneous fat.[63] However, hypertension is not an invariable consequence of obesity-related increases of neural activity, with neural activity to the kidney and skeletal muscle being elevated in obese normotensive as well as in obese hypertensive human subjects.[65] There may be multiple causes for activation of the sympathetic nervous system in obesity. Suggested mechanisms include hyperinsulinemia and/or insulin resistance; leptin or other adipokines; renin–angiotensin; lifestyle factors.

15.2.1.2 Sodium Retention

Obesity-related hypertension is associated with renal sodium retention and impaired pressure natriuresis.[66] Increased renal tubular reabsorption of sodium has been attributed to increased sympathetic outflow to the kidney. In addition increased pressures caused by fat surrounding the kidneys and increased abdominal pressure associated with visceral obesity may be responsible for impaired natriuresis. Also increased mineralcorticoid activity may cause impaired pressure-natriuresis, with plasma aldosterone being associated with blood pressure, BMI, waist circumference, and insulin resistance.[67,68]

15.2.1.3 Increased Plasma Renin–Angiotensin

Several reports indicate that plasma renin activity and plasma angiotensin II levels are increased in obesity, possibly as a consequence of increased sympathetic outflow to the kidney.[69,70] In obese hypertensive patients, pharmacologic blockade with angiotensin-converting enzymes (ACEs) or angiotensin II receptor blockers ameliorate hypertension and associated metabolic derangements, and reduce the incidence of type 2 diabetes.[71]

15.2.1.4 Increased Adipose Renin–Angiotensin

Adipose tissue expresses all the components of the renin–angiotensin system. Activation of an adipose renin–angiotensin system is associated with high blood pressure in a model of visceral obesity[72,73] and in adipose tissue from obese hypertensive patients.[74]

15.2.1.5 Impaired Vascular Endothelial Function

Vascular endothelial dysfunction is associated with a number of cardiovascular risk factors, including obesity, insulin resistance, and hypertension.[59,60] Visceral fat is independently linked to impaired vasoreactivity. Weight loss alone improves endothelial function in overweight/obese adults.[75] High fat diets increase palmitate levels in the blood and organs. As obesity increases the body finds alternate methods to store fat, such as ceramide. High ceramide levels increase death from coronary artery disease.[76] Ceramide and its glyco-metabolites are toxic to the kidneys in animals and humans.[77,78] This kidney toxicity can increase blood pressure. Ceramide increases NADPH oxidase activity and reactive oxygen species generation as part of its mechanism of toxicity.[79] Ceramide also causes enothelial nitric oxide synthase dysfunction, decreases NO and causes endothelial dysfunction leading to hypertension.[80]

15.2.1.6 Other Vascular Mechanisms

15.2.1.6.1 Adipokines. Blood pressure and leptin are modestly correlated in normotensive and hypertensive individuals after adjustment for fat mass.[81] Two prospective studies have reported that plasma leptin concentration independently predicts the onset of hypertension.[82,83] Leptin activates the sympathetic nervous system both by centrally mediated effects on the hypothalamus and by local peripheral actions.[84] Increased sympathetic outflow is a possible mechanism by which leptin may increase arterial pressure.

Other adipokines may also affect arterial pressure. Plasma adiponectin levels are lowered in obesity-induced insulin resistance.[85] Some studies suggest that adiponectin may protect against hypertension *via* an endothelial-dependent mechanism.[86] Angiotensinogen is an adipokine that increases blood pressure and smooth muscle cell proliferation in the arteries.

15.2.1.6.2 Insulin Resistance. Insulin resistance may be a link between obesity and hypertension. Obesity is associated with resistance to insulin-stimulated glucose uptake and hyperinsulinemia, and weight loss increases insulin sensitivity.[87] In several rat models of experimental hypertension, hypertension can be ameliorated or prevented by chemical agents that increase insulin sensitivity or have a primary lipid lowering effect (*e.g.* thiazolidinediones, metformin, clofibrate, lovostatin).[88] Possible mechanisms by which insulin resistance and/or hyperinsulinemia may increase blood pressure include an antinatriuretic effect of insulin, increased sympathetic nervous system activity, increased responses to endogenous vasoconstrictors, altered vascular membrane cation transport, impaired endothelium-dependent vasodilatation, and stimulation of vascular smooth muscle growth by insulin.

15.2.2 Mechanisms Linking Obesity to Cardiovascular Disease

Putative mechanisms include the following.

15.2.2.1 Insulin Resistance

Several of the possible mechanisms linking obesity to cardiovascular disease are believed to be related to insulin resistance. In individuals with normal insulin sensitivity, lipolysis of adipose tissue is well regulated, with the release of non-esterified (*i.e.* unsaturated, free) fatty acids (NEFAs) relative to the energy requirement of the different tissues. Insulin resistance in hypertrophic adipocytes leads to increased lipolysis and the release of NEFAs. Other mechanisms could also lead to increased NEFA levels, such as decreased fatty acid oxidation and low levels of adiponectin, which normally activates fatty acid oxidation by turning on AMP-activated protein kinase. Increased levels of NEFAs might affect nitric oxide production, thereby impairing endothelium-dependent vasodilation. They may also increase myocardial oxygen requirements and therefore give rise to ischemia, decrease myocardial contractility and induce cardiac arrhythmias.[89] High NEFA levels may increase reactive oxygen species (ROS) generation in mononuclear cells, and induce insulin resistance in myocytes and hepatocytes.

15.2.2.2 Adipokines

Several adipokines such as leptin, adiponectin, tumor necrosis factor-α (TNF-α), interleukin-6 (IL-6), resistin, visfatin, and retinol binding protein 4 (RBP4),[90] have been suggested to be associated with insulin resistance. Leptin has atherogenic and growth properties. Obesity in humans is associated with elevated leptin levels. There may be a direct link between hyperleptinemia and increased cardiovascular disease risk. Leptin may enhance platelet aggregation and arterial thrombosis,[91] promote angiogenesis, impair arterial distensibility and induce proliferation and migration of vascular smooth muscle cells (VSMCs).[92,93]

Adiponectin has important anti-atherogenic, anti-diabetic and anti-inflammatory properties, and is decreased in obesity, diabetes and other insulin-resistant states. Co-culture with visceral fat cells inhibits adiponectin secretion from subcutaneous adipocytes, suggesting that some inhibiting factors for adiponectin synthesis or secretion are secreted from visceral adipose tissue. The negative correlation between visceral adiposity and adiponectin levels might be explained by the increased secretion of TNF-α from accumulated visceral fat. Adiponectin specifically increases the tissue inhibitor of metalloproteinase-1 in macrophages though IL-10 expression[94] and selectively suppresses endothelial cell apoptosis.[95] Adiponectin may protect plaque rupture by the inhibition of matrix metalloproteinase function. Local adiponectin treatment suppresses the development of atherosclerosis in part by attenuating the expression of adhesion molecules, vascular cell adhesion molecule-1 (VCAM-1) and intercellular adhesion molecule-1 (ICAM-1) in vascular walls.[96]

A number of adipokines might originate from non-adipose cells in adipose tissue, in particular macrophages, and these might include atherogenic cytokines.[97] Different factors could induce macrophage infiltration and activation in white adipose tissue, such as adipocyte hypertrophy and hyperplasia, secretion of monocyte chemoattracant protein-1 (MCP-1), or local action of leptin or adiponectin.[98] Macrophages are found more frequently in visceral compared to subcutaneous adipose tissue[99] and are presumably recruited through the secretion of chemotactic molecules by adipose tissue.

15.2.2.3 Oxidative Stress

Oxidative stress is a potential pathogenic mechanism linking obesity and insulin resistance with endothelial dysfunction.[100] Several insulin resistance syndrome components predict high levels of oxidized LDL. Elevated oxidized LDL has been shown to predict myocardial infarction in well-functioning elderly individuals, even after adjusting for age, gender, race, smoking, metabolic syndrome.[101] A decrease in oxidative stress occurred in obese subjects after dietary restriction and weight loss.[102]

Oxidative stress is considered to be the common factor underlying insulin resistance, type 2 diabetes and cardiovascular disease, and may explain the presence of inflammation in all of these conditions. Release of IL-6, mainly from abdominal adipocytes, might have a pivotal role in the relationship between oxidative stress and endothelial dysfunction.

IL-6 not only contributes to C-reactive protein (CRP) elevation and low grade inflammation, but also has a close relationship with coagulation, insulin resistance, dyslipidemia and endothelial dysfunction. Proinflammatory cytokines such as TNF-α and IL-6 are acutely increased by hyperglycemia[103] and by NEFAs.[104]

15.3 Molecular Mechanisms of Obesity-associated Diabetes

15.3.1 Mechanisms Linking Obesity to Diabetes

The most critical factor in the development of metabolic disease is obesity. Adipose tissue modulates metabolism by releasing NEFAs and glycerol, hormones including leptin and adiponectin, and proinflammatory cytokines.[105,106] Retinol binding protein-4 (RBP4) induces insulin resistance through reduced phosophatidylinositol-3-OH kinase (PI3K) signaling in muscle and increased expression of the gluconeogenic enzyme phosphoenolpyruvate carboxylase in the liver through a retinol-dependent mechanism.[107] By contrast, adiponectin acts as an insulin sensitizer, stimulating fatty acid oxidation in an AMP-activated protein kinase (AMPK) and peroxisome proliferator activated receptor-α (PPAR-α)-dependent manner.[108] In addition to adipocyte-derived factors, increased release of TNF-α, IL-6, monocyte chemoattractant protein-1 (MCP-1) and additional products of macrophages and other cells present in adipose tissue might also have an effect in the development of insulin resistance.[109,110]

The release of NEFAs may be the most critical factor in modulating insulin sensitivity. Increased NEFA levels are observed in obesity and type 2 diabetes and are associated with the insulin resistance observed in both.[111,112] Insulin resistance develops within hours of an acute increase in plasma NEFA levels in human subjects.[113] Conversely, insulin-mediated glucose uptake and glucose tolerance improve with an acute decrease in NEFA levels after treatment with the antilipolytic agent Acipimox.[114] The delivery of NEFAs to the tissues might be modulated by their source. Intra-abdominal fat is more lipolytic than subcutaneous fat and is also less sensitive to the anti-lipolytic effect of insulin.[115] Acipimox is limited by liver toxicity and the induction of gout. When palmitate levels increase, ceramide production increases. As described previously, ceramide causes endothelial nitric oxide synthase dysfunction and increases NADPH oxidase activity. Both of these mechanisms increase reactive oxygen species generation. Increased muscle ceramide in humans is associated with increased insulin resistance.[116] Ceramide causes hepatic insulin resistance through a cannabinoid-1 receptor mechanism.[117] Ceramide in the pancreas increases apoptosis of β-cells and insulin resistance.[118]

Insulin sensitivity modulates β-cell function and is almost always decreased in obesity. Insulin-resistant individuals, whether lean or obese, have greater insulin responses and lower hepatic insulin clearance than insulin-sensitive individuals. In healthy individuals there is a feedback loop between the insulin-sensitive tissues and the β-cells, with β-cells increasing insulin supply in response to demand by the liver, muscles and adipose tissue.[5] For glucose tolerance to remain unchanged, changes in insulin sensitivity must be matched by a proportionate yet opposite change in circulating insulin levels. Failure of this feedback loop results in impaired glucose tolerance and underlies the development of diabetes.

NEFAs are important for normal β-cell function and potentiate insulin release in response to glucose and non-glucose secretagogues.[119] NEFAs regulate insulin secretion from β-cells by binding to the G-protein-coupled receptor GPR40 on the cell membrane, resulting in the activation of intracellular signaling and a subsequent increase in intracellular calcium and secretory granule exocytosis.[120] NEFAs may enhance the production of fatty acyl-CoA which increases insulin release both by directly stimulating secretory granule exocytosis and by PKC activation.[121]

Oxidative stress plays a key role in the pathogenesis of insulin resistance and β-cell dysfunction.[122] There is a bulk of evidence demonstrating that mitochondrial reactive oxygen species (ROS) (predominantly superoxide anion) overproduction is involved in diabetes and diabetic complications.[123] Earlier work suggested that glucose can directly stimulate ROS overproduction but it was later shown that high glucose activates various enzymatic cascades in mitochondria, including activation of NADPH oxidase, uncoupling of NO synthases and stimulation of xanthine oxidase.[124–126] Glycated proteins can also be the promoters of ROS formation.[127] Under physiological conditions, insulin signals through a signaling cascade that includes insulin or insulin-like growth factor (IGF)-1, insulin receptor (IR), insulin receptor substrate (IRS)-1 and phosphatidylinositol-3 kinase (PI3K)/Akt or ERK kinases. ROS overproduction can disturb this process at different stages, resulting in insulin resistance or increased insulin signal.

Oxidative stress with ensuing glucotoxicity and lipotoxicity are diabetes-related phenomena that have been involved in the pathogenesis of β-cell dysfunction.[128] Thus, hyperglycemia and hyperlipidemia that follow the primary pathogenic process of diabetes may exert additional toxic effects on β-cells. Hyperglycemia may be a prerequisite for the negative effects of lipotoxicity.

Mitochondria can be a source of TNF-α-induced ROS production in cells and this may contribute to the pathogenesis of TNF-α-induced insulin resistance. Increased production of TNF-α has been widely associated with obesity-related insulin resistance and abnormal vascular reactivity, the vasculature being an important target of TNF-α.[129] Circulating TNF-α may impair vascular function by altering the balance between endothelial-derived vasodilator and vasoconstrictor substances because it downregulates the expression of eNOS and upregulates endothelin-1 production in endothelial cells.[130,131] TNF-α may also directly activate NAD(P)H oxidase and increase the production of reactive oxygen species (ROS) in endothelial and vascular smooth muscle cells.[132] In addition, TNF-α might induce vascular dysfunction indirectly through stimulation of lipolysis, hence resulting in increased release of NEFAs. In healthy individuals, infusion of TNF-α inhibits the stimulating effect of insulin on glucose uptake and endothelium-dependent vasodilation.[133] Weight loss in healthy premenopausal obese women over one year results in a significant reduction in circulating TNF-α and IL-6 levels with parallel amelioration of endothelial dysfunction.[134]

15.4 Molecular Mechanisms in Exercise-induced Amelioration of Disease

Exercise is well known to alter blood levels of adipokines. Even in ultra-endurance athletes, resistin and leptin levels change after races.[135] Exercise alone in obese patients has been shown to increase adiponectin and visfatin, decrease blood levels of leptin, retinol binding protein-4, chemerin, interleukin-6, melanin concentrating hormone, ghrelin, C-reactive protein and resistin.[136–142] The meaning of these changes is difficult to decipher. The point is that exercise helps the body to alter its production of adipokines,[143] perhaps for the better.

Weight loss, with or without exercise, also alters blood adipokine levels. The levels of adiponectin increase, whereas decreases are found in leptin, interleukin-15, interleukin-18, monocyte chemotactic protein-1, macrophage inflammatory protein-1alpha, plasminogen activator inhibitor-1, E-selectin and apelin.[144–147] These changes probably demonstrate that weight loss and exercise help the body to alter its production of adipokines in order to improve health.

One of the major unanswered questions in all of this is the role of muscle in health during obesity and during recovery from obesity. Myokines are produced by muscle and are not well characterized. Irisin appears to be a myokine and is also an adipokine.[148,149] Irisin can increase the ability of subcutaneous fat to generate heat. Muscle expression of interleukin-6 increases with diet and exercise.[147] The ability of muscle to respond to adiponectin and insulin decreases during high fat diets and obesity.[150]

Muscle is known to be a major site of clearance of blood insulin. High fat diets decrease insulin clearance in the muscle. However, moderate diets and exercise programs do not routinely improve insulin sensitivity in obese patients.

It is known that prolonged regimens of diet and exercise improve insulin sensitivity, type 2 diabetes symptoms, heart disease symptoms, decrease the progression of arthritis and improve survival of cancer.[151] As stated in Chapter 4, when the body is in balance, the body can heal itself. There are still many questions left unanswered about the mechanisms involved in how the body heals itself.

15.5 Development of New Drugs and Therapies for Treatment of Obesity and Obesity-linked Diseases

15.5.1 Increasing Energy Expenditure through Thermogenesis

For a treatment strategy to have a significant impact on body weight, it must reduce energy intake, increase energy expenditure, or both. There is mounting evidence that mature white adipocytes, including human fat cells, can be converted into brown fat-like adipocytes, and that the typical fatty acid storage phenotype of white adipocyte can be altered towards a fat utilization phenotype. These findings open up new opportunities for the development of drugs for obesity and its metabolic and cardiovascular complications.[152,153] The main function of brown adipose tissue (BAT) is to burn lipids to release heat through the action of uncoupling protein-1 (UCP-1).[154] BAT is specialized in adaptive thermogenesis, the part of energy expenditure induced by cold exposure or diet.[155] Cold exposure of healthy human volunteers increases circulating irisin and fibroblast growth factor 21 (FGF21). The induction of irisin secretion is proportional to shivering intensity, in magnitude similar to exercise-stimulated secretion.[156] Irisin is an exercise-induced myokine that is secreted into the circulation following proteolytic cleavage from its cellular form, fibronectin-type III domain-containing 5 (FNDC5).[157] It reverses diet-induced obesity and diabetes by stimulating thermogenesis in rodents through increasing brown adipocyte-like cell abundance within white fat.[158] FGF21 is a recently identified brown adipokine that stimulates a nonshivering thermogenesis response in human individuals.[159,160] Cold-induced thermogenesis comprises nonshivering thermogenesis and shivering thermogenesis, the latter being stimulated by irisin. FNDC5 and/or FGF21 treatment upregulates human adipocyte brown fat gene/protein expression and thermogenesis in a depot-specific manner.[156] FGF21 increases expression of UCP-1 and other thermogenic genes in fat tissues.[161] The use of peroxisome proliferator-activated receptor γ (PPARγ) activators is also an efficient means of inducing 'browning' in white fat depots and even in human white adipocytes. Treatment of white adipocytes with the potent PPARγ-ligand rosiglitazone, besides promoting adipocyte differentiation, also leads to marked UCP-1 gene expression.[162]

The β_3-adrenergic receptor (β_3-AR) is expressed on the surface of both white and brown adipocytes,[163] and plays a significant role in regulating lipolysis and thermogenesis in rodent and human adipocyte tissues.[164] Like other β-ARs, β_3-AR belongs to the G-protein-coupled receptor (GPCR) family. It is reported that the stimulation of β_3-AR would induce a variety of pharmacological effects such as an increase in fat oxidation, enhancement of energy expenditure, and improvement of insulin-mediated glucose uptake. For these reasons, β_3-AR has been developed as a therapeutic candidate target for obesity and type 2 diabetes.[165,166]

15.5.2 Activating PPARγ in Adipose Tissue and Upregulating Adiponectin

Adipokines and adipokine receptors represent potential therapeutic targets to combat obesity-linked diseases characterized by inflammation and insulin resistance. Most adipokines are proinflammatory, such as leptin, TNF-α, and IL-6. One prominent exception is adiponectin, an anti-inflammatory adipokine that promotes insulin sensitization and protects cardiovascular tissue from ischemic injury.[167,168] Therapeutic strategies may include the upregulation of circulating adiponectin levels, the upregulation of adiponectin receptors, or the development of adiponectin receptor agonists.[108] Circulating adiponectin levels are upregulated by TZDs.[169,170] TZDs improve insulin sensitivity in animal models of insulin resistance and diabetes by enhancing glucose disposal in skeletal muscle and suppressing gluconeogenesis in the liver. It is proposed that TZDs ameliorate insulin resistance by binding to and activating PPARγ in adipose tissue, thereby promoting adipocyte differentiation and increasing the number of small adipocytes that are sensitive to insulin.[171,172] TZDs may upregulate adiponectin by generating small adipocytes that abundantly express and secrete adiponectin[171,173] and/or directly activating adiponectin gene transcription.[174] Since adiponectin is an insulin-sensitizing adipokine, it seems likely that the action whereby TZDs increase insulin sensitivity is mediated, at least in part, by increased circulating adiponectin levels. Although TZDs are a widely used class of anti-diabetic drugs, a significant proportion of all patients fail to show improvements in insulin sensitivity.[175] However, adiponectin increases in the majority of patients given TZDs, suggesting a disconnection between the induction of adiponectin and improvements in insulin sensitivity. It appears that the HMW form of adiponectin may be responsible for the insulin sensitizing effects of TZDs whereas the LMW form may even act as an antagonist.[176] TZDs increase body weight and heart disease.[177] Rosiglitazone was removed from the European market in 2010 for this reason.

15.5.3 Inhibiting Adipocyte Inflammation by PAR-2 Antagonist

During obesity, elevated concentrations of prothrombin factors such as plasminogen activator inhibitor-1 (PAI-1), tissue factor, proteases including mast cell tryptase, cathepsins, caspases and kallikreins, are implicated in metabolic and cardiovascular complications.[178–181] In addition, multiple G-protein-coupled receptors (GPCRs) regulate obesity, immune cell-dependent inflammation in adipose tissue, and metabolic functions in organs and cells.[182] Extracellular serine proteases, such as trypsin, tryptase, tissue factor VIIa, and kallikreins, activate protease-activated receptor 2 (PAR2) which belongs to a unique class of GPCRs that are N-terminally truncated by proteases to enable coupling and activation of intracellular G-protein signaling cascades.[183]

PAR2 activation is generally proinflammatory.[183] The expression of PAR2 is increased in adipose tissue of obese individuals, is stimulated in macrophages by dietary fatty acids, and is inhibited by a PAR2 antagonist, GB88 (5-isoxazoyl-Cha-Ile-spiroindene-1,4-piperidine).[184,185] Oral treatment of obese rats with GB88 (10 mg kg^{-1} d^{-1}) attenuates PAR2 signaling in adipose tissue and inhibits adipose inflammation, insulin resistance, diet-induced obesity, and cardiovascular abnormalities.[185] PAR2 may be a valuable new biomarker for metabolic dysfunction and PAR2 antagonism can be an effective intervention for treating obesity and metabolic dysfunction.

15.5.4 Blocking Interaction of Rab5 Family GTPases to $p75^{NTR}$

Glucose transporter 4 (GLUT4) is the principal insulin-stimulated glucose transporter expressed primarily in adipose tissue and skeletal muscle.[186] Insulin stimulates glucose uptake by inducing GLUT4 translocation to the plasma membrane.[187] GLUT4 trafficking from intracellular compartments to the plasma membrane is controlled by a number of small guanosine triphosphate hydrolases (GTPases), including Rab5 and its family member Rab31.[187,188] p75 Neurotrophin receptor ($p75^{NTR}$) is a member of the TNF receptor superfamily.[189] $p75^{NTR}$ is expressed in the nervous system and in adult nonneuronal tissues such as white adipose tissue and skeletal muscle.[190,191] Because $p75^{NTR}$ is expressed in white adipose tissue and skeletal muscle and regulates GTPase activity, it is hypothesized that $p75^{NTR}$ might participate in the regulation of glucose metabolism. Adipocytes from $p75^{NTR}$ knockout mice show increased Rab5 and decreased Rab31 activities, and dominant negative Rab5 rescues the increase in glucose uptake seen in $p75^{NTR}$ knockout adipocytes.[192] It is suggested that signaling from $p75^{NTR}$ to Rab5 family GTPases may represent a unique therapeutic target for insulin resistance and diabetes. Peptide inhibitors may be designed to block the interaction of the Rab5 family GTPases to helix 4 of $p75^{NTR}$ and thus increase glucose uptake as novel therapeutic approaches for insulin resistance.[192]

15.6 Future Directions

15.6.1 Healthy Lifestyle and Diet

The primary causes of obesity and its inflammatory complications are generally nutritional and lifestyle factors such as overeating and physical inactivity, and treatment of obesity and obesity-related diseases should first be directed at correcting these factors. In both developed and developing countries, insufficient daily physical exercise and an unhealthy choice of foods especially from fast food restaurants are important factors giving rise to an increased prevalence of obesity and metabolic disease.

Sedentary behaviours which include television (TV) watching, video game play, and recreational computer use have been assessed for their effects on acute energy intake from food and drink in children, adolescents and adults.[193] Energy intake in children was greater during TV watching than computer use, with no difference in energy intake between computer use and video game play.[194,195]

A high content of sugar in soft drinks and energy drinks is a major contributor to childhood obesity. The World Health Organization (WHO) has recommended that adults should eat less than the equivalent of six teaspoons of sugar a day if they are to avoid health risks such as weight gain and tooth decay linked to sugary diets. WHO has reiterated a strong recommendation that sugar should account for no more than 10% of total energy intake. It is estimated that obesity affects 0.5 billion people worldwide.[196]

Just as special programmes have been set up to reduce smoking and drug addiction, special initiatives are required to curb overeating and promote healthy food choices and daily exercise. This requires special educational initiatives aimed at children in schools and adults and families at home, with access to psychological and social support services.

15.6.2 Preclinical and Clinical Studies

Future studies using knockout animals/cells or transgenic animals will be directed towards identifying molecular pathways to be targeted by therapeutic intervention for obesity and metabolic disease.

Translational and clinical research will lead the development of more effective therapies having greater patient acceptance and compliance. It will be important to identify the genetic phenotype for obesity, cardiovascular disease and diabetes, so that children with genetic susceptibility to obesity and metabolic disease can be identified early and treatment initiated. Recent advances in genetic and molecular biology techniques heralded the promise of better biomarkers for disease diagnosis and treatment decisions, but translation of these research techniques to clinical practice has been slow. The scientific evidence base for pharmacogenetics continues to expand and the use of such information in the clinical management of patients and drug design/development becomes increasingly important. Genetic information provides accurate and affordable diagnosis and prognosis, targeted treatments, and monitoring solutions. The US Food and Drug Administration (FDA) has advised that the incorporation of genetic testing into clinical trials should be a routine consideration at the start of a drug development program.[197] Treatment of patients with early indications of disease (*e.g.* prediabetes) should be undertaken as soon as possible to retard or prevent disease progression.

15.6.3 Modifying Circulating Levels of Adipokines

Adiponectin is an adipokine that exerts a potent insulin-sensitizing effect by binding to its receptors AdipoR1 and AdipoR2, leading to the activation of 5^{I} AMP-activated protein kinase. Circulating levels of adiponectin, especially HMW adiponectin, are positively correlated with insulin sensitivity and altered by various genetic and environmental factors, pathological conditions and medications.[108] Monitoring the levels of HMW adiponectin is a good predictable marker for type 2 diabetes and the metabolic syndrome, which is the coexistence of obesity, hypertension, coronary artery disease, and type 2 diabetes. Methods to increase adiponectin levels such as TZD administration are effective for treatment of these diseases. Enhancing or mimicking adiponectin action through modulation of expression and/or function of adiponectin receptors may be a novel and promising therapeutic approach for insulin resistance, type 2 diabetes, and the metabolic syndrome.[108]

Omega-3 fatty acid dietary supplementation is known to increase blood adiponectin levels.[198–200] These studies show that resistin levels decrease after omega-3 fatty acid supplements. Leptin levels decrease in thin people and increase in obese people after supplementation.

Only rare cases of severe early childhood obesity have been associated with leptin deficiency.[201] The remainder of the obese population typically have elevated leptin levels.[202] It has been suggested that it might be possible to control the amount of bioavailable circulating leptin, and hence prevent leptin-induced inflammation, by means of a soluble, high affinity leptin-binding molecule analogous to the soluble TNF receptors used to treat rheumatoid arthritis.[203] A proposal of nanobodies (a unique form of antibodies characterized by a single antigen-binding domain that generally do not cross the blood–brain barrier) may lead to an antagonist that could selectively inhibit peripheral actions of leptin.[204] This form of leptin antagonist could be clinically useful as it could target peripheral adverse effect of leptin without inducing central weight gain.[202]

Electroacupuncture treatment of women with BMI 30 to 40 kg m^{-2} performed three times a week at selected ear and body acupuncture points was an effective therapy for obesity with a 4.8% reduction in body weight following 20 days of therapy.[205] In a study using high fat diet-induced obese rats, electroacupuncture given three times a week for four weeks at selected body acupuncture points led to a reduction of body weight, a decrease in plasma leptin levels, and an increase in leptin receptor expression in the hypothalamus.[206] Moreover a recent study of repeated application of electroacupuncture using selected body acupuncture points to obese rats with type 2-like diabetes was effective in reducing blood glucose levels[207] and in lowering leptin levels in blood serum and white adipose tissue.[208] While it did not alter adiponectin level in white adipose tissue, the ratio of adiponectin : leptin in adipose tissue was increased.[208] Thus, regulating the expression of leptin might be one of the molecular mechanisms underlying the reduction of body weight in diet-induced obese rats and ameliorating hyperglycemia in obese diabetic rats.

References

1. G. C. Kennedy, *Proc. R. Soc. London, Ser. B*, 1953, **140**, 578.
2. J. Gibbs, R. C. Young and G. P. Smith, *J. Comp. Physiol. Psychol.*, 1973, **84**, 488.
3. Y. Zhang, R. Proenca, M. Maffe, M. Barone, L. Leopold and J. M. Friedman, *Nature*, 1994, **372**, 425.
4. A. J. Sipols, D. G. Bsaskin and M. W. Schwartz, *Diabetes*, 1995, **44**, 147.
5. S. E. Kahn, R. L. Prigeon, D.K. McCulloch, E. J. Boyko, R. N. Bergman, M.W. Schwartz, J. L. Neifing, W. K. Ward, J. C. Beard, J.P. Palmer and D. Porte, *Diabetes*, 1993, **42**, 1663.
6. W. M. Mueller, F. M. Gregoire, K. L. Stanhope, C. V. Mobbs, T. M. Mizuno, C. H. Warden, J. S. Stern and P. J. Havel, *Endocrinology*, 1998, **139**, 551.
7. R. V. Considine, M. K. Sinha, M. L. Heiman, A. Kriauciunas, T. W. Stephens, M. R. Nyce, J. P. Ohannesian, C. C. Marco, L. J. McKee, T. L. Bauer *et al.*, *N. Engl. J. Med.*, 1996, **334**, 292.
8. L. A. Campfield, F. J. Smith, Y. Guisez, R. Devos and P. Burn, *Science*, 1995, **269**, 546.
9. S. C. Chua, W. K. Chung, X. S. Wu-Peng, Y. Zhang, S. M. Liu, L. Tartaglia and R. L. Leibel, *Science*, 1996, **271**, 994.
10. N. Zarjevski, I. Cusin, R. Vettor, F. Rohner-Jeanrenaud and B. Jeanrenaud, *Endocrinology*, 1993, **133**, 1753.
11. S. C. Woods, R. J. Seeley, D. Porte and M. W. Schwartz, *Science*, 1998, **280**, 1378.
12. D. P. Figlewicz, L. J. Stein, D. West, D. Porte and S. C. Woods, *Am. J. Physiol.*, 1986, **250**, R856.
13. S. Rittter, T. Dinh and M. Friedman, *Brain Res.*, 1994, **646**, 53.
14. J. B. Travers, S. P. Travers and R. Norgren, *Annu. Rev. Neurosci.*, 1987, **10**, 595.
15. M. I. Friedman, R. B. Harris, H. Ji, I. Ramirez and M. G. Tordoff, *Am. J. Physiol.*, 1999, **276**, R1046.
16. T. H. Moran and G. J. Schwartz, *Crit. Rev. Neurobiol.*, 1994, **9**, 1.
17. C. A. Matson and R. C. Ritter, *Am. J. Physiol.: Regul., Integr. Comp. Physiol.*, 1999, **276**, R1038.

18. M. Emond, G. J. Schwartz, E. E. Ladenheim and T. H. Moran, *Am. J. Physiol.*, 1999, **276**, R1545.
19. M. W. Schwartz, S. C. Woods, D. Porte, R. J. Seeley and D. G. Baskin, *Nature*, 2000, **404**, 661.
20. M. Lutter and E. J. Nestler, *J. Nutr.*, 2009, **139**, 629.
21. P. J. Kenny, *Neuron*, 2011, **69**, 664.
22. B. Swinburn, G. Sacks and E. Ravussin, *Am. J. Clin. Nutr.*, 2009, **90**, 1453.
23. G. J. Wang, N. D. Volkow, F. Telang, M. Jayne, J. Ma, M. Rao, W. Zhu, C. T. Wong, N. R. Pappas, A. Geliebter and J. S. Fowler, *NeuroImage*, 2004, **21**, 1790.
24. V. Bragulat, M. Dzemidzic, C. Bruno, C. A. Cox, T. Talavage, R. V. Considine and D. A. Kareken, *Obesity*, 2010, **18**, 1566.
25. G. Sescousse, J. Redoute and J. C. Dreher, *J. Neurosci.*, 2010, **30**, 13095.
26. M. S. Man, H. F. Clarke and A. C. Roberts, *Cereb. Cortex*, 2009, **19**, 899.
27. E. T. Rolls, *Acta Physiol. Hung.*, 2008, **95**, 131.
28. K. S. LaBar, D. R. Gitelman, T. B. Parrish, Y. H. Kim, A. C. Nobre and M. M. Mesulam, *Behav. Neurosci.*, 2001, **115**, 493.
29. A. P. Goldstone, C. G. Prechtl de Hernandez, J. D. Beaver, K. Muhammed, C. Croese, G. Bell, G. Durighel, E. Hughes, A. D. Waldman, G. Frost and J. D. Bell, *Eur. J. Neurosci.*, 2009, **30**, 1625.
30. M. A. Cornier, A. K. Salzberg, D. C. Endly, D. H. Bessesen, D. C. Rojas and J .R. Tregellas, *PLoS One*, 2009, **4**, e6310.
31. D. P. Figlewicz, S. B. Evans, J. Murphy, M. Hoen and D. G. Baskin, *Brain Res.*, 2003, **964**, 107.
32. J. D. Hommel, R. Trinko, R. M. Sears, D. Georgescu, Z. W. Liu, X. B. Gao, J. J. Thurmon, M. Martinelli and R. J. DiLeone, *Neuron*, 2006, **51**, 801.
33. S. Malik, F. McGlone, D. Bedrossian and A. Dagher, *Cell Metab.*, 2008, 7, 400.
34. A. Abizaid, Z. W. Liu, Z. B. Andrews, M. Shanabrough, E. Borok, J. D. Elsworth, R. H. Roth, M. W. Sleeman, M. R. Picciotto, M. H. Tschop, X. B. Gao and T. L. Horvath, *J. Clin. Invest.*, 2006, **116**, 3229.
35. E. Jerlhag, E. Egecioglu, S. L. Dickson, A. Douhan, L. Svensson and J. A. Engel, *Addict. Biol.*, 2007, **12**, 6.
36. E. Stice, S. Spoor, C. Bohon and D. M. Small, *Science*, 2008, **322**, 449.
37. E. Stice, S. Yokum, C. Bohon, N. Marti and A. Smolen, *NeuroImage*, 2010, **50**, 1618.
38. R. D. Levitan, M. Masellis, R. W. Lam, P. Muglia, V. S. Basile, U. Jain, A. S. Kaplan, S. Tharmalingam, S. H. Kennedy and J. L. Kennedy, *Neuropsychopharmacology*, 2004, **29**, 179.
39. P. P. Silveira, A. K. Portella, J. L. Kennedy, H. Gaudreau, C. Davis, M. Steiner, C. N. Soares, S. G. Matthews, M. B. Sokolowski, L. Dube *et al.*, *Appetite*, 2014, **73**, 15.
40. B. J. Everitt, D. Belin, D. Economidou, Y. Pelloux, J. W. Dalley and T. W. Robbins, *Philos. Trans. R. Soc., B*, 2008, **363**, 3125.

41. N. D. Volkow and R. A. Wise, *Nat. Neurosci.*, 2005, **8**, 555.
42. E. D. Rosen and B. M. Spiegelman, *Nature*, 2006, **444**, 847.
43. A. H. Mokdad, E. S. Ford, B. A. Bowman, W.H. Dietz, F. Vinicor, V. S. Bales and J. S. Marks, *JAMA, J. Am. Med. Assoc.*, 2003, **289**, 76.
44. M. W. Schwartz, R. J. Seeley, L. A. Campfield, P. Burn and D. G. Baskin, *J. Clin. Invest.*, 1996, **98**, 1101.
45. C. Bjorbaek and B. B. Kahn, *Recent Prog. Horm. Res.*, 2004, **59**, 305.
46. E. S. Bachman, H. Dhillon, C. Y. Zhang, S. Cinti, A. C. Bianco, B. K. Kobilka and B. B. Lowell, *Science*, 2002, **297**, 843.
47. B. B. Lowell, V. S-Susulic, A. Hamann, J. A. Lawitts, J. Himms-Hagen, B. B. Boyer, L. P. Kozak and J. S. Flier, *Nature*, 1993, **366**, 740.
48. I. Shimomura, R. E. Hammer, S. Ikemoto, M. S. Brown and J. L. Goldstein, *Nature*, 1999, **401**, 73.
49. E. A. Oral, V. Simha, E. Ruiz, A. Andewelt, A. Premkumar, P. Snell, A. J. Wagner, A. M. DePaoli, M. L. Reitman, S. I. Taylor, P. Gorden and A. Garg, *N. Engl. J. Med.*, 2002, **346**, 570.
50. M. S. Winzell, R. Nogueiras, C. Dieguez and B. Ahren, *Biochem. Biophys. Res. Commun.*, 2004, **321**, 154.
51. N. Kubota, Y. Terauchi, T. Kubota, H. Kumagai, S. Itoh, H. Satoh, W. Yano, H. Ogata, K. Tokuyama, I. Takamoto *et al.*, *J. Biol. Chem.*, 2006, **281**, 8748.
52. N. Maeda, M. Takahashi, T. Funahashi, S. Kihara, H. Nishizawa, K. Kishida, H. Nagaretani, M. Matsuda, R. Komuro, N. Ouchi *et al.*, *Diabetes*, 2001, **50**, 2094.
53. G. S. Hotamisligil, *J. Intern. Med.*, 1999, **245**, 621.
54. H. Xu, G. T. Barnes, Q. Yang, G. Tan, D. Yang, C. J. Chou, J. Sole, A. Nichols, J. S. Ross, L. A. Tartaglia and H. Chen, *J. Clin. Invest.*, 2003, **112**, 1821.
55. A. Must, J. Spadano, E. H. Coakley, A. E. Field, G. Colditz and W. H. Dietz, *JAMA, J. Am. Med. Assoc.*, 1999, **282**, 1523.
56. J. Sorof and S. Daniels, *Hypertension*, 2002, **40**, 441.
57. D. Blair, J. P. Habicht, E. A. Sims, D. Sylwester and S. Abraham, *Am. J. Epidemiol.*, 1984, **119**, 526.
58. H. Kanai, Y. Matsuzawa, K. Kotani, Y. Keno, T. Kobatake, Y. Nagai, S. Fujioka, S. Tokunaga and S. Tarui, *Hypertension*, 1990, **16**, 484.
59. H. O. Steinberg, H. Chaker, R. Leaming, A. Johnson, G. Brechtel and A. D. Baron, *J. Clin. Invest.*, 1996, **97**, 2601.
60. M. R. Meyers and N. Gokce, *Curr. Opin. Endocrinol., Diabetes Obes.*, 2007, **14**, 365.
61. Z. Pausova, M. Jomphe, L. Houde, H. Vezina, S. N. Orlov, F. Gossard, D. Gaudet, J. Tremblay, T. A. Kotchen, A. W. Cowley, G. Bouchard and P. Hamet, *Obes. Res.*, 2002, **10**, 463.
62. P. Hamet, E. Merlo, O. Seda, U. Broeckel, J. Tremblay, M. Kaldunski, D. Gaudet, G. Bouchard, B. Deslauriers, F. Gagnon *et al.*, *Am. J. Hum. Genet.*, 2005, **76**, 815.
63. G. E. Alvarez, S. D. Beske, T. P. Ballard and K. P. Davy, *Circulation*, 2002, **106**, 2533.

64. K. P. Davy, *Curr. Hypertens. Rep.*, 2004, **6**, 241.
65. M. Vaz, G. Jennings, A. Turner, H. Cox, G. Lambert and M. Esler, *Circulation*, 1997, **96**, 3423.
66. J. E. Hall, *Hypertension*, 2003, **41**, 625.
67. T. L. Goodfriend, B. M. Egan and D. E. Kelley, *Endocr. Res.*, 1998, **24**, 789.
68. A. P. Rocchini, V. L. Katch, R. Grekin, C. Moorhead and J. Anderson, *Am. J. Cardiol.*, 1986, **57**, 613.
69. A. M. Sharma, *Hypertension*, 2004, **44**, 12.
70. S. Engeli, J. Bohnke, K. Gorzelniak, J. Janke, P. Schling, M. Bader, F. C. Luft and A. M. Sharma, *Hypertension*, 2005, **45**, 356.
71. S. Yusuf, P. Sleight, J. Pogue, J. Bosch, R. Davies and G. Dagenais, *N. Engl. J. Med.*, 2000, **342**, 145.
72. K. Rahmouni, M. L. Correia, W. G. Haynes and A. L. Mark, *Hypertension*, 2005, **45**, 9.
73. S. Yasue, H. Masuzaki, S. Okada, T. Ishii, C. Kozuka, T. Tanaka, J. Fujikura, K. Ebihara, K. Hosoda, A. Katsurada *et al.*, *Am. J. Hypertens.*, 2010, **23**, 425.
74. S. Thatcher, F. Yiannikouris, M. Gupte and L. Cassis, *Mol. Cell. Endocrinol.*, 2009, **302**, 111.
75. G. L. Pierce, S. D. Beske, B. R. Lawson, K. L. Southall, F. J. Benay, A. J. Donato and D. R. Seals, *Hypertension*, 2008, **52**, 72.
76. V. Tarasov, K. Ekroos, M. Suoniemi, D. Kauhanen, T. Sylvanne, R. Hurme, I. Gouni-berthold, H. Berthold, M. Kleber, R. Laaksonen and W. Marz, *J. Clin. Endocrinol. Metab.*, 2014, **99**, E45.
77. K Boini, C. Zhang, M. Xia, J. Poklis and P. Li, *J. Pharmacol. Exp. Ther.*, 2010, **334**, 839.
78. A. Alayoubi, J. Wang, B. Au, S. Carpentier, V. Garcia, S. Sworski, S. El-Ghamrasni, K. Kirouac, M. Exertier, Z. Xiong *et al.*, *EMBO Mol. Med.*, 2013, **5**, 827.
79. P. Li and Y. Zhang, *Handb. Exp. Pharmacol.*, 2013, **216**, 17.
80. J. D. Symons and E. D. Abel, *Rev. Endocr. Metab. Disord.*, 2013, **14**, 59.
81. M. L. Correia and W. G. Haynes, *Curr. Hypertens. Rep.*, 2004, **6**, 230.
82. F. Galletti, L. D'Elia, G. Barba, A. Siani, F.P. Cappuccio, E. Farinaro, R. Iacone, O. Russo, D. De Palma, R. Ippolito and P. Strazzullo, *J. Clin. Endocrinol. Metab.*, 2008, **93**, 3922.
83. C. Asferg, R. Mogelvang, A. Flyvbjerg, J. Frystyk, J. S. Jensen, J. L. Marott, M. Appleyard, G. B. Jensen and J. Jeppesen, *Am. J. Hypertens.*, 2010, **23**, 327.
84. A. L. Mark, K. Agassandian, D. A. Morgan, X. Liu, M. D. Cassell and K. Rahmouni, *Hypertension*, 2009, **53**, 375.
85. N. Rasouli and P. A. Kern, *J. Clin. Endocrinol. Metab.*, 2008, **93**(11 Suppl 1), S64.
86. F. Yiannikouris, M. Gupte, K. Putnam and L. Cassis, *Curr. Opin. Nephrol. Hypertens.*, 2010, **19**, 195.

87. I. M. O'Shaughnessy and T. A. Kotchen, *Curr. Opin. Cardiol.*, 1993, **8**, 757.
88. T. A. Kotchen, *Drug Dev. Res.*, 1994, **32**, 100.
89. S. R. Smith and P. W. Wilson, *J. Clin. Endocrinol. Metab.*, 2006, **91**, 2506.
90. T. E. Graham, Q. Yang, M. Bluher, A. Hammarstedt, T. P. Ciaraldi, R. R. Henry, C. J. Wason, A. Oberbach, P. A. Jansson, U. Smith and B. B. Kahn, *N. Engl. J. Med.*, 2006, **354**, 2552.
91. S. Konstantinides, K. Schafer, S. Koschnick and D. J. Loskutoff, *J. Clin. Invest.*, 2001, **108**, 1533.
92. M. R. Sierra-Honigmann, A. K. Nath, C. Murakami, G. Garcia-Cardena, A. Papapetropoulos, W. C. Sessa, L.A. Madge, J. S. Schechner, M. B. Schwabb, P. J. Polverini and J. R. Flores-Riveros, *Science*, 1998, **281**, 1683.
93. A. Singhal, S. Farooqi, T. J. Cole, S. O'Rahilly, M. Fewtrell, M. Kattenhorn, A. Lucas and J. Deanfield, *Circulation*, 2002, **106**, 1919.
94. M. Kumada, S. Kihara, N. Ouchi, H. Kobayashi, Y. Okamoto, K. Ohashi, K. Maeda, H. Nagaretani, K. Kishida, N. Maeda *et al.*, *Circulation*, 2004, **109**, 2046.
95. H. Kobayashi, N. Ouchi, S. Kihara, K. Walsh, M. Kumada, Y. Abe, T. Funahashi and Y. Matsuzawa, *Circ. Res.*, 2004, **94**, e27.
96. C. J. Li, H. W. Sun, F. L. Zhu, L. Chen, Y. Y. Rong, Y. Zhang and M. Zhang, *J. Endocrinol.*, 2007, **193**, 137.
97. S. P. Weisberg, D. McCann, M. Desai, M. Rosenbaum, R. L. Leibel and A. W. Ferrante, *J. Clin. Invest.*, 2003, **112**, 1796.
98. R. Cancello and K. Clement, *BJOG*, 2006, **113**, 1141.
99. S. R. Bornstein, M. Abu-Asab, A. Glasow, G. Path, H. Hauner, M. Tsokos, G. P. Chrousos and W. A. Scherbaum, *Diabetes*, 2000, **49**, 532.
100. A. Ceriello and E. Motz, *Arterioscler., Thromb., Vasc. Biol.*, 2004, **24**, 816.
101. P. Holvoet, S. B. Kritchevsky, R. P. Tracy, A. Mertens, S. M. Rubin, J. Butler, B. Goodpaster and T. B. Harris, *Diabetes*, 2004, **53**, 1068.
102. E. Tumova, W. Sun, P. H. Jones, M. Vrablik, C. M. Ballantyne and R. C. Hoogeveen, *J. Obes.*, 2013, 729515.
103. K. Esposito, F. Nappo, R. Marfella, G. Giugliano, F. Giugliano, M. Ciotola, L. Quagliaro, A. Ceriello and D. Giugliano, *Circulation*, 2002, **106**, 2067.
104. D. Tripathy, P. Mohanty, S. Dhindsa, T. Syed, H. Ghanim, A. Aljada and P. Dandona, *Diabetes*, 2003, **52**, 2882.
105. P. E. Scherer, *Diabetes*, 2006, **55**, 1537.
106. S. E. Shoelson, J. Lee and A. B. Goldfine, *J. Clin. Invest.*, 2006, **116**, 1793.
107. Q. Yang, T. E. Graham, N. Mody, F. Preitner, O. D. Peroni, J. M. Zabolotny, K. Kotani, L Quadro and B. B. Kahn, *Nature*, 2005, **436**, 356.
108. T. Kadowaki, T. Yamauchi, N. Kubota, K. Hara, K. Ueki and K. Tobe, *J. Clin. Invest.*, 2006, **116**, 1784.
109. K. E. Wellen and G. S. Hotamisligil, *J. Clin. Invest.*, 2005, **115**, 111.

110. J. N. Fain, A. K. Madan, M. L. Hiler, P. Cheema and S. W. Bahouth, *Endocrinology*, 2004, **145**, 2273.
111. G. M. Reaven, C. B. Hollenbeck, C. Y. Jeng, M. S. Wu and Y. D. Chen, *Diabetes*, 1988, **37**, 1020.
112. G. Boden, *Diabetes*, 1997, **46**, 3.
113. M. Roden, T. B. Price, G. Perseghin, K. F. Petersen, D. L. Rothman, G. W. Cline and G. I. Shulman, *Clin. Invest.*, 1996, **97**, 2859.
114. A. T. Santomauro, G. Boden, M. E. Silva, D. M. Rocha, R. F. Santos, M. J. Ursich, P. G. Strassmann and B. L. Wajchenberg, *Diabetes*, 1999, **48**, 1836.
115. C. T. Montague and S. O'Rahilly, *Diabetes*, 2000, **49**, 883.
116. I. Samjoo, A. Safdar, M. Hamadeh, A. Gover, N. Mocellin, J. Santana, J. Little, G. Steinberg, S. Raha and M. Tarnopolsky, *PLoS One*, 2013, **8**, e66322.
117. R. Cinar, G. Godlewski, J. Liu, J. Tam, T. Jourdan, B. Mukhopadhyay, J. Harvey-White and G. Kunos, *Hepatology*, 2014, **59**, 143.
118. S. Galadari, A. Rahman, S. Palichankandy, A. Galadari and F. Thayyullathil, *Lipids Health Dis.*, 2013, **12**, 98.
119. R. L. Dobbins, M. W. Chester, B. E. Stevenson, M. B. Daniels, D. T. Stein and J. D. McGarry, *J. Clin. Invest.*, 1998, **101**, 2370.
120. Y. Itoh, Y. Kawamata, M. Harada, M. Kobayashi, R. Fujii, S. Fukusumi, K. Ogi, M. Hosoya, Y. Tanaka, H. Uejima *et al.*, *Nature*, 2003, **422**, 173.
121. M. Prentki, E. Joly, W. El-Assaad and R. Roduit, *Diabetes*, 2002, **51**(Suppl 3), S405.
122. D. Pitocco, M. Tesauro, R. Alessandro, G. Ghirlanda and C. Cardillo, *Int. J. Mol. Sci.*, 2013, **14**, 21525.
123. X. Du, K. Stockklauser-Farber and P. Rosen, *Free Radical Biol. Med.*, 1999, **27**, 752.
124. S. Gupta, E. Chough, J. Daley, P. Oates, K. Tornheim, N. B. Ruderman and J. F. Keaney, *Am. J. Physiol.: Cell Physiol.*, 2002, **282**, C560.
125. T. Nishikawa, D. Edelstein, X. L. Du, S. Yamagishi, T. Matsumura, Y. Kaneda, M. A. Yorek, D. Beebe, P. J. Oates, H. P. Hammes, I. Giardino and M. Brownlee, *Nature*, 2000, **404**, 787.
126. X. L. Du, D. Edelstein, L. Rossetti, I. G. Fantus, H. Goldberg, F. Ziyadeh, J. Wu and M. Brownlee, *Proc. Natl. Acad. Sci. U. S. A.*, 2000, **97**, 1222.
127. C. J. Mullarkey, D. Edelstein and M. Brownlee, *Biochem. Biophys. Res. Commun.*, 1990, **173**, 932.
128. V. Poitout and R. P. Robertson, *Endocr. Rev.*, 2008, **29**, 351.
129. L. Zhang, C. M. Wheatley, S. M. Richards, E. J. Barrett, M. G. Clark and S. Rattigan, *Am. J. Physiol.: Endocrinol. Metab.*, 2003, **285**, E654.
130. M. Yoshizumi, M. A. Perrella, J. C. Burnett and M. E. Lee, *Circ. Res.*, 1993, **73**, 205.
131. F. Mohamed, J. C. Monge, A. Gordon, P. Cernacek, D. Blais and D. J. Stewart, *Arterioscler., Thromb., Vasc. Biol.*, 1995, **15**, 52.

132. G. W. De Keulenaer, R. W. Alexander, M. Ushio-Fukai, N. Ishizaka and K. K. Griendling, *Biochem. J.*, 1998, **329**, 653.
133. C. Rask-Madsen, H. Dominguez, N. Ihlemann, T. Hermann, L. Kober and C. Torp-Pedersen, *Circulation*, 2003, **108**, 1815.
134. P. Ziccardi, F. Nappo, G. Giugliano, K. Esposito, R. Marfella, M. Cioffi, F. D'Andrea, A. M. Molinari and D. Giugliano, *Circulation*, 2002, **105**, 804.
135. N. D. Roupas, I. Mamali, S. Maragkos, L. Leonidou, A. K. Armeni, G. K. Markantes, A. Tsekouras, G. C. Sakellaropoulos, K. B. Markou and N. A. Georgopoulos, *Hormones*, 2013, **12**, 275.
136. I. G. Fatouros, A. Chatzinikolaou, S. Tournis, M. G. Nikolaidis, A. Z. Jamurtas, I. I. Douroudos, I. Papassotiriou, P. M. Thomakos, K. Taxildaris, G. Mastorakos and A. Mitrakou, *Diabetes Care*, 2009, **32**, 2161.
137. K. A. Varady, S. Bhutani, E. C. Church and S. A. Phillips, *Med. Sci. Sports Exercise*, 2010, **42**, 456.
138. E. Riesco, S. Choquette, M. Audet, J. Lebon, D. Tessier and I. J. Dionne, *Metabolism*, 2012, **61**, 273.
139. R. Plinta, M. Olszanecka-Glinianowicz, A. Drosdzol-Cop, J. Chudek and V. Skrzypulec-Plinta, *J. Endocrinol. Invest.*, 2012, **35**, 595.
140. A. de Piano, M. T. de Mello, Pde. L. Sanches, P. L. da Silva, R. M. Campos, J. Carnier, F. Corgosinho, D. Foschini, D. L. Masquio, L. Tock *et al.*, *Eur. J. Gastroenterol. Hepatol.*, 2012, **24**, 1313.
141. S. Numao, H. Sasai, Y. Nomata, T. Matsuo, M. Eto, T. Tsujimoto and K. Tanaka, *Obes. Facts*, 2012, **5**, 845.
142. M. Venojarvi, N. Wasenius, S. Manderoos, O. Heinonen, M. Hernelahti, H. Lindholm, J. Surakka, J. Lindstrom, S. Aunola, M. Atalay and J. G. Eriksson, *Ann. Med.*, 2013, **45**, 162.
143. H. J. Yoo, S. Y. Hwang, G. J. Cho, H. C. Hong, H. Y. Choi, T. G. Hwang, S. M. Kim, M. Bluher, B. S. Youn, S. H. Baik and K. M. Choi, *J. Clin. Endocrinol. Metab.*, 2013, **98**, 2897.
144. J. Krist, K. Wider, N. Klotting, A. Oberbach, S. Kralisch, T. Wiesner, M. R. Schon, D. Gartner, A. Dietrich, E. Shang *et al.*, *Obes. Facts*, 2013, **6**, 57.
145. P. Koncsos, I. Seres, M. Harangi, D. Pall, L. Jozsa, L. Bajnok, E. V. Nagy and G. Paragh, *J. Am. Coll. Nutr.*, 2011, **30**, 333.
146. D. C. Masquio, A. de Piano, P. L. Sanches, F. C. Corgosinho, R. M. Campos, J. Carnier, P. L. da Silva, D. A. Caranti, L. Tock, L. M. Oyama *et al.*, *Clin. Endocrinol.*, 2013, **79**, 55.
147. T. Christiansen, S. K. Paulsen, J. M. Bruun, S. B. Pedersen and B. Richelsen, *Am. J. Physiol.: Endocrinol. Metab.*, 2010, **298**, E824.
148. U. Ebelt, T. Hofmann and A. Stengel, *Curr. Opin. Clin. Nutr. Metab. Care*, 2013, **16**, 541.
149. A. Roca-Rivada, C. Castelao, L. L. Senin, M. O. Landrove, J. Baltar, A. Belen Crujeiras, L. M. Seoane, F. F. Casanueva and M. Pardo, *PLoS One*, 2013, **8**, e60563.

150. R. A. Gulli, J. M. Tishinsky, T. MacDonald, L. E. Robinson, D. C. Wright and D. J. Dyck, *Am. J. Physiol.: Regul., Integr. Comp. Physiol.*, 2012, **303**, R1062.
151. J. D. Adams, *World J. Pharmacol.*, 2013, **2**, 73.
152. D. Langin, *Biochim. Biophys. Acta*, 2010, **1801**, 372.
153. P. Lee, M. M. Swarbrick and K. K. Ho, *Endocr. Rev.*, 2013, **34**, 413.
154. B. Cannon and J. Nedergaard, *Physiol. Rev.*, 2004, **84**, 277.
155. J. Himms-Hagen, *FASEB J.*, 1990, **4**, 2890.
156. P. Lee, J. D. Linderman, S. Smith, R. J. Brychta, J. Wang, C. Idelson, R. M. Perron, C. D. Werner, G. Q. Phan, U. S. Kammula *et al.*, *Cell Metab.*, 2014, **19**, 302.
157. P. Bostrom, J. Wu, M. P. Jedrychowski, A. Korde, L. Ye, J. C. Lo, K. A. Rasbach, E. A. Bostrom, J. H. Choi, J. Z. Long *et al.*, *Nature*, 2012, **481**, 463.
158. J. Wu, P. Bostrom, L. M. Sparks, L. Ye, J. H. Choi, A. H. Giang, M. Khandekar, K. A. Virtanen, P. Nuutila, G. Schaart *et al.*, *Cell*, 2012, **150**, 366.
159. P. Lee, R. J. Brychta, J. Linderman, S. Smith, K. Y. Chen and F. S. Celi, *J. Clin. Endocrinol. Metab.*, 2013, **98**, E98.
160. P. Lee, C. D. Werner, E. Kebebew and F. S. Celi, *Int. J. Obes.*, 2014, **38**, 170.
161. F. M. Fisher, S. Kleiner, N. Douris, E. C. Fox, R. J. Mepani, F. Verdeguer, J. Wu, A. Kharitonenkov, J. S. Flier, E. Maratos-Flier and B. M. Spiegelman, *Genes Dev.*, 2012, **26**, 271.
162. N. Petrovic, T. B. Walden, I. G. Shabalina, J. A. Timmons, B. Cannon and J. Nedergaard, *J. Biol. Chem.*, 2010, **285**, 7153.
163. P. Muzzin, J. P. Revelli, F. Kuhne, J. D. Gocayne, W. R. McCombie, J. C. Venter, J. P. Giacobino and C. M. Fraser, *J. Biol. Chem.*, 1991, **266**, 24053.
164. J. R. Arch, A. T. Ainsworth, M. A. Cawthorne, V. Piercy, M. V. Sennitt, V. E. Thody, C. Wilson and S. Wilson, *Nature*, 1984, **309**, 163.
165. C. P. Kordik and A. B. Reitz, *J. Med. Chem.*, 1999, **42**, 181.
166. A. D. Strosberg, *Trends Pharmacol. Sci.*, 1997, **18**, 449.
167. N. Ouchi, S. Kihara, T. Funahashi, Y. Matsuzawa and K. Walsh, *Curr. Opin. Lipidol.*, 2003, **14**, 561.
168. R. Shibata, K. Sato, D. R. Pimentel, Y. Takemura, S. Kihara, K. Ohashi, T. Funahashi, N. Ouchi and K. Walsh, *Nat. Med.*, 2005, **11**, 1096.
169. J. G. Yu, S. Javorschi, A. L. Hevener, Y. T. Kruszynska, R. A. Norman, M. Sinha and J. M. Olefsky, *Diabetes*, 2002, **51**, 2968.
170. H. Hirose, T. Kawai, Y. Yamamoto, M. Taniyama, M. Tomita, K. Matsubara, Y. Okazaki, T. Ishii, Y. Oguma, I. Takei and T. Saruta, *Metabolism*, 2002, **51**, 314.
171. A. Okuno, H. Tamemoto, K. Tobe, K. Ueki, Y. Mori, K. Iwamoto, K. Umesono, Y. Akanuma, T. Fujiwara, H. Horikoshi, Y. Yazaki and T. Kadowaki, *J. Clin. Invest.*, 1998, **101**, 1354.
172. J. M. Olefsky and A. R. Saltiel, *Trends Endocrinol. Metab.*, 2000, **11**, 362.

173. T. Yamauchi, J. Kamon, H. Waki, K. Murakami, K. Motojima, K. Komeda, T. Ide, N. Kubota, Y. Terauchi, K. Tobe *et al.*, *J. Biol. Chem.*, 2001, **276**, 41245.
174. M. Iwaki, M. Matsuda, N. Maeda, T. Funahashi, Y. Matsuzawa, M. Makishima and I. Shimomura, *Diabetes*, 2003, **52**, 1655.
175. J. M. Olefsky, *J. Clin. Invest.*, 2000, **106**, 467.
176. U. B. Pajvani, M. Hawkins, T. P. Combs, M. W. Rajala, T. Doebber, J. P. Berger, J. A. Wagner, M. Wu, A. Knopps, A. H. Xiang *et al.*, *J. Biol. Chem.*, 2004, **279**, 12152.
177. Z. T. Bloomgarden, *Diabetes Care*, 2005, **28**, 488.
178. I. Shimomura, T. Funahashi, M. Takahashi, K. Maeda, K. Kotani, T. Nakamura, S. Yamashita, M. Miura, Y. Fukuda, K. Takemura, K. Tokunaga and Y. Matsuzawa, *Nat. Med.*, 1996, **2**, 800.
179. R. Stienstra, L. A. Joosten, T. Koenen, B. van Tits, J. A. van Diepen, S. A. van den Berg, P. C. Rensen, P. J. Voshol, G. Fantuzzi, A. Hijmans *et al.*, *Cell Metab.*, 2010, **12**, 593.
180. N. Naour, C. Rouault, S. Fellahi, M. E. Lavoie, C. Poitou, M. Keophiphath, D. Eberle, S. Shoelson, S. Rizkalla, J. P. Bastard *et al.*, *J. Clin. Endocrinol. Metab.*, 2010, **95**, 1861.
181. F. Samad, M. Pandey and D. J. Loskutoff, *Blood*, 2001, **98**, 3353.
182. D. Y. Oh, S. Talukdar, E. J. Bae, T. Imamura, H. Morinaga, W. Fan, P. Li, W. J. Lu, S. M. Watkins and J. M. Olefsky, *Cell*, 2010, **142**, 687.
183. M. N. Adams, R. Ramachandran, M. K. Yau, J. Y. Suen, D. P. Fairlie, M. D. Hollenberg and J. D. Hooper, *Pharmacol. Ther.*, 2011, **130**, 248.
184. J. Y. Suen, G. D. Barry, R. J. Lohman, M. A. Halili, A. J. Cotterell, G. T. Le and D. P. Fairlie, *Br. J. Pharmacol.*, 2012, **165**, 1413.
185. J. Lim, A. Iyer, L. Liu, J. Y. Suen, R. J. Lohman, V. Seow, M. K. Yau, L. Brown and D. P. Fairlie, *FASEB J.*, 2013, **27**, 4757.
186. M. A. Herman and B. B. Kahn, *J. Clin. Invest.*, 2006, **116**, 1767.
187. J. C. Hou and J. E. Pessin, *Curr. Opin. Cell Biol.*, 2007, **19**, 466.
188. I. J. Lodhi, S. H. Chiang, L. Chang, D. Vollenweider, R. T. Watson, M. Inoue, J. E. Pessin and A. R. Saltiel, *Cell Metab.*, 2007, **5**, 59.
189. M. V. Chao, *Nat. Rev. Neurosci.*, 2003, **4**, 299.
190. M. R. Peeraully, J. R. Jenkins and P. Trayhurn, *Am. J. Physiol.: Endocrinol. Metab.*, 2004, **287**, E331.
191. D. Deponti, R. Buono, G. Catanzaro, C. De Palma, R. Longhi, R. Meneveri, N. Bresolin, M. T. Bassi, G. Cossu, E. Clementi and S. Brunelli, *Mol. Biol. Cell*, 2009, **20**, 3620.
192. B. Baeza-Raja, P. Li, N. Le Moan, B. D. Sachs, C. Schachtrup, D. Davalos, E. Vagena, D. Bridges, C. Kim, A. R. Saltiel, J. M. Olefsky and K. Akassoglou, *Proc. Natl. Acad. Sci. U. S. A.*, 2012, **109**, 5838.
193. J. L. Temple, A. M. Giacomelli, K. M. Kent, J. N. Roemmich and L. H. Epstein, *Am. J. Clin. Nutr.*, 2007, **85**, 355.
194. S. Marsh, C. Ni Mhurchu and R. Maddison, *Appetite*, 2013, **71**, 259.

195. S. Marsh, C. Ni Mhurchu, Y. Jiang and R. Maddison, *Appetite*, 2014, **77**, 13.
196. http://www.foodnavigator.com/Legislation/WHO-recommends-halving-sugar-intake-advice.
197. S. W. Junod, 2008, http://www.fda.gov/AboutFDA/WhatWeDo/History/Overviews/ucm304485.htm.
198. E. Mohammadi, M. Rafraf, L. Farzadi, M. Asghari-Jafarabadi and S. Sabour, *Asia Pac. J. Clin. Nutr.*, 2012, **21**, 511.
199. B. Gray, F. Steyn, P. S. Davies and L. Vitetta, *Eur. J. Clin. Nutr.*, 2013, **67**, 1234.
200. M. Mostowik, G. Gajos, J. Zalewski, J. Nessler and A. Undas, *Cardiovasc. Drugs Ther.*, 2013, **27**, 289.
201. C. T. Montague, I. S. Farooqi, J. P. Whitehead, M. A. Soos, H. Rau, N. J. Wareham, C.P. Sewter, J. E. Digby, S. N. Mohammed, J. A. Hurst *et al.*, *Nature*, 1997, **387**, 903.
202. R. Yang and L. A. Barouch, *Circ. Res.*, 2007, **101**, 545.
203. F. Lago, C. Dieguez, J. Gomez-Reino and O. Gualillo, *Nat. Clin. Pract. Rheumatol.*, 2007, **3**, 716.
204. A. Gertier, *Trends Endocrinol. Metab.*, 2006, **17**, 372.
205. M. T. Cabioglu and N. Ergene, *Int. J. Neurosci.*, 2007, **117**, 579.
206. M. Gong, X. Wang, Z. Mao, Q. Shao, X. Xiang and B. Xu, *Am. J. Chin. Med.*, 2012, **40**, 511.
207. P. V. Peplow and S. M. Han, *J. Acupunct. Meridian Stud.*, 2014, 7, 1.
208. P. V. Peplow, *J. Acupunct, Meridian Stud.*, 2014, DOI: 10.1016/j.jams.2014.06.013.

Subject Index

Page numbers in *italics* refer to tables or figures.